Fungal Infections in Immunocompromised Hosts

Fungal Infections in Immunocompromised Hosts

Special Issue Editors

Dimitrios P. Kontoyiannis
Monica Slavin

MDPI • Basel • Beijing • Wuhan • Barcelona • Belgrade

MDPI

Special Issue Editors
Dimitrios P. Kontoyiannis
The University of Texas MD Anderson Cancer Center
USA

Monica Slavin
Department of Infectious Diseases and Infection Prevention, Peter MacCallum Cancer Centre
Australia

Editorial Office
MDPI
St. Alban-Anlage 66
4052 Basel, Switzerland

This is a reprint of articles from the Special Issue published online in the open access journal *Journal of Fungi* (ISSN 2309-608X) from 2018 to 2019 (available at: https://www.mdpi.com/journal/jof/special_issues/ICHS2018).

For citation purposes, cite each article independently as indicated on the article page online and as indicated below:

LastName, A.A.; LastName, B.B.; LastName, C.C. Article Title. *Journal Name* **Year**, *Article Number*, Page Range.

ISBN 978-3-03897-716-2 (Pbk)
ISBN 978-3-03897-717-9 (PDF)

Cover image courtesy of Dimitrios P. Kontoyiannis and Monica Slavin.

Contents

About the Special Issue Editors

Dimitrios P. Kontoyiannis is the Texas 4000 Distinguished Endowed Professor and Deputy Head in the Division of Internal Medicine at MD Anderson Cancer Center in Houston, TX. Dr. Kontoyiannis has authored over 550 peer-reviewed manuscripts and has given over 330 lectures in international conferences and institutions in US and abroad. He is considered a leading mycology expert worldwide with an H index of 98 and over 42,000 citations. His research group is credited for many and sustained contributions to clinical, translational and experimental mycology. He has been the recipient of many national and international awards and was the president elect of Immunocompromised Host Society (2016–2018).

Monica Slavin is Director of the Department of Infectious Diseases at Peter MacCallum Cancer Centre and a National Health and Medical Research Council funded Centre of Research Excellence in Infections in cancer. She is Professor, in the Department of Medicine, the University of Melbourne. Her research focuses on identifying risk factors, improving early diagnosis, prevention, and treatment of infection, particularly fungal infection in hematology and transplant patients. She was a founding member of the Australasian Society for Infectious Diseases Mycoses Interest Group and is the current president of the International Immunocompromised Host Society.

Preface to "Fungal Infections in Immunocompromised Hosts"

First convened in 1980, the International Immunocompromised Host Society (ICHS) symposium is the premier, international, multi-disciplinary forum for scientific and clinical exchange to improve understanding and management of infections in an immunocompromised host. Since its inception, mycology has been a key theme throughout the bi-annual meetings. The recent meeting in June 2018 in Athens, Greece similarly had a cutting edge mycology program delivered by leading international experts addressing current and future innovations in mycology.

Worldwide deaths from Candida, Aspergillus, Pneumocystis, and Cryptococcus infections are estimated to exceed 1.4 million annually with individual patient mortality rates in excess of 30%. Invasive fungal diseases (IFDs) are occurring in expanding populations at risk such as those with liver disease and receiving novel treatments such as biologics and small molecule protein kinase inhibitors, providing insight into fungal pathogenesis. Outbreaks occurring in healthcare or after natural disasters are increasingly identified and require specialized investigation. The emerging global problem of antifungal resistance has prompted an increasing focus on antifungal drug discovery as well as optimizing PK/PD of antifungals, and on fungal biofilms that may select resistance. Approaches such as harnessing the mycobiome and immunotherapy may offer new options for managing IFD.

The development and integration of improved diagnostic tests for IFD including MALDI-TOF, nucleic acid and biomarker detection, and immunogenetic assays offers promise for implementing more precise individual risk profiles and early intervention to prevent or treat IFD. At a population level, these tests offer important antifungal stewardship tools to support the judicious use of antifungals. In this unique supplement, we have compiled several state-of-the-art topics that are based on lectures delivered by eminent mycology experts during the 37th ICHS meeting. We hope that the esteemed audience of the *Journal of Fungi* will enjoy and appreciate the ever-evolving and complex field of fungal infections in vulnerable hosts.

Dimitrios P. Kontoyiannis, Monica Slavin
Special Issue Editors

Journal of
Fungi

MDPI

Review

From the Clinical Mycology Laboratory: New Species and Changes in Fungal Taxonomy and Nomenclature

Nathan P. Wiederhold * and Connie F. C. Gibas

Fungus Testing Laboratory, Department of Pathology and Laboratory Medicine, University of Texas Health Science Center at San Antonio, San Antonio, TX 78229, USA; gibas@uthscsa.edu
* Correspondence: wiederholdn@uthscsa.edu

Received: 29 October 2018; Accepted: 13 December 2018; Published: 16 December 2018

Abstract: Fungal taxonomy is the branch of mycology by which we classify and group fungi based on similarities or differences. Historically, this was done by morphologic characteristics and other phenotypic traits. However, with the advent of the molecular age in mycology, phylogenetic analysis based on DNA sequences has replaced these classic means for grouping related species. This, along with the abandonment of the dual nomenclature system, has led to a marked increase in the number of new species and reclassification of known species. Although these evaluations and changes are necessary to move the field forward, there is concern among medical mycologists that the rapidity by which fungal nomenclature is changing could cause confusion in the clinical literature. Thus, there is a proposal to allow medical mycologists to adopt changes in taxonomy and nomenclature at a slower pace. In this review, changes in the taxonomy and nomenclature of medically relevant fungi will be discussed along with the impact this may have on clinicians and patient care. Specific examples of changes and current controversies will also be given.

Keywords: taxonomy; fungal nomenclature; phylogenetics; species complex

1. Introduction

Kingdom Fungi is a large and diverse group of organisms for which our knowledge is rapidly expanding. This kingdom includes numerous species that are capable of causing disease in humans, animals and plants. Infections caused by fungi are highly prevalent in humans, as it is estimated that greater than 1 billion people worldwide have infections caused by these organisms [1,2]. However, the full extent of fungi capable of causing infections in humans remains unknown. Although only several hundred species have been reported to cause disease in humans [3], it is estimated that there are between 1.5 million to 5 million fungal species and only approximately 100,000 species have been identified [4,5]. The potential clinical relevance of yet to be discovered species is highlighted by the nearly 10-fold increase in reports of newly described fungal pathogens in plants, animals and humans since 1995 [6], as well as by outbreaks of infections caused by fungi previously not associated with severe disease in humans [7–10]. Those that are capable of causing systemic infections in humans often have key attributes that make this possible (e.g., growth at 37 °C, penetrate or circumvent host barriers, digest and absorb components of human tissue, withstand immune responses of host) [11]. Many are also capable of persisting in the environment due to saprobic potential (i.e., the ability to grow on dead or decaying material) [12]. In addition, many species may be generalist pathogens with little host specificity and have dynamic genomes allowing for rapid adaption and evolution [11–13]. Thus, the number of fungal species that are etiologic agents of human infections will continue to grow. As the number of pathogenic species continues to grow, many of which are opportunists, new classifications and nomenclature will be introduced. In addition, revisions to current taxonomy will continue to be made based on our increased understanding of the diversity of this kingdom. In this review, changes in taxonomy and nomenclature of clinically relevant fungi will be discussed as will the challenges posed to clinicians and clinical microbiology laboratories by these changes.

2. Changes in Fungal Taxonomy and Nomenclature

Over the last several years significant changes have occurred in fungal taxonomy and nomenclature, as new fungi are discovered and the relationships of individual species to others and within larger taxonomic groups have been re-evaluated and redefined. Although the discovery of new fungal species and their classification has been a continuous process since the advent of the field of mycology, the pace of discovery and re-evaluation of taxonomic status has increased with the introduction of molecular and proteomic tools. Historically, morphologic characteristics and other phenotypic traits (e.g., growth on different media at different temperature, biochemical analysis) have been used for both taxonomic evaluation and species identification in clinical settings. However, the phenotypic traits that are observed may vary under different conditions and are thus subjective. Errors in species identification may occur because of this. DNA sequence analysis is now considered the gold standard for fungal species identification and has been a driving force for the increased pace of the discovery of new species and changes in fungal taxonomy and nomenclature [14–16]. Phylogenetic analysis based on the sequences of multiple loci within fungal DNA is often used for taxonomic designation of new species and in the re-evaluation of previous classifications that had been based solely on phenotypic characteristics. An advantage of phylogenetic analysis for taxonomic purposes is that close relatives become grouped together regardless of differences in morphology and these relationships may be useful for predicting pathogenicity and susceptibility to antifungal drugs [17]. These methods have led to the discovery of numerous cryptic species, which are indistinguishable from closely related species based on morphologic characteristics but can be identified by molecular means [18]. However, the use of phylogenetic analysis for taxonomic re-evaluation is not without its flaws, as the relationships created may be subject to change with increased understanding of fungal diversity since phylogenetic trees are highly subject to sampling effects [17]. In addition, no delimitation criteria exist above the species level [19]. Newer technologies, such as matrix assisted laser desorption ionization time-of-flight mass spectrometry (MALDI-TOF MS), are also being used with increased frequency for rapid species identification in clinical settings as well as for the taxonomic evaluation of fungi [20–23]. It should be noted that clinical laboratories may need to exercise caution in the adoption of these technologies for the identification of all fungal isolates until appropriately validated in the literature. Some examples of new and clinically relevant fungal species are listed in Table 1. Clearly, the description and recognition of new species helps to advance the field of medical mycology by increasing our understanding of the epidemiology of various fungal infections, the geographic distribution of species that cause these infections and how infections caused by different species may respond differently to treatment [24–28].

In addition to new tools for fungal identification and taxonomic re-evaluation, changes in fungal nomenclature have also been brought about by the elimination of the dual nomenclature system. When fungal taxonomy was based solely on morphologic characteristics, many fungi were forced to have multiple names describing either their sexual (teleomorph) or asexual (anamorph) life cycle stages under Article 59 of the International Code for Botanical Nomenclature. However, this dual nomenclature system became obsolete with the introduction of molecular tools since different morphologic stages are identical at the genetic level [17,19,29]. Thus, the system was abolished under the newly named International Code of Nomenclature of algae, fungi and plants in which fungi are now only to have one name [30]. However, decisions regarding which names to use have not always been straightforward. Some examples of clinically relevant changes in nomenclature for yeasts and molds are shown in Table 2.

Table 1. Examples of recently described and medically relevant fungi.

Species	Family	Order	Sites & Infections in Humans	Reference
Apophysomyces mexicanus	Saksenaeaceae	Mucorales	Necrotizing fasciitis	[31]
Aspergillus citrinoterreus	Aspergillaceae	Eurotiales	Pulmonary infection	[32]
Aspergillus suttoniae	Aspergillaceae	Eurotiales	Human sputum	[33]
Aspergillus tanneri	Aspergillaceae	Eurotiales	Lung, gastric abscess	[34]
Candida auris	Incertae sedis	Saccharomycetales	Various sites, candidemia	[35]
Curvularia americana	Pleosporaceae	Pleosporales	Nasal sinus, bone marrow	[36]
Curvularia chlamydospora	Pleosporaceae	Pleosporales	Nasal sinus, nail	[36]
Emergomyces canadensis	Ajellomycetaceae	Onygenales	Pneumonia, fungemia	[37,38]
Exophiala polymorpha	Herpotrichiellaceae	Chaetothyriales	Subcutaneous & cutaneous infections	[39]
Paracoccidioides lutzi	Ajellomycetaceae	Onygenales	Various	[40]
Rasamsonia aegroticola	Aspergillaceae	Eurotiales	Pulmonary infections	[41,42]
Spiromastigoides albida	Spiromastigaceae	Onygenales	Lung biopsy	[43]

Table 2. Examples of fungal nomenclature changes in medically relevant fungi.

New Name	Previous Name	Family	Order	Reference
Yeasts				
Apiotrichum mycotoxinivorans	*Trichosporon mycotoxinivorans*	Trichosporonaceae	Trichosporonales	[44]
Candida duobushaemulonii	*Candida haemulonii* group II	Incertae sedis	Saccharomycetales	[45]
Kluyveromyces marxianus	*Candida kefyr*	Saccharomycetaceae	Saccharomycetales	[46]
Magnusiomyces capitatus	*Blastoschizomyces capitatus/Geotrichum capitatum*	Dipodascaceae	Saccharomycetales	[47]
Meyerozyma guilliermondii	*Candida guilliermondii*	Debaryomycetaceae	Saccharomycetales	[48]
Moulds				
Blastomyces helicus	*Emmonsia helica*	Onygenaceae	Onygenalses	[33]
Blastomyces parvus	*Emmonsia parva*	Onygenaceae	Onygenales	[33]
Curvularia australiensis	*Bipolaris australiensis*	Pleosporaceae	Pleosporales	[49]
Curvularia hawaiiensis	*Bipolaris hawaiiensis*	Pleosporaceae	Pleosporales	[49]
Curvularia spicifera	*Bipolaris spicifera*	Pleosporaceae	Pleosporales	[49]
Lichtheimia corymbifera	*Absidia corymbifera*	Lichtheimiaceae	Mucorales	[50]
Neocosmospora solani	*Fusarium solani*	Nectriaceae	Hypocreales	[51]
Purpureocillium lilacinum	*Paecilomyces lilacinus*	Ophiocordycipitaceae	Hypocreales	[52]
Aspergillus thermomutatus	*Neosartorya pseudofischeri*	Aspergillaceae	Eurotiales	[53,54]
Aspergillus udagawae	*Neosartorya udagawae*	Aspergillaceae	Eurotiales	[54]
Rasamsonia argillacea	*Geosmithia argillacea*	Aspergillaceae	Eurotiales	[37]
Scedosporium boydii	*Pseudallescheria boydii*	Microascaceae	Microascales	[55]
Verruconis gallopava	*Ochroconis gallopava*	Sympoventuriaceae	Venturiales	[56]
Talaromyces marneffei	*Penicillium marneffei*	Aspergillaceae	Eurotiales	[57]

3. Implications of Changes in Nomenclature for Medical Mycology

The abolishment of the dual nomenclature system and the introduction of molecular tools for species identification have implications for medical mycology. There is concern that these changes may lead to confusion in the clinical literature regarding the names of the organisms or the diseases they cause among clinicians who do not closely follow taxonomic changes but are still responsible for navigating the medical publications to find clinically useful information regarding invasive mycoses and their etiologic agents in order to optimize patient care [17]. In addition, there is no single source that can be used to stay abreast of changes in fungal taxonomy and literature, as descriptions of new species or revised classifications are published in various scientific journals [58], many of which lack clinical scope. Websites that serve as useful online repositories include Mycobank (http://www.mycobank.org) and Index Fungorum (http://www.indexfungorum.org). Other useful resources include the Westerdijk Fungal Biodiversity Institute (http://www.westerdijkinstitute.nl/), the Atlas of Clinical Fungi, (http://www.clinicalfungi.org/), The Yeasts website (http://theyeasts.org/) and the International Commission of *Penicillium* and *Aspergillus* (https://www.aspergilluspenicillium.org/).

4. Recommendations for Nomenclature Changes in Medical Mycology

The relevance of nomenclature changes to medical mycology is often unknown at first and only later once cryptic or sibling species have been further evaluated in in vitro studies, animal models, or with the publications of case reports, does the clinical significance, or lack thereof, become better understood [17,19]. Because of this and the confusion that may be present in the literature due to differences in fungal nomenclature used between the clinical and purely mycologic literature, the International Society for Human and Animal Mycology Working Group on Nomenclature of Medical Fungi has made recommendations on the adoption of new fungal names. In general, this group has proposed that the clinical arena be allowed to follow and adopt changes in nomenclature at a slower pace [17,19]. At the genus level and higher, the taxa with similar medical attributes/characteristics would be maintained and changes should be made once validated and a consensus is reached regarding new classifications and nomenclature. Taxa should not be too large as this could possibly conceal phenotypic differences of clinical importance. Conversely, taxa should not be too small as this would reduce the distinction between genus and species; however, monotypic genera do exist (e.g., *Epidermophyton* and *Lophophyton*) [19,59].

At the species level, the term species complex should be used to cover the name used in medical practice for a group of similar organisms when there is a lack of evidence of the clinical relevance of cryptic species. Once the significance of the cryptic species becomes known to the medical community, the new name can be adopted and used by clinical laboratories and medical mycologists.

5. Species Complexes

In the mycology literature there has been an increased use of the term species complex. However, there is no clear taxonomic definition/statute for this term and various authors have used it in different contexts [60]. Some have used it to describe a selected group of organisms that are difficult to differentiate between based on standard diagnostic means, including classic morphologic and other phenotypic characteristics and in some cases DNA barcode analysis using single targets [60]. An example of this is the *Aspergillus viridinutans* species complex within *Aspergillus* section *Fumigati*, which includes 10 closely related species, including the human and animal pathogens *A. udagawae*, *A. felis*, *A. pseudofelis*, *A. parafelis*, *A. pseudoviridinutans and A. wyomingensis* [61].

In contrast, others have used the term species complex as a substitute for the subgenus term section. Examples of this use can be found within the *Fusarium, Aspergillus and Trichoderma* genera [62,63]. Still others have used species complexes to group together well-described species for which there are no known or insignificant differences in clinical parameters. An example of this is the *Aspergillus niger* species complex, in which there is a lack in differences in antifungal susceptibility profiles between the various species [64]. Other examples of the use of this term in this fashion may include the *Coccidioides immitis* species complex [65,66], which is now recognized to consist of the separate species *C. immitis* and *C. posadasii* [67], the *Candida albicans* species complex, which known to consist of *C. albicans*, *C. africana* and *C. stellatoidea* [66,68–71] and the *Candida glabrata* species complex, consisting of *C. glabrata*, *C. nivariensis* and *C. bracarensis* [66,71–74]. Although *C. immitis* and *C. posadasii* may differ in their geographic distributions [75,76], no clinically relevant differences appear to exist between these two species. Interestingly, the term species complex has been used in the literature for the examples listed above, even when it is known that these consist of distinct species [60,66,71,74].

Lastly, species complex has also been employed to group together species when the taxonomy is unsettled or under debate in the literature. This has been proposed for seven separate *Cryptococcus* species (*Cryptococcus neoformans* species complex) [23], although this is still under debate and different groups have different opinions as to the lumping or splitting of these species [66,77]. Although there may be important phenotypic differences among the species, the clinical relevance of these is not fully understood. Another example where key phenotypic differences may exist but the clinical relevance is not fully known is the *Candida parapsilosis* species complex (*C. parapsilosis*, *C. metapsilosis* and *C. orthopsilosis*). The original species, *C. parapsilosis*, is known to have reduced in vitro susceptibility to the echinocandins [78], although patients often respond well to therapy [79–81]. It is now known that *C. metapsilosis* and *C. orthopsilosis* are hybrids and this may be of clinical relevance [82,83].

One way that clinical laboratories can use species complexes is in the reporting of preliminary microbiologic test results. If a preliminary identification of an isolate can be reported to a clinician at the species complex level, this information may be useful in making treatment decisions while further studies are performed to identify the exact species. Once the species is known, the final results should then be provided. However, clinicians should also be made aware that all species within a particular complex may not have the same antifungal susceptibility profiles. Thus, a full species identification should be provided if available. This information will then be available for clinicians, epidemiologists and other mycologists for further study.

6. Clinically Relevant Changes in Fungal Nomenclature and Current Controversies

Acute invasive aspergillosis and chronic pulmonary aspergillosis are primarily caused by the species *A. fumigatus*, *A. flavus*, *A. nidulans*, *A. niger* and *A. terreus* [84–86]. However, surveillance studies that have used molecular means of species identification have reported higher rates of cryptic species than previously appreciated. In the TRANSNET study, which included solid organ and hematopoietic stem cell transplant recipients in U.S. centers, 11% of the 218 *Aspergillus* species isolated were found to be cryptic species, including *A. lentulus* (1.8%) and *A. udagawae* (1.4%) from section *Fumigati*, *A. tubingensis* (2.8%) from section *Nigri* and *A. calidoustus* (2.8%) from section *Usti* [87]. Similarly, in the FILPOP study, a population-based survey study conducted in Spain, 14.5% of the *Aspergillus* isolates were considered to be cryptic species [86]. This may be of clinical importance as several cryptic species have reduced susceptibility to the azoles or multiple classes of clinically available antifungals. For example, section *Fumigati*, there are currently at least 63 phylogenetically distinct species, of which at least 19 have been reported to cause disease in humans and animals [88–90]. This includes several that were previously known as *Neosartorya* species, including *A. fischeri* (formerly *N. fischeri*), *A. hiratsukae* (formerly *N. hiratsukae*), *A. thermomutatus* (formerly *N. pseudofischeri*) and *A. udagawae* (formerly *N. udagawae*) [91–95]. The previously discussed *A. viridinutans* species complex also falls into section *Fumigati*. Although the importance of distinguishing between members of this complex in the clinical setting is unknown, it is important to know that the species causing infection falls within this complex as these species are often associated with chronic infections as well as reduced antifungal susceptibility and thus may be refractory to therapy [61].

Another group of clinically important fungi that has undergone major taxonomic and nomenclature changes over the last decade is that of *Scedosporium*. Previously, *Pseudallescheria boydii* and *Scedosporium apiospermum* were considered to be the same species and were identified by morphology in clinical microbiology laboratories as *P. boydii* (teleomorph) or *S. apiospermum* (anamorph) based on their ability to develop sexual structures on routine culture media. This changed when it was determined that *P. boydii* (anamorph *Scedosporium boydii*) and *Pseudallescheria apiosperma* (anamorph *S. apiospermum)* were separate species based on phylogenetic analysis [96,97]. Subsequently, other species that are morphologically identical but genetically different have been discovered through the use of molecular phylogenetics [55,96,98]. The *Scedosporium apiospermum* species complex is composed of *S. apiospermum*, *S. boydii* and *Pseudallescheria angusta* [60]. However, other *Scedosporium* species, including *S. aurantiacum, S. dehoogii* and *S. minutisporum,* have not been placed within this species complex due to clear phylogenetic differences among the species and those that comprise this group, as well as differences in antifungal susceptibility patterns [60,99]. The morphologically distinct species previously known as *Scedosporium prolificans* has been renamed *Lomentospora prolificans* based on significant phylogenetic differences [55]. As *L. prolificans* is highly resistant to multiple antifungals [99–104] and infections caused by this organism are extremely difficult to treat [100], the distinction between this species and those in the genus *Scedosporium* species is clinically relevant.

Recently, a revision to the taxonomy of *Cryptococcus* species that frequently cause disease in humans was proposed. In a study that included 115 isolates, *Cryptococcus neoformans* var. *grubii* and *Cryptococcus neoformans* var. *neoformans* were split into the separate species *Cryptococcus neoformans* and *Cryptococcus deneoformans*, respectively [23], while *Cryptococcus gattii* was proposed to be split into 5 distinct species (Table 3). This was based on the results from multi-locus sequence typing (MLST) based phylogenetic analysis using 11 different loci, differences in phenotypic characteristics and other means. Phenotypic characteristics that were evaluated in this study and others have included temperature, melanin content, virulence in a *Drosophila melanogaster* model, sensitivity to mycophenolic acid and growth on L-canavanine glycine bromothymol blue (CGB) agar and creatinine dextrose bromothymol blue thymine (CDBT) agar [23,66]. The authors also evaluated MALDI-TOF MS and reported that this technology could also readily distinguish between the different *Cryptococcus* species.

Table 3. Proposed names for *Cryptococcus neoformans* and *C. gattii* species [23].

Current Name	Molecular Type	Proposed Name
Cryptococcus neoformans var. *grubii*	VNI, VNII, VNB	*Cryptococcus neoformans*
Cryptococcus neoformans var. *neoformans*	VNIV	*Cryptococcus deneoformans*
Cryptococcus gattii	VGI	*Cryptococcus gattii*
	VGIII	*Cryptococcus bacillisporus*
	VGII	*Cryptococcus deuterogattii*
	VGIV	*Cryptococcus tetragattii*
	VGIV/VGIIIc	*Cryptococcus decagattii*
Serotypes AD hybrid	VNIII	*Cryptococcus neoformans* x *Cryptococcus deneoformans* hybrid
Serotypes DB hybrid	AFLP8	*Cryptococcus deneoformans* x *Cryptococcus gattii* hybrid
Serotypes AB hybrid	AFLP9	*Cryptococcus neoformans* x *Cryptococcus gattii* hybrid
Serotypes AB hybrid	AFLP11	*Cryptococcus neoformans* x *Cryptococcus deuterogattii* hybrid

This proposal to divide the *Cryptococcus neoformans/gattii* species complex into different species has not been without criticism. In an editorial, Kwon-Chung et al. argued that the proposed division was premature as an insufficient number of isolates were used to make this taxonomic change [77]. A previous, larger analysis including over 2000 isolates, had showed greater genetic diversity and the possibility of even more species [77]. In addition, since loci from only 6 of the 14 chromosomes in *Cryptococcus* were used in the MLST-based phylogenetic analysis, the true extent of diversity and recombination events remains unknown. It was also argued that the proposed division is impractical for routine use in clinical microbiology laboratories. Eleven concatenated loci were used in the phylogenetic analysis that supported separating the species and even the most commonly used MLST scheme of seven concatenated loci recommended by the ISHAM Genotyping Working Group of *C. neoformans* and *C. gattii* is too complicated for clinical microbiology laboratories and even reference laboratories, especially since the loci commonly used for molecular identification of fungal species (i.e., ITS and D1/D2) are not included [77,105]. In addition, the MALDI-TOF MS score threshold used was somewhat different than the usual score cutoff value for species recognition [23,77,106] and the newly proposed species are not currently available in databases cleared by regulatory agencies for use in clinical microbiology laboratories. It should be noted that other studies have reported that lower score thresholds can be used to reliably identify fungal species [107–109]. However, many clinical microbiology laboratories may be reluctant to use lower score thresholds without internal validation studies. Concern was also raised regarding the possible creation of confusion between the taxonomic and clinical literature. Specifically, under the proposed nomenclature the Vancouver *C. gattii* epidemic reference strain R265 would no longer be *C. gattii* but instead would be reclassified as *C. deuterogattii*. Although it was recognized that the designation of seven separate species would be an important step for the formal recognition of the biodiversity of pathogenic *Cryptococcus* species, Kwon-Chung et al. instead proposed the use of *Cryptococcus neoformans* species complex and *Cryptococcus gattii* species complex based on these issues and our current insufficient understanding of the clinical differences among the various proposed *Cryptococcus* species. In a rebuttal, Hagen et al. defended the nomenclature changes and noted that the main advantage will be the advancement of the field through stimulation of further studies to assess for similarities and differences between the recognized species [66]. Additional work has subsequently reported that the newly proposed *Cryptococcus* species may indeed have clinically significant differences [23,66,110,111]. However, many clinical microbiology laboratories may not be able to adapt to the new nomenclature into their routine workflow in the near future, as the new species are not yet incorporated into commercially available assays and databases cleared for clinical use by regulatory agencies for diagnosis or species identification.

Fusarium species are significant causes of invasive infections in highly immunocompromised hosts [112,113]. In addition, infections including keratitis and onychomycosis, can also occur in immunocompetent patients [27,114]. Human infections can be caused by species grouped within 8 different species complexes, including: *Fusarium solani* species complex, *Fusarium oxysporum* species complex, *Fusarium fujikuroi* species complex, *Fusarium chlamydosporum* species complex, *Fusarium dimerum* species complex, *Fusarium incarnatum-equiseti* species complex, *Fusarium sambucinum* species complex and *Fusarium tricinctum* species complex [25], although most infections are caused by members of the *F. solani* and *F. oxysporum* species complexes [112]. The *F. solani* species complex encompass at least 60 phylogenetically distinct species and in addition to causing disease in humans and animals, also includes a number of important agricultural pathogens [115]. Traditionally, clinical microbiology laboratories have identified and reported these isolates as *F. solani* species complex and some reference laboratories also report the specific haplotype based on MLST of the translation elongation factor 1α and RNA polymerase II gene. Although some members of this complex have received formal species names (e.g., *F. petroliphilum* [halplotype 1], *F. keratoplasticum* [haplotype 2], *F. falciforme* [haplotype 3+4] and *F. solani* [haplotype 5]) many others have not [116,117].

Recently, it has been proposed that members of the *F. solani* species complex be moved to the genus *Neocosmospora* based on the results of phylogenetic analysis [53] and new species previously classified only as haplotypes have been described [118]. This includes the species *N. petroliphila* (*F. petroliphilum*), *N. keratoplastica* (*F. keratoplasticum*), *N. falciformis* (*F. falciforme*) and *N. solani* (*F. solani*), along with the new species *N. gamsii* (haplotype 7), *N. suttoniana* (haplotype 20) and *N. catenata* (haplotype 43). Others have argued against renaming members of the *F. solani* species complex based on the long-standing, historical concept of this genus [119]. In order to provide up-to-date information to both clinicians and clinical microbiologists, as well as facilitate their ability to find relevant information in the medical literature, the reports generated by our reference mycology laboratory provide both the name commonly used in the medical literature as well as the new nomenclature. For example, for a recent species identification of an isolate cultured from the cornea of a patient, the name frequently found in the clinical literature, *Fusarium falciforme*, was provided along with a statement that the species is now known as *Neocosmospora falciformis*.

7. Summary

The field of medical mycology is rapidly changing due to the introduction of new molecular and proteomic technologies. New species are rapidly being discovered in the environment, as are new etiologic agents of disease in humans and animals. The adoption of these technologies, along with the abandonment of the dual nomenclature system, has led to marked changes in fungal taxonomy and nomenclature, as organisms previously thought to be unrelated are now recognized as being genetically similar. Conversely, we are now learning that species that were previously considered to be related are in fact very different from each other. The rapidity of these changes has caused concern among some medical mycologists and clinicians that the nomenclature changes may lead to negative clinical consequences, as the ever-changing literature could cause confusion among those who are responsible direct patient care. To mitigate this possibility, it has been proposed that medical mycology, specifically clinicians and clinical microbiology laboratories, may need to adopt changes in fungal nomenclature and taxonomy at a more measured pace. In addition, clinical microbiology and reference laboratories should provide useful information that will aide clinicians in this endeavor. This includes keeping abreast of changes in taxonomy and nomenclature, serving as a resource for clinicians as to previous names that may be published in the literature, as well as the clinical significance of the new classifications. An example of how the clinical laboratories may provide up-to-date information as well as species names that are prevalent in the clinical literature is provided above (*Neocosmospora falciformis* and *Fusarium falciforme*). The discovery of new fungal species capable of causing disease in humans and animals and the reclassification of various groups will continue as our knowledge of fungal diversity increases. However, this should not impede clinicians in their treatment of patients with fungal infections.

Author Contributions: Writing—original draft preparation, N.P.W; review and editing, N.P.W and C.F.C.G.

Funding: The authors received no funding for this work.

Conflicts of Interest: NPW has received research support to the UT Health San Antonio from Astellas, bioMerieux, Cidara, F2G, Merck, Pfizer and Viamet and has served on advisory boards for Astellas and Mayne Pharma and as a speaker for Gilead.

References

1. Bongomin, F.; Gago, S.; Oladele, R.O.; Denning, D.W. Global and multi-national prevalence of fungal diseases-estimate precision. *J. Fungi* **2017**, *3*, 57. [CrossRef] [PubMed]
2. Havlickova, B.; Czaika, V.A.; Friedrich, M. Epidemiological trends in skin mycoses worldwide. *Mycoses* **2008**, *51* (Suppl. 4), 2–15. [CrossRef] [PubMed]
3. Taylor, L.H.; Latham, S.M.; Woolhouse, M.E. Risk factors for human disease emergence. *Philos. Trans. R. Soc. Lond. B Biol. Sci.* **2001**, *356*, 983–989. [CrossRef] [PubMed]
4. Jones, N. Planetary disasters: It could happen one night. *Nature* **2013**, *493*, 154–156. [CrossRef] [PubMed]
5. O'Brien, H.E.; Parrent, J.L.; Jackson, J.A.; Moncalvo, J.M.; Vilgalys, R. Fungal community analysis by large-scale sequencing of environmental samples. *Appl. Environ. Microbiol.* **2005**, *71*, 5544–5550. [CrossRef] [PubMed]
6. Fisher, M.C.; Henk, D.A.; Briggs, C.J.; Brownstein, J.S.; Madoff, L.C.; McCraw, S.L.; Gurr, S.J. Emerging fungal threats to animal, plant and ecosystem health. *Nature* **2012**, *484*, 186–194. [CrossRef]
7. Kauffman, C.A.; Pappas, P.G.; Patterson, T.F. Fungal infections associated with contaminated methylprednisolone injections. *N. Engl. J. Med.* **2013**, *368*, 2495–2500. [CrossRef]

8. Nucci, M.; Akiti, T.; Barreiros, G.; Silveira, F.; Revankar, S.G.; Sutton, D.A.; Patterson, T.F. Nosocomial fungemia due to *Exophiala jeanselmei* var. *jeanselmei and a Rhinocladiella species: Newly described causes of bloodstream infection. J. Clin. Microbiol.* **2001**, *39*, 514–518.
9. Nucci, M.; Akiti, T.; Barreiros, G.; Silveira, F.; Revankar, S.G.; Wickes, B.L.; Sutton, D.A.; Patterson, T.F. Nosocomial outbreak of Exophiala jeanselmei fungemia associated with contamination of hospital water. *Clin. Infect. Dis.* **2002**, *34*, 1475–1480. [CrossRef]
10. Saracli, M.A.; Mutlu, F.M.; Yildiran, S.T.; Kurekci, A.E.; Gonlum, A.; Uysal, Y.; Erdem, U.; Basustaoglu, A.C.; Sutton, D.A. Clustering of invasive *Aspergillus ustus* eye infections in a tertiary care hospital: A molecular epidemiologic study of an uncommon species. *Med. Mycol.* **2007**, *45*, 377–384. [CrossRef]
11. Kohler, J.R.; Casadevall, A.; Perfect, J. The spectrum of fungi that infects humans. *Cold Spring Harb. Perspect. Med.* **2014**, *5*, a019273. [CrossRef] [PubMed]
12. Casadevall, A. Determinants of virulence in the pathogenic fungi. *Fungal Biol. Rev.* **2007**, *21*, 130–132. [CrossRef] [PubMed]
13. Odds, F.C. Ecology and epidemiology of Candida species. *Zentralbl. Bakteriol. Mikrobiol. Hyg. A* **1984**, *257*, 207–212. [PubMed]
14. Petti, C.A. Detection and identification of microorganisms by gene amplification and sequencing. *Clin. Infect. Dis.* **2007**, *44*, 1108–1114. [PubMed]
15. Schoch, C.L.; Seifert, K.A.; Huhndorf, S.; Robert, V.; Spouge, J.L.; Levesque, C.A.; Chen, W.; Bolchacova, E.; Voigt, K.; Crous, P.W.; et al. Nuclear ribosomal internal transcribed spacer (ITS) region as a universal DNA barcode marker for Fungi. *Proc. Natl. Acad. Sci. USA* **2012**, *109*, 6241–6246. [CrossRef] [PubMed]
16. Seifert, K.A. Progress towards DNA barcoding of fungi. *Mol. Ecol. Resour.* **2009**, *9* (Suppl. 1), 83–89. [CrossRef] [PubMed]
17. De Hoog, G.S.; Chaturvedi, V.; Denning, D.W.; Dyer, P.S.; Frisvad, J.C.; Geiser, D.; Graser, Y.; Guarro, J.; Haase, G.; Kwon-Chung, K.J.; et al. Name changes in medically important fungi and their implications for clinical practice. *J. Clin. Microbiol.* **2015**, *53*, 1056–1062. [CrossRef]
18. Howard, S.J. Multi-resistant aspergillosis due to cryptic species. *Mycopathologia* **2014**, *178*, 435–439. [CrossRef]
19. De Hoog, G.S.; Haase, G.; Chaturvedi, V.; Walsh, T.J.; Meyer, W.; Lackner, M. Taxonomy of medically important fungi in the molecular era. *Lancet Infect. Dis.* **2013**, *13*, 385–386. [CrossRef]
20. Rychert, J.; Slechta, E.S.; Barker, A.P.; Miranda, E.; Babady, N.E.; Tang, Y.W.; Gibas, C.; Wiederhold, N.; Sutton, D.; Hanson, K.E. Multicenter Evaluation of the Vitek MS v3.0 System for the Identification of Filamentous Fungi. *J. Clin. Microbiol.* **2018**, *56*, e01353-17. [CrossRef]
21. Lau, A.F.; Drake, S.K.; Calhoun, L.B.; Henderson, C.M.; Zelazny, A.M. Development of a clinically comprehensive database and a simple procedure for identification of molds from solid media by matrix-assisted laser desorption ionization-time of flight mass spectrometry. *J. Clin. Microbiol.* **2013**, *51*, 828–834. [CrossRef] [PubMed]
22. Brun, S.; Madrid, H.; Gerrits Van Den Ende, B.; Andersen, B.; Marinach-Patrice, C.; Mazier, D.; De Hoog, G.S. Multilocus phylogeny and MALDI-TOF analysis of the plant pathogenic species *Alternaria dauci* and relatives. *Fungal Biol.* **2013**, *117*, 32–40. [CrossRef] [PubMed]
23. Hagen, F.; Khayhan, K.; Theelen, B.; Kolecka, A.; Polacheck, I.; Sionov, E.; Falk, R.; Parnmen, S.; Lumbsch, H.T.; Boekhout, T. Recognition of seven species in the *Cryptococcus gattii/Cryptococcus neoformans* species complex. *Fungal Genet. Biol.* **2015**, *78*, 16–48. [CrossRef] [PubMed]
24. Zhang, Y.; Hagen, F.; Stielow, B.; Rodrigues, A.M.; Samerpitak, K.; Zhou, X.; Feng, P.; Yang, L.; Chen, M.; Deng, S.; et al. Phylogeography and evolutionary patterns in *Sporothrix* spanning more than 14000 human and animal case reports. *Persoonia* **2015**, *35*, 1–20. [CrossRef] [PubMed]
25. O'Donnell, K.; Sarver, B.A.; Brandt, M.; Chang, D.C.; Noble-Wang, J.; Park, B.J.; Sutton, D.A.; Benjamin, L.; Lindsley, M.; Padhye, A.; et al. Phylogenetic diversity and microsphere array-based genotyping of human pathogenic Fusaria, including isolates from the multistate contact lens-associated U.S. keratitis outbreaks of 2005 and 2006. *J. Clin. Microbiol.* **2007**, *45*, 2235–2248. [CrossRef] [PubMed]
26. Short, D.P.; O'Donnell, K.; Thrane, U.; Nielsen, K.F.; Zhang, N.; Juba, J.H.; Geiser, D.M. Phylogenetic relationships among members of the *Fusarium solani* species complex in human infections and the descriptions of *F. keratoplasticum* sp. nov. and *F. petroliphilum* stat. nov. *Fungal Genet. Biol.* **2013**, *53*, 59–70. [CrossRef] [PubMed]

27. Chang, D.C.; Grant, G.B.; O'Donnell, K.; Wannemuehler, K.A.; Noble-Wang, J.; Rao, C.Y.; Jacobson, L.M.; Crowell, C.S.; Sneed, R.S.; Lewis, F.M.; et al. Multistate outbreak of Fusarium keratitis associated with use of a contact lens solution. *JAMA* **2006**, *296*, 953–963. [CrossRef]

28. Marimon, R.; Cano, J.; Gene, J.; Sutton, D.A.; Kawasaki, M.; Guarro, J. *Sporothrix brasiliensis, S. globosa,* and *S. mexicana,* three new *Sporothrix* species of clinical interest. *J. Clin. Microbiol.* **2007**, *45*, 3198–3206. [CrossRef]

29. Hawksworth, D.L. A new dawn for the naming of fungi: Impacts of decisions made in Melbourne in July 2011 on the future publication and regulation of fungal names. *IMA Fungus* **2011**, *2*, 155–162. [CrossRef]

30. Norvell, L.L. Melbourne approves a new CODE. *Mycotaxon* **2011**, *116*, 481–490. [CrossRef]

31. Bonifaz, A.; Stchigel, A.M.; Guarro, J.; Guevara, E.; Pintos, L.; Sanchis, M.; Cano-Lira, J.F. Primary cutaneous mucormycosis produced by the new species *Apophysomyces mexicanus. J. Clin. Microbiol.* **2014**, *52*, 4428–4431. [CrossRef] [PubMed]

32. Guinea, J.; Sandoval-Denis, M.; Escribano, P.; Pelaez, T.; Guarro, J.; Bouza, E. Aspergillus citrinoterreus, a new species of section Terrei isolated from samples of patients with nonhematological predisposing conditions. *J. Clin. Microbiol.* **2015**, *53*, 611–617. [CrossRef] [PubMed]

33. Siqueira, J.P.Z.; Wiederhold, N.; Gene, J.; Garcia, D.; Almeida, M.T.G.; Guarro, J. Cryptic *Aspergillus* from clinical samples in the USA and description of a new species in section *Flavipedes. Mycoses* **2018**, *61*, 814–825. [CrossRef] [PubMed]

34. Sugui, J.A.; Peterson, S.W.; Clark, L.P.; Nardone, G.; Folio, L.; Riedlinger, G.; Zerbe, C.S.; Shea, Y.; Henderson, C.M.; Zelazny, A.M.; et al. *Aspergillus tanneri* sp. nov., a new pathogen that causes invasive disease refractory to antifungal therapy. *J. Clin. Microbiol.* **2012**, *50*, 3309–3317. [CrossRef] [PubMed]

35. Satoh, K.; Makimura, K.; Hasumi, Y.; Nishiyama, Y.; Uchida, K.; Yamaguchi, H. *Candida auris* sp. nov., a novel ascomycetous yeast isolated from the external ear canal of an inpatient in a Japanese hospital. *Microbiol. Immunol.* **2009**, *53*, 41–44. [CrossRef]

36. Madrid, H.; da Cunha, K.C.; Gene, J.; Dijksterhuis, J.; Cano, J.; Sutton, D.A.; Guarro, J.; Crous, P.W. Novel Curvularia species from clinical specimens. *Persoonia* **2014**, *33*, 48–60. [CrossRef]

37. Jiang, Y.; Dukik, K.; Muñoz, J.F.; Sigler, L.; Schwartz, I.S.; Govender, N.P.; Kenyon, C.; Feng, P.; van den Ende, B.G.; Stielow, J.B.; et al. Phylogeny, ecology and taxonomy of systemic pathogens and their relatives in *Ajellomycetaceae* (Onygenales): *Blastomyces, Emergomyces, Emmonsia, Emmonsiellopsis. Fungal Divers.* **2018**, *90*, 245–291. [CrossRef]

38. Schwartz, I.S.; Sanche, S.; Wiederhold, N.P.; Patterson, T.F.; Sigler, L. *Emergomyces canadensis,* a Dimorphic Fungus Causing Fatal Systemic Human Disease in North America. *Emerg. Infect. Dis.* **2018**, *24*, 758–761. [CrossRef]

39. Yong, L.K.; Wiederhold, N.P.; Sutton, D.A.; Sandoval-Denis, M.; Lindner, J.R.; Fan, H.; Sanders, C.; Guarro, J. Morphological and Molecular Characterization of *Exophiala polymorpha* sp. nov. Isolated from Sporotrichoid Lymphocutaneous Lesions in a Patient with Myasthenia Gravis. *J. Clin. Microbiol.* **2015**, *53*, 2816–2822. [CrossRef]

40. Teixeira, M.M.; Theodoro, R.C.; Nino-Vega, G.; Bagagli, E.; Felipe, M.S. Paracoccidioides species complex: Ecology, phylogeny, sexual reproduction, and virulence. *PLoS Pathog.* **2014**, *10*, e1004397. [CrossRef]

41. Houbraken, J.; Giraud, S.; Meijer, M.; Bertout, S.; Frisvad, J.C.; Meis, J.F.; Bouchara, J.P.; Samson, R.A. Taxonomy and antifungal susceptibility of clinically important Rasamsonia species. *J. Clin. Microbiol.* **2013**, *51*, 22–30. [CrossRef] [PubMed]

42. Hong, G.; White, M.; Lechtzin, N.; West, N.E.; Avery, R.; Miller, H.; Lee, R.; Lovari, R.J.; Massire, C.; Blyn, L.B.; et al. Fatal disseminated Rasamsonia infection in cystic fibrosis post-lung transplantation. *J. Cyst. Fibros* **2017**, *16*, e3–e7. [CrossRef] [PubMed]

43. Stchigel, A.M.; Sutton, D.A.; Cano-Lira, J.F.; Wiederhold, N.; Guarro, J. New Species Spiromastigoides albida from a Lung Biopsy. *Mycopathologia* **2017**, *182*, 967–978. [CrossRef] [PubMed]

44. Liu, X.Z.; Wang, Q.M.; Goker, M.; Groenewald, M.; Kachalkin, A.V.; Lumbsch, H.T.; Millanes, A.M.; Wedin, M.; Yurkov, A.M.; Boekhout, T.; et al. Towards an integrated phylogenetic classification of the Tremellomycetes. *Stud. Mycol.* **2015**, *81*, 85–147. [CrossRef]

45. Cendejas-Bueno, E.; Kolecka, A.; Alastruey-Izquierdo, A.; Theelen, B.; Groenewald, M.; Kostrzewa, M.; Cuenca-Estrella, M.; Gomez-Lopez, A.; Boekhout, T. Reclassification of the *Candida haemulonii* complex as *Candida haemulonii* (*C. haemulonii* group I), *C. duobushaemulonii* sp. nov. (*C. haemulonii* group II), and *C. haemulonii* var. vulnera var. nov.: Three multiresistant human pathogenic yeasts. *J. Clin. Microbiol.* **2012**, *50*, 3641–3651. [CrossRef]

46. Van der Walt, J.P. The emendation of the genus *Kluyveromyces* v. d. Walt. *Antonie Van Leeuwenhoek* **1965**, *31*, 341–348. [CrossRef]

47. De Hoog, G.S.; Smith, M.T. Ribsomal gene phylogeny and species delimitation in Geotrichum and its teleomorphs. *Stud. Mycol.* **2004**, *50*, 489–515.

48. Kurtzman, C.P.; Suzuki, M. Phylogenetic analysis of ascomycete yeasts that form coenzyme Q-9 and the proposal of the new genera *Babjeviella*, *Meyerozyma*, *Millerozyma*, *Priceomyces*, and *Scheffersomyces*. *Mycoscience* **2010**, *51*, 2–14. [CrossRef]

49. Manamgoda, D.S.; Cai, L.; McKenzie, H.C.; Crous, P.W.; Madrid, H.; Chikeatirote, E.; Shivas, R.G.; Tan, Y.P.; Hyde, K.D. A phylogenetic and taxonomic re-evaluation of the *Bipolaris-Cochliobolus-Curvularia* complex. *Fungal Divers.* **2012**, *56*, 131–144. [CrossRef]

50. Alastruey-Izquierdo, A.; Hoffmann, K.; de Hoog, G.S.; Rodriguez-Tudela, J.L.; Voigt, K.; Bibashi, E.; Walther, G. Species recognition and clinical relevance of the zygomycetous genus *Lichtheimia* (syn. *Absidia pro parte, Mycocladus)*. *J. Clin. Microbiol.* **2010**, *48*, 2154–2170. [CrossRef]

51. Lombard, L.; van der Merwe, N.A.; Groenewald, J.Z.; Crous, P.W. Generic concepts in Nectriaceae. *Stud. Mycol.* **2015**, *80*, 189–245. [CrossRef] [PubMed]

52. Luangsa-Ard, J.; Houbraken, J.; van Doorn, T.; Hong, S.B.; Borman, A.M.; Hywel-Jones, N.L.; Samson, R.A. Purpureocillium, a new genus for the medically important *Paecilomyces lilacinus*. *FEMS Microbiol. Lett.* **2011**, *321*, 141–149. [CrossRef] [PubMed]

53. Peterson, S.W. *Neosartorya pseudofischeri* sp. nov. and its relationship to other species in *Aspergillus* section *Fumigati*. *Mycol. Res.* **1992**, *96*, 547–554. [CrossRef]

54. Samson, R.A.; Visagie, C.M.; Houbraken, J.; Hong, S.B.; Hubka, V.; Klaassen, C.H.; Perrone, G.; Seifert, K.A.; Susca, A.; Tanney, J.B.; et al. Phylogeny, identification and nomenclature of the genus *Aspergillus*. *Stud. Mycol.* **2014**, *78*, 141–173. [CrossRef] [PubMed]

55. Lackner, M.; de Hoog, G.S.; Yang, L.; Moreno, L.F.; Ahmed, S.A.; Andreas, F.; Kaltseis, J.; Nagl, M.; Lass-Florl, C. Proposed nomenclature for *Pseudallescheria, Scedosporium* and related genera. *Fungal Divers.* **2014**, *67*, 1–10. [CrossRef]

56. Samerpitak, K.; Van der Linde, E.; Choi, H.J.; Van den Ende, B.G.; Machouart, M.; Gueidan, C.; De Hoog, G.S. Taxonomy of Ochroconis, genus including opportunistic pathogens on humans and animals. *Fungal Divers.* **2014**, *65*, 89–126. [CrossRef]

57. Samson, R.A.; Yilmaz, N.; Houbraken, J.; Spierenburg, H.; Seifert, K.A.; Peterson, S.W.; Varga, J.; Frisvad, J.C. Phylogeny and nomenclature of the genus *Talaromyces* and taxa accommodated in *Penicillium* subgenus *Biverticillium*. *Stud. Mycol.* **2011**, *70*, 159–183. [CrossRef]

58. Warnock, D.W. Name Changes for Fungi of Medical Importance, 2012 to 2015. *J. Clin. Microbiol.* **2017**, *55*, 53–59. [CrossRef]

59. De Hoog, G.S.; Dukik, K.; Monod, M.; Packeu, A.; Stubbe, D.; Hendrickx, M.; Kupsch, C.; Stielow, J.B.; Freeke, J.; Goker, M.; et al. Toward a Novel Multilocus Phylogenetic Taxonomy for the Dermatophytes. *Mycopathologia* **2017**, *182*, 5–31. [CrossRef]

60. Chen, M.; Zeng, J.; De Hoog, G.S.; Stielow, B.; Gerrits Van Den Ende, A.H.; Liao, W.; Lackner, M. The 'species complex' issue in clinically relevant fungi: A case study in *Scedosporium apiospermum*. *Fungal Biol.* **2016**, *120*, 137–146. [CrossRef]

61. Talbot, J.J.; Barrs, V.R. One-health pathogens in the *Aspergillus viridinutans* complex. *Med. Mycol.* **2018**, *56*, 1–12. [CrossRef]

62. Balajee, S.A.; Houbraken, J.; Verweij, P.E.; Hong, S.B.; Yaghuchi, T.; Varga, J.; Samson, R.A. *Aspergillus* species identification in the clinical setting. *Stud. Mycol.* **2007**, *59*, 39–46. [CrossRef]

63. Chaverri, P.; Branco-Rocha, F.; Jaklitsch, W.; Gazis, R.; Degenkolb, T.; Samuels, G.J. Systematics of the *Trichoderma harzianum* species complex and the re-identification of commercial biocontrol strains. *Mycologia* **2015**, *107*, 558–590. [CrossRef] [PubMed]

64. Howard, S.J.; Harrison, E.; Bowyer, P.; Varga, J.; Denning, D.W. Cryptic species and azole resistance in the *Aspergillus niger* complex. *Antimicrob. Agents Chemother.* **2011**, *55*, 4802–4809. [CrossRef]

65. Koufopanou, V.; Burt, A.; Szaro, T.; Taylor, J.W. Gene genealogies, cryptic species, and molecular evolution in the human pathogen *Coccidioides immitis* and relatives (Ascomycota, Onygenales). *Mol. Biol. Evol.* **2001**, *18*, 1246–1258. [CrossRef] [PubMed]

66. Hagen, F.; Lumbsch, H.T.; Arsic Arsenijevic, V.; Badali, H.; Bertout, S.; Billmyre, R.B.; Bragulat, M.R.; Cabanes, F.J.; Carbia, M.; Chakrabarti, A.; et al. Importance of Resolving Fungal Nomenclature: The Case of Multiple Pathogenic Species in the *Cryptococcus* Genus. *mSphere* **2017**, *2*, e00238-17. [CrossRef] [PubMed]

67. Fisher, M.C.; Koenig, G.L.; White, T.J.; Taylor, J.W. Molecular and phenotypic description of *Coccidioides posadasii* sp. nov., previously recognized as the non-California population of *Coccidioides immitis*. *Mycologia* **2002**, *94*, 73–84. [CrossRef] [PubMed]

68. Chowdhary, A.; Hagen, F.; Sharma, C.; Al-Hatmi, A.M.S.; Giuffre, L.; Giosa, D.; Fan, S.; Badali, H.; Felice, M.R.; de Hoog, S.; et al. Whole Genome-Based Amplified Fragment Length Polymorphism Analysis Reveals Genetic Diversity in *Candida africana*. *Front. Microbiol.* **2017**, *8*, 556. [CrossRef] [PubMed]

69. Tietz, H.J.; Hopp, M.; Schmalreck, A.; Sterry, W.; Czaika, V. *Candida africana* sp. nov., a new human pathogen or a variant of *Candida albicans*? *Mycoses* **2001**, *44*, 437–445. [CrossRef]

70. Romeo, O.; Criseo, G. First molecular method for discriminating between *Candida africana*, *Candida albicans*, and *Candida dubliniensis* by using *hwp1* gene. *Diagn. Microbiol. Infect. Dis.* **2008**, *62*, 230–233. [CrossRef]

71. Arastehfar, A.; Fang, W.; Pan, W.; Liao, W.; Yan, L.; Boekhout, T. Identification of nine cryptic species of *Candida albicans*, *C. glabrata*, and *C. parapsilosis* complexes using one-step multiplex PCR. *BMC Infect. Dis.* **2018**, *18*, 480. [CrossRef] [PubMed]

72. Alcoba-Florez, J.; Mendez-Alvarez, S.; Cano, J.; Guarro, J.; Perez-Roth, E.; del Pilar Arevalo, M. Phenotypic and molecular characterization of *Candida nivariensis* sp. nov., a possible new opportunistic fungus. *J. Clin. Microbiol.* **2005**, *43*, 4107–4111. [CrossRef] [PubMed]

73. Correia, A.; Sampaio, P.; James, S.; Pais, C. *Candida bracarensis* sp. nov., a novel anamorphic yeast species phenotypically similar to *Candida glabrata*. *Int. J. Syst. Evol. Microbiol.* **2006**, *56*, 313–317. [CrossRef] [PubMed]

74. Lockhart, S.R.; Messer, S.A.; Gherna, M.; Bishop, J.A.; Merz, W.G.; Pfaller, M.A.; Diekema, D.J. Identification of *Candida nivariensis* and *Candida bracarensis* in a large global collection of *Candida glabrata* isolates: Comparison to the literature. *J. Clin. Microbiol.* **2009**, *47*, 1216–1217. [CrossRef] [PubMed]

75. Hirschmann, J.V. The early history of coccidioidomycosis: 1892–1945. *Clin. Infect. Dis.* **2007**, *44*, 1202–1207. [CrossRef] [PubMed]

76. Brown, J.; Benedict, K.; Park, B.J.; Thompson, G.R., III. Coccidioidomycosis: Epidemiology. *Clin. Epidemiol.* **2013**, *5*, 185–197.

77. Kwon-Chung, K.J.; Bennett, J.E.; Wickes, B.L.; Meyer, W.; Cuomo, C.A.; Wollenburg, K.R.; Bicanic, T.A.; Castaneda, E.; Chang, Y.C.; Chen, J.; et al. The Case for Adopting the "Species Complex" Nomenclature for the Etiologic Agents of Cryptococcosis. *mSphere* **2017**, *2*, e00357-16. [CrossRef]

78. Trevino-Rangel Rde, J.; Garza-Gonzalez, E.; Gonzalez, J.G.; Bocanegra-Garcia, V.; Llaca, J.M.; Gonzalez, G.M. Molecular characterization and antifungal susceptibility of the *Candida parapsilosis* species complex of clinical isolates from Monterrey, Mexico. *Med. Mycol.* **2012**, *50*, 781–784. [CrossRef]

79. Mora-Duarte, J.; Betts, R.; Rotstein, C.; Colombo, A.L.; Thompson-Moya, L.; Smietana, J.; Lupinacci, R.; Sable, C.; Kartsonis, N.; Perfect, J.; et al. Comparison of caspofungin and amphotericin B for invasive candidiasis. *N. Engl. J. Med.* **2002**, *347*, 2020–2029. [CrossRef]

80. Kuse, E.R.; Chetchotisakd, P.; da Cunha, C.A.; Ruhnke, M.; Barrios, C.; Raghunadharao, D.; Sekhon, J.S.; Freire, A.; Ramasubramanian, V.; Demeyer, I.; et al. Micafungin versus liposomal amphotericin B for candidaemia and invasive candidosis: A phase III randomised double-blind trial. *Lancet* **2007**, *369*, 1519–1527. [CrossRef]

81. Reboli, A.C.; Rotstein, C.; Pappas, P.G.; Chapman, S.W.; Kett, D.H.; Kumar, D.; Betts, R.; Wible, M.; Goldstein, B.P.; Schranz, J.; et al. Anidulafungin versus fluconazole for invasive candidiasis. *N. Engl. J. Med.* **2007**, *356*, 2472–2482. [CrossRef]

82. Pryszcz, L.P.; Nemeth, T.; Gacser, A.; Gabaldon, T. Genome comparison of *Candida orthopsilosis* clinical strains reveals the existence of hybrids between two distinct subspecies. *Genome Biol. Evol.* **2014**, *6*, 1069–1078. [CrossRef] [PubMed]

83. Pryszcz, L.P.; Nemeth, T.; Saus, E.; Ksiezopolska, E.; Hegedusova, E.; Nosek, J.; Wolfe, K.H.; Gacser, A.; Gabaldon, T. The Genomic Aftermath of Hybridization in the Opportunistic Pathogen *Candida metapsilosis*. *PLoS Genet.* **2015**, *11*, e1005626. [CrossRef]

84. Pappas, P.G.; Alexander, B.D.; Andes, D.R.; Hadley, S.; Kauffman, C.A.; Freifeld, A.; Anaissie, E.J.; Brumble, L.M.; Herwaldt, L.; Ito, J.; et al. Invasive fungal infections among organ transplant recipients: Results of the Transplant-Associated Infection Surveillance Network (TRANSNET). *Clin. Infect. Dis.* **2010**, *50*, 1101–1111. [CrossRef]

85. Kontoyiannis, D.P.; Marr, K.A.; Park, B.J.; Alexander, B.D.; Anaissie, E.J.; Walsh, T.J.; Ito, J.; Andes, D.R.; Baddley, J.W.; Brown, J.M.; et al. Prospective surveillance for invasive fungal infections in hematopoietic stem cell transplant recipients, 2001–2006: Overview of the Transplant-Associated Infection Surveillance Network (TRANSNET) Database. *Clin. Infect. Dis.* **2010**, *50*, 1091–1100. [CrossRef] [PubMed]

86. Alastruey-Izquierdo, A.; Mellado, E.; Pelaez, T.; Peman, J.; Zapico, S.; Alvarez, M.; Rodriguez-Tudela, J.L.; Cuenca-Estrella, M.; Group, F.S. Population-based survey of filamentous fungi and antifungal resistance in Spain (FILPOP Study). *Antimicrob. Agents Chemother.* **2013**, *57*, 3380–3387. [CrossRef]

87. Balajee, S.A.; Kano, R.; Baddley, J.W.; Moser, S.A.; Marr, K.A.; Alexander, B.D.; Andes, D.; Kontoyiannis, D.P.; Perrone, G.; Peterson, S.; et al. Molecular identification of *Aspergillus* species collected for the Transplant-Associated Infection Surveillance Network. *J. Clin. Microbiol.* **2009**, *47*, 3138–3141. [CrossRef]

88. Frisvad, J.C.; Larsen, T.O. Extrolites of *Aspergillus fumigatus* and Other Pathogenic Species in *Aspergillus* Section *Fumigati*. *Front. Microbiol.* **2015**, *6*, 1485. [CrossRef] [PubMed]

89. Barrs, V.R.; Beatty, J.A.; Dhand, N.K.; Talbot, J.J.; Bell, E.; Abraham, L.A.; Chapman, P.; Bennett, S.; van Doorn, T.; Makara, M. Computed tomographic features of feline sino-nasal and sino-orbital aspergillosis. *Vet. J.* **2014**, *201*, 215–222. [CrossRef] [PubMed]

90. Sugui, J.A.; Peterson, S.W.; Figat, A.; Hansen, B.; Samson, R.A.; Mellado, E.; Cuenca-Estrella, M.; Kwon-Chung, K.J. Genetic relatedness versus biological compatibility between *Aspergillus fumigatus* and related species. *J. Clin. Microbiol.* **2014**, *52*, 3707–3721. [CrossRef] [PubMed]

91. Balajee, S.A.; Gribskov, J.; Brandt, M.; Ito, J.; Fothergill, A.; Marr, K.A. Mistaken identity: *Neosartorya pseudofischeri* and its anamorph masquerading as *Aspergillus fumigatus*. *J. Clin. Microbiol.* **2005**, *43*, 5996–5999. [CrossRef] [PubMed]

92. Matsumoto, N.; Shiraga, H.; Takahashi, K.; Kikuchi, K.; Ito, K. Successful treatment of *Aspergillus* peritonitis in a peritoneal dialysis patient. *Pediatr. Nephrol.* **2002**, *17*, 243–245. [CrossRef]

93. Jarv, H.; Lehtmaa, J.; Summerbell, R.C.; Hoekstra, E.S.; Samson, R.A.; Naaber, P. Isolation of *Neosartorya pseudofischeri* from blood: First hint of pulmonary Aspergillosis. *J. Clin. Microbiol.* **2004**, *42*, 925–928. [CrossRef] [PubMed]

94. Zbinden, A.; Imhof, A.; Wilhelm, M.J.; Ruschitzka, F.; Wild, P.; Bloemberg, G.V.; Mueller, N.J. Fatal outcome after heart transplantation caused by *Aspergillus lentulus*. *Transpl. Infect. Dis. Off. J. Transplant. Soc.* **2012**, *14*, E60–E63. [CrossRef] [PubMed]

95. Ghebremedhin, B.; Bluemel, A.; Neumann, K.H.; Koenig, B.; Koenig, W. Peritonitis due to *Neosartorya pseudofischeri* in an elderly patient undergoing peritoneal dialysis successfully treated with voriconazole. *J. Med. Microbiol.* **2009**, *58*, 678–682. [CrossRef] [PubMed]

96. Gilgado, F.; Cano, J.; Gene, J.; Sutton, D.A.; Guarro, J. Molecular and phenotypic data supporting distinct species statuses for *Scedosporium apiospermum* and *Pseudallescheria boydii* and the proposed new species *Scedosporium dehoogii*. *J. Clin. Microbiol.* **2008**, *46*, 766–771. [CrossRef]

97. Gilgado, F.; Gene, J.; Cano, J.; Guarro, J. Heterothallism in *Scedosporium apiospermum* and description of its teleomorph *Pseudallescheria apiosperma* sp. nov. *Med. Mycol.* **2010**, *48*, 122–128. [CrossRef] [PubMed]

98. Gilgado, F.; Cano, J.; Gene, J.; Guarro, J. Molecular phylogeny of the *Pseudallescheria boydii* species complex: Proposal of two new species. *J. Clin. Microbiol.* **2005**, *43*, 4930–4942. [CrossRef] [PubMed]

99. Lackner, M.; de Hoog, G.S.; Verweij, P.E.; Najafzadeh, M.J.; Curfs-Breuker, I.; Klaassen, C.H.; Meis, J.F. Species-specific antifungal susceptibility patterns of *Scedosporium* and *Pseudallescheria* species. *Antimicrob. Agents Chemother.* **2012**, *56*, 2635–2642. [CrossRef] [PubMed]

100. Cortez, K.J.; Roilides, E.; Quiroz-Telles, F.; Meletiadis, J.; Antachopoulos, C.; Knudsen, T.; Buchanan, W.; Milanovich, J.; Sutton, D.A.; Fothergill, A.; et al. Infections caused by *Scedosporium* spp. *Clin. Microbiol. Rev.* **2008**, *21*, 157–197. [CrossRef]

101. Walsh, T.J.; Groll, A.; Hiemenz, J.; Fleming, R.; Roilides, E.; Anaissie, E. Infections due to emerging and uncommon medically important fungal pathogens. *Clin. Microbiol. Infect. Off. Publ. Eur. Soc. Clin. Microbiol. Infect. Dis.* **2004**, *10* (Suppl. 1), 48–66. [CrossRef]

102. Lackner, M.; Hagen, F.; Meis, J.F.; Gerrits van den Ende, A.H.; Vu, D.; Robert, V.; Fritz, J.; Moussa, T.A.; de Hoog, G.S. Susceptibility and diversity in the therapy-refractory genus scedosporium. *Antimicrob. Agents Chemother.* **2014**, *58*, 5877–5885. [CrossRef] [PubMed]

103. Lewis, R.E.; Wiederhold, N.P.; Klepser, M.E. In vitro pharmacodynamics of amphotericin B, itraconazole, and voriconazole against *Aspergillus*, *Fusarium*, and *Scedosporium* spp. *Antimicrob. Agents Chemother.* **2005**, *49*, 945–951. [CrossRef] [PubMed]

104. Wiederhold, N.P.; Lewis, R.E. Antifungal activity against *Scedosporium* species and novel assays to assess antifungal pharmacodynamics against filamentous fungi. *Med. Mycol.* **2009**, *47*, 422–432. [CrossRef] [PubMed]

105. Meyer, W.; Aanensen, D.M.; Boekhout, T.; Cogliati, M.; Diaz, M.R.; Esposto, M.C.; Fisher, M.; Gilgado, F.; Hagen, F.; Kaocharoen, S.; et al. Consensus multi-locus sequence typing scheme for *Cryptococcus neoformans* and *Cryptococcus gattii*. *Med. Mycol.* **2009**, *47*, 561–570. [CrossRef]

106. Firacative, C.; Trilles, L.; Meyer, W. MALDI-TOF MS enables the rapid identification of the major molecular types within the *Cryptococcus neoformans/C. gattii* species complex. *PLoS ONE* **2012**, *7*, e37566. [CrossRef] [PubMed]

107. Normand, A.C.; Cassagne, C.; Gautier, M.; Becker, P.; Ranque, S.; Hendrickx, M.; Piarroux, R. Decision criteria for MALDI-TOF MS-based identification of filamentous fungi using commercial and in-house reference databases. *BMC Microbiol.* **2017**, *17*, 25. [CrossRef]

108. Van Herendael, B.H.; Bruynseels, P.; Bensaid, M.; Boekhout, T.; De Baere, T.; Surmont, I.; Mertens, A.H. Validation of a modified algorithm for the identification of yeast isolates using matrix-assisted laser desorption/ionisation time-of-flight mass spectrometry (MALDI-TOF MS). *Eur. J. Clin. Microbiol. Infect. Dis.* **2012**, *31*, 841–848. [CrossRef]

109. Vlek, A.; Kolecka, A.; Khayhan, K.; Theelen, B.; Groenewald, M.; Boel, E.; Multicenter Study, G.; Boekhout, T. Interlaboratory comparison of sample preparation methods, database expansions, and cutoff values for identification of yeasts by matrix-assisted laser desorption ionization-time of flight mass spectrometry using a yeast test panel. *J. Clin. Microbiol.* **2014**, *52*, 3023–3029. [CrossRef]

110. Herkert, P.F.; Dos Santos, J.C.; Hagen, F.; Ribeiro-Dias, F.; Queiroz-Telles, F.; Netea, M.G.; Meis, J.F.; Joosten, L.A.B. Differential In Vitro Cytokine Induction by the Species of *Cryptococcus gattii* Complex. *Infect. Immun.* **2018**, *86*. [CrossRef]

111. Nyazika, T.K.; Hagen, F.; Meis, J.F.; Robertson, V.J. *Cryptococcus tetragattii* as a major cause of cryptococcal meningitis among HIV-infected individuals in Harare, Zimbabwe. *J. Infect.* **2016**, *72*, 745–752. [CrossRef] [PubMed]

112. Nucci, M.; Marr, K.A.; Queiroz-Telles, F.; Martins, C.A.; Trabasso, P.; Costa, S.; Voltarelli, J.C.; Colombo, A.L.; Imhof, A.; Pasquini, R.; et al. *Fusarium* infection in hematopoietic stem cell transplant recipients. *Clin. Infect. Dis.* **2004**, *38*, 1237–1242. [CrossRef] [PubMed]

113. Nucci, M.; Garnica, M.; Gloria, A.B.; Lehugeur, D.S.; Dias, V.C.; Palma, L.C.; Cappellano, P.; Fertrin, K.Y.; Carlesse, F.; Simoes, B.; et al. Invasive fungal diseases in haematopoietic cell transplant recipients and in patients with acute myeloid leukaemia or myelodysplasia in Brazil. *Clin. Microbiol. Infect. Off. Publ. Eur. Soc. Clin. Microbiol. Infect. Dis.* **2013**, *19*, 745–751. [CrossRef] [PubMed]

114. Nucci, M.; Anaissie, E. Cutaneous infection by *Fusarium* species in healthy and immunocompromised hosts: Implications for diagnosis and management. *Clin. Infect. Dis.* **2002**, *35*, 909–920. [CrossRef] [PubMed]

115. Aoki, T.; O'Donnell, K.; Homma, Y.; Lattanzi, A.R. Sudden-death syndrome of soybean is caused by two morphologically and phylogenetically distinct species within the *Fusarium solani* species complex-*F. virguliforme* in North America and *F. tucumaniae* in South America. *Mycologia* **2003**, *95*, 660–684. [PubMed]

116. O'Donnell, K.; Rooney, A.P.; Proctor, R.H.; Brown, D.W.; McCormick, S.P.; Ward, T.J.; Frandsen, R.J.; Lysoe, E.; Rehner, S.A.; Aoki, T.; et al. Phylogenetic analyses of *RPB1* and *RPB2* support a middle Cretaceous origin for a clade comprising all agriculturally and medically important fusaria. *Fungal Genet. Biol.* **2013**, *52*, 20–31. [CrossRef] [PubMed]

J. Fungi **2018**, *4*, 138

117. O'Donnell, K.; Humber, R.A.; Geiser, D.M.; Kang, S.; Park, B.; Robert, V.A.; Crous, P.W.; Johnston, P.R.; Aoki, T.; Rooney, A.P.; et al. Phylogenetic diversity of insecticolous fusaria inferred from multilocus DNA sequence data and their molecular identification via FUSARIUM-ID and Fusarium MLST. *Mycologia* **2012**, *104*, 427–445. [CrossRef] [PubMed]

118. Sandoval-Denis, M.; Crous, P.W. Removing chaos from confusion: Assigning names to common human and animal pathogens in Neocosmospora. *Persoonia* **2018**, *41*, 109–129. [CrossRef]

119. Geiser, D.M.; Aoki, T.; Bacon, C.W.; Baker, S.E.; Bhattacharyya, M.K.; Brandt, M.E.; Brown, D.W.; Burgess, L.W.; Chulze, S.; Coleman, J.J.; et al. One fungus, one name: Defining the genus *Fusarium* in a scientifically robust way that preserves longstanding use. *Phytopathology* **2013**, *103*, 400–408. [CrossRef] [PubMed]

Journal of
Fungi

MDPI

Review

New Concepts in Diagnostics for Invasive Mycoses: Non-Culture-Based Methodologies

Thomas F. Patterson * and J. Peter Donnelly

Division of Infectious Diseases, San Antonio Center for Medical Mycology, The University of Texas Health Science Center at San Antonio and the South Texas Veterans Health Care System,
7703 Floyd Curl Drive-MSC 7881, San Antonio, TX 78229-3900, USA; p.donnelly@usa.net
* Correspondence: patterson@uthscsa.edu

Received: 11 December 2018; Accepted: 15 January 2019; Published: 17 January 2019

Abstract: Non-culture-based diagnostics have been developed to help establish an early diagnosis of invasive fungal infection. Studies have shown that these tests can significantly impact the diagnosis of infection in high risk patients. *Aspergillus* galactomannan EIA testing is well-recognized as an important adjunct to the diagnosis of invasive aspergillosis and can be detected in serum, bronchoalveolar lavage and other fluids. Galactomannan testing used along with PCR testing has been shown to be effective when integrated into care paths for high risk patients for both diagnoses and as a surrogate marker for outcome when used in serial testing. Beta-D-glucan assays are non-specific for several fungal genera including *Aspergillus* and *Candida* and in high risk patients have been an important tool to augment the diagnosis. Lateral flow technology using monoclonal antibodies to *Aspergillus* are available that allow rapid testing of clinical samples. While standard PCR for *Candida* remains investigational, T2 magnetic resonance allows for the rapid diagnosis of *Candida* species from blood cultures. *Aspergillus* PCR has been extensively validated with standardized approaches established for these methods and will be included in the diagnostic criteria in the revised European Organization for Research and Treatment of Cancer/Mycoses Study Group (EORTC-MSG) definitions. Finally, these non-culture-based tests can be used in combination to significantly increase the detection of invasive mycoses with the ultimate aim of establishing an early diagnosis of infection.

Keywords: invasive fungal infection; non-culture-based diagnostics; aspergillosis; candidiasis; *Aspergillus* PCR; galactomannan; lateral flow; beta-D-glucan; T2 *Candida*

1. Introduction

Invasive fungal infections remain a significant cause of morbidity and mortality in immunocompromised patients. The diagnosis of these infections is delayed due to lack of positive cultures from blood or from tissues, which require invasive procedures to obtain and are often difficult to perform in these critically ill patients. Non-culture-based diagnostics have been developed to help establish an early diagnosis of infection with the aim of allowing prompt initiation of antifungal therapy and improving patient outcomes.

Non-culture-based diagnostics have been developed for both *Aspergillus* and *Candida* along with other opportunistic fungal pathogens [1]. These assays have been largely focused on *Aspergillus* due to its prominence as the most common mold in immunocompromised hosts and for *Candida*, to augment diagnosis in the setting of negative or delayed positive blood cultures [2]. Assays are being developed for opportunistic pathogens including mucorales but are less widely available in clinical settings [3,4]. Additionally, for endemic fungi including *Coccidioides*, *Histoplasma*, and *Blastomyces* as well as *Cryptococcus*, non-culture-based methods for diagnosis are available but beyond the scope of this review.

Non-culture-based tests include galactomannan, which can be used in serum, bronchoalveolar lavage fluid and other samples; beta-D-glucan, a non-specific assay for *Aspergillus, Candida,* and other mycoses; lateral flow technology using an *Aspergillus* monoclonal antibody; and others including *Candida* PCR and T2 magnetic resonance. *Aspergillus* PCR has been extensively validated for standardized methodologies and is now included in the recent EORTC/MSG definition updates. In this review, the data supporting the use of clinically available non-culture-based methods for *Aspergillus* and *Candida* will be discussed and their utility alone and in combination will be summarized.

2. Risk Factors and Impact of Diagnostics

When approaching the use of these assays in the clinical setting, it is important to recognize the risk factors associated with invasive fungal infection, in order to improve the utility of their performance. The risk factors for invasive fungal infections have been extensively evaluated, as they significantly impact the incidence of invasive fungal infections and thus the performance of diagnostic assays. Herbrecht and colleagues outlined host factors for 'high risk patients', including those with allogeneic stem cell transplants, acute myelogenous leukemia/myelodysplastic syndrome, chronic granulomatous disease and others; those at 'intermediate' risk, including solid organ transplant recipients, other haematological malignancies, uncontrolled HIV infection, and others; while 'low risk' includes patients with autologous stem cell transplants, kidney transplant, solid tumors and others [5]. Additional risk factors influence the host condition, including innate immune defects; underlying conditions (neutropenia, graft vs. host disease, corticosteroid use, other biological agents, chemotherapy, etc); environmental factors and exposures; and other co-morbidities (diabetes, respiratory diseases and others) [5].

Fleming and colleagues established a risk stratification for patients with hematological malignancies. High risk patients are those with >10% incidence of invasive fungal disease, that is, patients with prolonged neutropenia (<0.1 $\times$ 10^9/L for > 3 weeks or <0.5 $\times$ 10^9/L for >5 weeks), unrelated, mismatched or cord blood donor SCT, graft vs. host disease (GVHD), high doses of corticosteroids, certain chemotherapeutic agents (high-dose cytarabine, fludarabine, alemtuzumab, and others), and certain hematological malignancies (acute myelogenous leukemia (AML) and acute lymphocytic leukemia (ALL)) [6]. An intermediate risk group with an incidence of invasive fungal disease of around 10% includes those with less profound neutropenia (0.1–0.5 $\times$ 10^9/L for 3–5 weeks or 0.1–0.5 $\times$ 10^9/L for <3 weeks with lymphopenia), while low risk patients (~2% incidence of invasive fungal disease) would include autologous SCT and lymphoma [6]. Clearly, these patients with hematological malignancies have significant differences in risk for fungal infection and it becomes critically important to consider these differences when interpreting the clinical utility of these non-culture-based diagnostic tests, based on the prior probability of disease.

It is also critical to recognize the impact that diagnostic tests can have on underestimates of infection and the impact that diagnosis has on outcomes. Ceesay and colleagues evaluated a series of 203 patients with hematological malignancies using a strict diagnostic algorithm including a pre-treatment computed tomography of the chest, twice weekly serum galactomannan, and beta-D-glucan with suspicion of infection and tissue for diagnosis [7]. The series showed that the incidence of established infection rose from 10.5% with galactomannan alone to almost 20% with a combination of galactomannan and beta-D-glucan, and was 21.1% when all tests were combined. Furthermore, at 45%, the survival of those with proven/probable infection was significantly lower than those with possible disease, at 66%. The survival rate was 87% for those without infection ($p < 0.001$), supporting the importance of using these tools to establish a diagnosis of invasive fungal disease.

3. Galactomannan

The detection of galactomannan by EIA is a well-established and extensively studied method for the diagnosis of invasive aspergillosis [2,8,9]. Monoclonal antibody EB-A2 is used in a double sandwich ELISA to detect an antigenic side chain of β-1,5-galactofuronosyl with a linear core of mannan with α1,2 and α1,6 linkages [10]. Early studies by Maertens and others prior to the availability of anti-mold prophylaxis showed a sensitivity and specificity of 89% and 98%, respectively [11]. Subsequently, other studies showed more limited sensitivity (43–70%) but with specificity of 70–93% and studies confirmed the validation of a galactomannan index (GMI) of 0.5 as the threshold for positivity [12,13]. It was also appreciated that there were several sources for false-positive results including weakly positive samples, cross-reactivity with other fungi and antibiotics (such as pipercillin-tazobactam, which has now been resolved) [14]. Other sources of false-positives such as dietary reactivity, laboratory contamination or fluids for bronchoalveolar lavage (such as plasmalyte which also appears to have resolved) may continue [14–19]. A number of other factors may affect the performance of the galactomannan assay, including biological factors (such as site of infection, *Aspergillus* species, prior use of antifungals, renal clearance, hepatic metabolism, underlying condition, storage of the samples, and others) and epidemiological factors (such as the prevalence of disease, sampling strategies and definitions for positive results) [17]. An important study by Duarte and colleagues showed the dramatic impact of antifungal prophylaxis on the strategy of serial sampling [20]. In this study the positive predictive value dropped to 11.8%, but was still useful with a positive predictive value of 89.6% when used for the diagnosis of invasive fungal disease on suspicion of disease [20].

Galactomannan testing in bronchoalveolar (BAL) lavage fluid is very useful for establishing a diagnosis of infection. This method is more sensitive than cytology, culture, transbronchial biopsy or serum galactomannan testing [21]. Galactomannan detection increased the sensitivity from serum from 47% to 85% from BAL, with a positive predictive value approaching 100% and with a BAL GMI of <0.5, indicating that it is useful to exclude the diagnosis in high risk patients with hematological malignancies [22,23]. On the other hand, in solid organ transplant patients, false positive results were more likely in lung transplant patients and in those colonized with *Aspergillus* [24]. The utility of galactomannan has been shown to improve in combination with PCR. Reinwald and colleagues showed that positivity from BAL of both galactomannan (at a GMI of >0.5) with positive PCR results highly supported the diagnosis [25]. Notably, the consensus regarding the cutoff value for a positive BAL galactomannan is still lacking, as performance will also vary in different patient settings (i.e., hematological malignancy, solid organ transplantation, intensive care units, etc.) so that a higher cutoff threshold (GMI > 1.0) may correlate with better diagnostic utility [26,27].

Galactomannan along with PCR was evaluated in a diagnostic vs. empirical therapy approach by Morrissey and colleagues [28]. In this study, empirical antifungal therapy was reduced from 32% in the standard diagnosis group to 15% in the biomarker diagnosis group, even though the rate of proven/probable disease was increased from 1% to 15% [28]. Mortality was 15% in the standard diagnosis group and 10% in the biomarker group [28]. In this study, 10/39 (26%) of the patients receiving empirical antifungal therapy would have been diagnosed with invasive aspergillosis a median of 4 days earlier and 5/6 (83%) of those who died from invasive aspergillosis would have been diagnosed a median of 7 days earlier.

Finally, serial galactomannan measurements can also be used for the assessment of outcomes. Chai and colleagues showed that a GMI reduction of >35% between baseline and week 1 predicted a satisfactory response in patients enrolled in the Global Voriconazole Aspergillosis trial [29]. Poorer responses occurred with increasing GMI after 2 weeks. Similarly, outcomes are better in patients who become GMI negative during their course of treatment [9,30].

4. Beta-D-Glucan

Testing with (1-3)-β-D-glucan activates *Limulus* amebocyte lysate through factor G initiation of the complement cascade [31]. The output can be measured using a chromogenic substrate (Fungitell (Associates of Cape Cod, Falmouth, MA) and others or by turbidity after gel clot (Wako Pure Chemical Industries, Osaka, Japan) [31]. These assays detect a number of important fungal genera including *Aspergillus*, *Candida*, *Trichosporon*, *Fusarium*, and *Exerohilum* but not mucorales or cryptococcosis [32]. Early studies showed the value of the assay in candidemia and in invasive fungal disease in patients with acute leukemia, which allowed regulatory clearance as well as inclusion in the European Organization for Research and Treatment of Cancer/Mycoses Study Group (EORTC-MSG) definitions for fungal infection [33,34].

More recent studies in invasive candidiasis have evaluated the role of beta-D-glucan with *Candida*, real-time PCR and blood cultures [35]. In this study, both PCR and beta-D-glucan were more sensitive for deep seated *Candida* infection: 88% and 62%, respectively, vs. 17% for blood cultures. Beta-D-glucan was shown to anticipate the diagnosis of blood culture-negative intraabdominal candidiasis and may be an important adjunct to that diagnosis [36].

Beta-D-glucan testing has also shown utility in specific clinical settings. In pneumocystis pneumonia, beta-D-glucan levels are frequently extremely elevated (>500 pg/mL) so that in a likely clinical setting more invasive testing might be obviated [37]. In addition, during an outbreak of fungal meningitis due to contaminated steroids, it was recognized that *Exserohilum* spp. (the etiological cause of the outbreak) produces high levels of beta-D-glucan [38]. It was subsequently shown that beta-D-glucan is highly sensitive for the diagnosis and is correlated with response to therapy [38].

5. Lateral Flow Technology in Invasive Aspergillosis

Lateral flow assays for invasive aspergillosis offer the potential for a rapid diagnosis, ease of performance and point of care use. A murine monoclonal antibody, JF4, binds to an extracellular glycoprotein antigen secreted during the growth of *Aspergillus* and distinguishes between hyphae and conidia [39]. Earlier and more consistent detection compared to beta-D-glucan or galactomannan was seen in pre-clinical models [39–41]. A prototype lateral flow device (LFD) was evaluated in high risk hematological malignancy patients and was shown to improve the diagnostic yield, especially when combined with galactomannan and PCR [42–44]. In BAL samples this assay had an overall sensitivity of 73% and a specificity of 90% [45,46]. A recent European conformity (CE)-marked LFD (OLM Diagnostics, Newcastle-on-Tyne, UK) showed similar sensitivity to the prototype device of 71%, but with an improved sensitivity of 100% [46]. Another lateral flow assay for *Aspergillus* (IMMY, Norman, OK, USA) was compared in a small study to the LFD in BAL fluid and showed similar sensitivity and specificities of 89% and 88% [47].

6. T2 Magnetic Resonance

While real-time PCR for *Candida* remains investigational, the use of T2 magnetic resonance using nanoparticles has been cleared for clinical use (T2 *Candida*, T2 Biosystems, Lexington, MA, USA) [48–50]. The assay requires a dedicated instrument and detects *Candida* species directly from blood samples, but unlike blood cultures it does not require viable organisms [51]. The assay detects five major *Candida* species that are grouped based on typical susceptibility patterns. Using spiked blood samples, it was shown that this method could detect *C. albicans/C. tropicalis*, *C. parapsilosis*, and *C. krusei/C. glabrata* at a sensitivity of 91.1% with a time to positivity of 4.4 hrs and a limit of detection of 1–3 CFU/mL [52]. In a follow-up study of patients with candidemia, follow-up blood cultures were compared with the T2 *Candida* assay and showed T2 *Candida* positivity in 45% of the follow-up blood cultures compared to 24% with standard culture techniques. These results suggest that this method could be used to detect candidemia in patients who are receiving empirical antifungal therapy and could be very useful in allowing empirical therapy to be discontinued in patients with negative results [51]. In addition, T2 shortened the time to positivity to <3 h and identified a bloodstream infection not detected by blood cultures, while retaining sensitivity during antifungal therapy.

7. *Aspergillus* PCR Development and Standardization

PCR for *Aspergillus* species has been evaluated for more than 25 years to fulfil the mycological criteria for the diagnosis of invasive fungal disease [53]. The EORTC/MSG clinical definitions were published in 2002 and revised in 2008 and provided important guidelines for criteria of proven, probable and possible invasive fungal disease, in order to facilitate clinical research including drug trials, epidemiology, and diagnostic tests [54,55]. These were not intended as a guide to clinical practice, but the elements of the definitions have been widely used as an adjunct to clinical management. The 2008 definitions defined proven disease when fungi are detected in specimens from a sterile body site or in a biopsy. In contrast, possible and probable disease both require a host factor and a clinical feature. Possible disease is assigned in the absence of any mycological criteria. Probable disease is assigned when the mycological criteria are met by direct mycology–cytology, direct microscopy or culture, or indirectly, by the detection of galactomannan and beta-D-glucan, but not PCR, due to the lack of standardization and validation [55]. Subsequently, the European *Aspergillus* PCR Initiative (EAPCRI) was established to develop a standard for *Aspergillus* PCR methodology so that PCR could be incorporated into future consensus definitions for invasive fungal disease [56].

Initial efforts included a systemic review published in 2009 of more than 10,000 blood, serum or plasma samples from 1618 patients at risk for *Aspergillus* and concluded at that time that two positive tests were required to confirm the diagnosis due to the specificity of these tests and a single PCR-negative result was sufficient to exclude a diagnosis of invasive aspergillosis [57]. However, it was also noted that there was a lack of homogeneity of the PCR methods so that subsequent collaboration aimed at a formal validation of that process.

Aspergillus standardization was evaluated through a collaboration of 21 European medical centers [56]. While specific protocols were not developed, *compliant* and *non-compliant* centers were noted. Twelve centers used 10 different DNA extraction protocols and nine different PCR amplification procedures that were compliant. Nine centers used seven extraction protocols and seven different PCR amplification procedures that were noncompliant. The sensitivity, specificity, and diagnostic odds ratio (DOR) of compliant centers were 88.7%, 91.6%, and 119.9, respectively, for non-compliant centers these were and 57.7%, 77.2%, and 8.9, respectively. These results were highly statistically significant for sensitivity ($p = 0.008$) and DOR ($p = 0.006$). The factors with the most influence on PCR included compliance with the extraction method, bead beating of the sample, and use of an internal control PCR. This led to recommendations for the whole blood PCR of >3 mL of blood, bead-beating to lyse fungal cells, a real time PCR platform with multi-copy targets with specific probes, internal control PCR and others [56]. Subsequent work showed that most of the *Aspergillus* protocols used to test serum generated satisfactory analytical performance and that the testing serum required less standardization [58]. Additional evaluation evaluated plasma vs. serum and showed improved sensitivity with plasma with PCR positivity occurring earlier, while maintaining methodological simplicity [59].

An extensive Cochrane study was performed for the diagnosis of invasive aspergillosis in immunocompromised people [60]. The studies included were those that compared the results of blood PCR tests with reference to the EORTC/MSG standard and included false positive, true positive, false negative and true negative results and that evaluated the test(s) prospectively in cohorts from patients at high risk for invasive aspergillosis. Overall, 1672 records were identified and 155 were screened to eventually include 18 studies for the meta-analysis [60]. For one single PCR specimen a sensitivity of 80.5% and a specificity of 78.5 were reported in 17 studies. For two or more PCR specimens the sensitivity decreased to 57.9% and the specificity increased to 96.2%. The authors concluded that PCR shows moderate diagnostic accuracy when used for screening and has a high negative predictive value (NPV) that allows the diagnosis of invasive aspergillosis to be ruled out. A poor positive predictive value (PPV) when the prevalence of disease is low, limits the ability to rule in a diagnosis. Since other non-culture-based methods (such as galactomannan) detect different aspects of the disease, combinations of both together are likely to be more useful. PCR used in combination with galactomannan has been shown to improve the detection of invasive aspergillosis before detection by CT findings in high risk hematological patients [61] and to improve the sensitivity and specificity of testing [62]. In a randomized controlled trial, a combined monitoring strategy based on serum galactomannan and *Aspergillus* PCR was associated with an earlier diagnosis of invasive aspergillosis [63]. Robust recommendations for plasma and serum are available that establish a 'standard for PCR' but not a 'standard method for PCR' so that *Aspergillus* PCR will be included in the revised EORTC/MSG consensus definitions [64].

7. Clinical Utility and Summary

The importance of assessing risk and using non-culture-based diagnostics for invasive fungal disease is clear. Several methods have been evaluated and validated for clinical use including galactomannan, beta-D-glucan, lateral flow technology, T2 magnetic resonance, PCR and others. Non-culture-based biomarkers provide more reliable negative than positive predictive values. When the prevalence of disease is higher than 15%, negative test results exclude the diagnosis, while positive test results include the diagnosis. Clinicians and laboratories need to consider when a test is being requested for screening (when a patient is at risk for invasive fungal disease) as opposed to diagnosis (in which there is a high clinical suspicion of an invasive fungal disease), which will have a substantially higher pre-test probability. Finally, combinations of these tests may provide the greatest benefit in establishing a diagnosis of invasive fungal disease.

Author Contributions: Conceptualization, writing—review and editing, T.F.P and J.P.D.

Funding: There was no financial support for this work.

Conflicts of Interest: For activities outside of the submitted work, T.F.P. has been a consultant for or served on advisory boards to Astellas, Basilea, Gilead, Merck, Pfizer, Scynexis, and Toyama. J.P.D is a member of the European *Aspergillus* PCR initiative (EAPCRI).

References

1. Mikulska, M.; Furfaro, E.; Viscoli, C. Non-cultural methods for the diagnosis of invasive fungal disease. *Expert Rev. Anti Infect. Ther.* **2015**, *13*, 103–117. [CrossRef] [PubMed]
2. Hope, W.W.; Walsh, T.J.; Denning, D.W. Laboratory diagnosis of invasive aspergillosis. *Lancet Infect. Dis.* **2005**, *5*, 609–622. [CrossRef]
3. Baldin, C.; Soliman, S.S.M.; Jeon, H.H.; Alkhazraji, S.; Gebremariam, T.; Gu, Y.; Bruno, V.M.; Cornely, O.A.; Leather, H.L.; Sugrue, M.W.; et al. Pcr-based approach targeting mucorales-specific gene family for diagnosis of mucormycosis. *J. Clin. Microbiol.* **2018**, *56*. [CrossRef] [PubMed]
4. Scherer, E.; Iriart, X.; Bellanger, A.P.; Dupont, D.; Guitard, J.; Gabriel, F.; Cassaing, S.; Charpentier, E.; Guenounou, S.; Cornet, M.; et al. Quantitative pcr (qpcr) detection of mucorales DNA in bronchoalveolar lavage fluid to diagnose pulmonary mucormycosis. *J. Clin. Microbiol.* **2018**, *56*. [CrossRef] [PubMed]
5. Herbrecht, R.; Bories, P.; Moulin, J.C.; Ledoux, M.P.; Letscher-Bru, V. Risk stratification for invasive aspergillosis in immunocompromised patients. *Ann. N. Y. Acad. Sci.* **2012**, *1272*, 23–30. [CrossRef]
6. Fleming, S.; Yannakou, C.K.; Haeusler, G.M.; Clark, J.; Grigg, A.; Heath, C.H.; Bajel, A.; van Hal, S.J.; Chen, S.C.; Milliken, S.T.; et al. Consensus guidelines for antifungal prophylaxis in haematological malignancy and haemopoietic stem cell transplantation, 2014. *Intern. Med. J.* **2014**, *44*, 1283–1297. [CrossRef] [PubMed]
7. Ceesay, M.M.; Desai, S.R.; Berry, L.; Cleverley, J.; Kibbler, C.C.; Pomplun, S.; Nicholson, A.G.; Douiri, A.; Wade, J.; Smith, M.; et al. A comprehensive diagnostic approach using galactomannan, targeted beta-d-glucan, baseline computerized tomography and biopsy yields a significant burden of invasive fungal disease in at risk haematology patients. *Br. J. Haematol.* **2015**, *168*, 219–229. [CrossRef] [PubMed]
8. Miceli, M.H.; Maertens, J. Role of non-culture-based tests, with an emphasis on galactomannan testing for the diagnosis of invasive aspergillosis. *Semin. Respir. Crit. Care Med.* **2015**, *36*, 650–661.
9. Mercier, T.; Guldentops, E.; Lagrou, K.; Maertens, J. Galactomannan, a surrogate marker for outcome in invasive aspergillosis: Finally coming of age. *Front. Microbiol.* **2018**, *9*, 661. [CrossRef]
10. Stynen, D.; Goris, A.; Sarfati, J.; Latge, J.P. A new sensitive sandwich enzyme-linked immunosorbent assay to detect galactofuran in patients with invasive aspergillosis. *J. Clin. Microbiol.* **1995**, *33*, 497–500.
11. Maertens, J.; Verhaegen, J.; Lagrou, K.; Van Eldere, J.; Boogaerts, M. Screening for circulating galactomannan as a noninvasive diagnostic tool for invasive aspergillosis in prolonged neutropenic patients and stem cell transplantation recipients: A prospective validation. *Blood* **2001**, *97*, 1604–1610. [CrossRef] [PubMed]

12. Marr, K.A.; Balajee, S.A.; McLaughlin, L.; Tabouret, M.; Bentsen, C.; Walsh, T.J. Detection of galactomannan antigenemia by enzyme immunoassay for the diagnosis of invasive aspergillosis: Variables that affect performance. *J. Infect. Dis.* **2004**, *190*, 641–649. [CrossRef] [PubMed]
13. Herbrecht, R.; Letscher-Bru, V.; Oprea, C.; Lioure, B.; Waller, J.; Campos, F.; Villard, O.; Liu, K.L.; Natarajan-Ame, S.; Lutz, P.; et al. Aspergillus galactomannan detection in the diagnosis of invasive aspergillosis in cancer patients. *J. Clin. Oncol.* **2002**, *20*, 1898–1906. [CrossRef] [PubMed]
14. Mikulska, M.; Furfaro, E.; Del Bono, V.; Raiola, A.M.; Ratto, S.; Bacigalupo, A.; Viscoli, C. Piperacillin/tazobactam (tazocin) seems to be no longer responsible for false-positive results of the galactomannan assay. *J. Antimicrob. Chemother.* **2012**, *67*, 1746–1748. [CrossRef] [PubMed]
15. Machetti, M.; Majabo, M.J.; Furfaro, E.; Solari, N.; Novelli, A.; Cafiero, F.; Viscoli, C. Kinetics of galactomannan in surgical patients receiving perioperative piperacillin/tazobactam prophylaxis. *J. Antimicrob. Chemother.* **2006**, *58*, 806–810. [CrossRef] [PubMed]
16. Maertens, J.; Theunissen, K.; Verhoef, G.; Van Eldere, J. False-positive aspergillus galactomannan antigen test results. *Clin. Infect. Dis.* **2004**, *39*, 289–290. [CrossRef]
17. Mennink-Kersten, M.A.; Donnelly, J.P.; Verweij, P.E. Detection of circulating galactomannan for the diagnosis and management of invasive aspergillosis. *Lancet Infect. Dis.* **2004**, *4*, 349–357. [CrossRef]
18. Spriet, I.; Lagrou, K.; Maertens, J.; Willems, L.; Wilmer, A.; Wauters, J. Plasmalyte: No longer a culprit in causing false-positive galactomannan test results. *J. Clin. Microbiol.* **2016**, *54*, 795–797. [CrossRef]
19. Viscoli, C.; Machetti, M.; Cappellano, P.; Bucci, B.; Bruzzi, P.; Van Lint, M.T.; Bacigalupo, A. False-positive galactomannan platelia aspergillus test results for patients receiving piperacillin-tazobactam. *Clin. Infect. Dis.* **2004**, *38*, 913–916. [CrossRef]
20. Duarte, R.F.; Sanchez-Ortega, I.; Cuesta, I.; Arnan, M.; Patino, B.; Fernandez de Sevilla, A.; Gudiol, C.; Ayats, J.; Cuenca-Estrella, M. Serum galactomannan-based early detection of invasive aspergillosis in hematology patients receiving effective antimold prophylaxis. *Clin. Infect. Dis.* **2014**, *59*, 1696–1702. [CrossRef]
21. Nguyen, M.H.; Leather, H.; Clancy, C.J.; Cline, C.; Jantz, M.A.; Kulkarni, V.; Wheat, L.J.; Wingard, J.R. Galactomannan testing in bronchoalveolar lavage fluid facilitates the diagnosis of invasive pulmonary aspergillosis in patients with hematologic malignancies and stem cell transplant recipients. *Biol. Blood Marrow Transplant.* **2011**, *17*, 1043–1050. [CrossRef] [PubMed]
22. Becker, M.J.; Lugtenburg, E.J.; Cornelissen, J.J.; Van Der Schee, C.; Hoogsteden, H.C.; De Marie, S. Galactomannan detection in computerized tomography-based broncho-alveolar lavage fluid and serum in haematological patients at risk for invasive pulmonary aspergillosis. *Br. J. Haematol.* **2003**, *121*, 448–457. [CrossRef] [PubMed]
23. D'Haese, J.; Theunissen, K.; Vermeulen, E.; Schoemans, H.; De Vlieger, G.; Lammertijn, L.; Meersseman, P.; Meersseman, W.; Lagrou, K.; Maertens, J. Detection of galactomannan in bronchoalveolar lavage fluid samples of patients at risk for invasive pulmonary aspergillosis: Analytical and clinical validity. *J. Clin. Microbiol.* **2012**, *50*, 1258–1263. [CrossRef] [PubMed]
24. Husain, S.; Paterson, D.L.; Studer, S.M.; Crespo, M.; Pilewski, J.; Durkin, M.; Wheat, J.L.; Johnson, B.; McLaughlin, L.; Bentsen, C.; et al. Aspergillus galactomannan antigen in the bronchoalveolar lavage fluid for the diagnosis of invasive aspergillosis in lung transplant recipients. *Transplantation* **2007**, *83*, 1330–1336. [CrossRef] [PubMed]
25. Reinwald, M.; Spiess, B.; Heinz, W.J.; Vehreschild, J.J.; Lass-Florl, C.; Kiehl, M.; Schultheis, B.; Krause, S.W.; Wolf, H.H.; Bertz, H.; et al. Diagnosing pulmonary aspergillosis in patients with hematological malignancies: A multicenter prospective evaluation of an aspergillus pcr assay and a galactomannan elisa in bronchoalveolar lavage samples. *Eur. J. Haematol.* **2012**, *89*, 120–127. [CrossRef] [PubMed]
26. Maertens, J.; Maertens, V.; Theunissen, K.; Meersseman, W.; Meersseman, P.; Meers, S.; Verbeken, E.; Verhoef, G.; Van Eldere, J.; Lagrou, K. Bronchoalveolar lavage fluid galactomannan for the diagnosis of invasive pulmonary aspergillosis in patients with hematologic diseases. *Clin. Infect. Dis.* **2009**, *49*, 1688–1693. [CrossRef] [PubMed]
27. Clancy, C.J.; Jaber, R.A.; Leather, H.L.; Wingard, J.R.; Staley, B.; Wheat, L.J.; Cline, C.L.; Rand, K.H.; Schain, D.; Baz, M.; et al. Bronchoalveolar lavage galactomannan in diagnosis of invasive pulmonary aspergillosis among solid-organ transplant recipients. *J. Clin. Microbiol.* **2007**, *45*, 1759–1765. [CrossRef]

28. Morrissey, C.O.; Chen, S.C.; Sorrell, T.C.; Milliken, S.; Bardy, P.G.; Bradstock, K.F.; Szer, J.; Halliday, C.L.; Gilroy, N.M.; Moore, J.; et al. Galactomannan and pcr versus culture and histology for directing use of antifungal treatment for invasive aspergillosis in high-risk haematology patients: A randomised controlled trial. *Lancet Infect. Dis.* **2013**, *13*, 519–528. [CrossRef]
29. Chai, L.Y.; Kullberg, B.J.; Johnson, E.M.; Teerenstra, S.; Khin, L.W.; Vonk, A.G.; Maertens, J.; Lortholary, O.; Donnelly, P.J.; Schlamm, H.T.; et al. Early serum galactomannan trend as a predictor of outcome of invasive aspergillosis. *J. Clin. Microbiol.* **2012**, *50*, 2330–2336. [CrossRef]
30. Maertens, J.; Buve, K.; Theunissen, K.; Meersseman, W.; Verbeken, E.; Verhoef, G.; Van Eldere, J.; Lagrou, K. Galactomannan serves as a surrogate endpoint for outcome of pulmonary invasive aspergillosis in neutropenic hematology patients. *Cancer* **2009**, *115*, 355–362. [CrossRef]
31. Obayashi, T.; Negishi, K.; Suzuki, T.; Funata, N. Reappraisal of the serum (1–>3)-beta-d-glucan assay for the diagnosis of invasive fungal infections–a study based on autopsy cases from 6 years. *Clin. Infect. Dis.* **2008**, *46*, 1864–1870. [CrossRef] [PubMed]
32. Odabasi, Z.; Paetznick, V.L.; Rodriguez, J.R.; Chen, E.; McGinnis, M.R.; Ostrosky-Zeichner, L. Differences in beta-glucan levels in culture supernatants of a variety of fungi. *Med. Mycol.* **2006**, *44*, 267–272. [CrossRef] [PubMed]
33. Odabasi, Z.; Mattiuzzi, G.; Estey, E.; Kantarjian, H.; Saeki, F.; Ridge, R.J.; Ketchum, P.A.; Finkelman, M.A.; Rex, J.H.; Ostrosky-Zeichner, L. Beta-d-glucan as a diagnostic adjunct for invasive fungal infections: Validation, cutoff development, and performance in patients with acute myelogenous leukemia and myelodysplastic syndrome. *Clin. Infect. Dis.* **2004**, *39*, 199–205. [CrossRef]
34. Ostrosky-Zeichner, L.; Alexander, B.D.; Kett, D.H.; Vazquez, J.; Pappas, P.G.; Saeki, F.; Ketchum, P.A.; Wingard, J.; Schiff, R.; Tamura, H.; et al. Multicenter clinical evaluation of the (1–>3) beta-d-glucan assay as an aid to diagnosis of fungal infections in humans. *Clin. Infect. Dis.* **2005**, *41*, 654–659. [CrossRef]
35. Nguyen, M.H.; Wissel, M.C.; Shields, R.K.; Salomoni, M.A.; Hao, B.; Press, E.G.; Shields, R.M.; Cheng, S.; Mitsani, D.; Vadnerkar, A.; et al. Performance of candida real-time polymerase chain reaction, beta-d-glucan assay, and blood cultures in the diagnosis of invasive candidiasis. *Clin. Infect. Dis.* **2012**, *54*, 1240–1248. [CrossRef] [PubMed]
36. Tissot, F.; Lamoth, F.; Hauser, P.M.; Orasch, C.; Fluckiger, U.; Siegemund, M.; Zimmerli, S.; Calandra, T.; Bille, J.; Eggimann, P.; et al. Beta-glucan antigenemia anticipates diagnosis of blood culture-negative intraabdominal candidiasis. *Am. J. Respir. Crit. Care Med.* **2013**, *188*, 1100–1109. [CrossRef] [PubMed]
37. Koo, S.; Baden, L.R.; Marty, F.M. Post-diagnostic kinetics of the (1 –> 3)-beta-d-glucan assay in invasive aspergillosis, invasive candidiasis and pneumocystis jirovecii pneumonia. *Clin. Microbiol. Infect.* **2012**, *18*, E122–E127. [CrossRef]
38. Litvintseva, A.P.; Lindsley, M.D.; Gade, L.; Smith, R.; Chiller, T.; Lyons, J.L.; Thakur, K.T.; Zhang, S.X.; Grgurich, D.E.; Kerkering, T.M.; et al. Utility of (1-3)-beta-d-glucan testing for diagnostics and monitoring response to treatment during the multistate outbreak of fungal meningitis and other infections. *Clin. Infect. Dis.* **2014**, *58*, 622–630. [CrossRef]
39. Thornton, C.R. Development of an immunochromatographic lateral-flow device for rapid serodiagnosis of invasive aspergillosis. *Clin. Vaccine Immunol.* **2008**, *15*, 1095–1105. [CrossRef]
40. Thornton, C.R. Breaking the mould—novel diagnostic and therapeutic strategies for invasive pulmonary aspergillosis in the immune deficient patient. *Expert Rev. Clin. Immunol.* **2014**, *10*, 771–780. [CrossRef]
41. Wiederhold, N.P.; Thornton, C.R.; Najvar, L.K.; Kirkpatrick, W.R.; Bocanegra, R.; Patterson, T.F. Comparison of lateral flow technology and galactomannan and (1->3)-beta-d-glucan assays for detection of invasive pulmonary aspergillosis. *Clin. Vaccine Immunol.* **2009**, *16*, 1844–1846. [CrossRef] [PubMed]
42. Eigl, S.; Prattes, J.; Lackner, M.; Willinger, B.; Spiess, B.; Reinwald, M.; Selitsch, B.; Meilinger, M.; Neumeister, P.; Reischies, F.; et al. Multicenter evaluation of a lateral-flow device test for diagnosing invasive pulmonary aspergillosis in icu patients. *Crit. Care* **2015**, *19*, 178. [CrossRef] [PubMed]
43. Hoenigl, M.; Prattes, J.; Spiess, B.; Wagner, J.; Prueller, F.; Raggam, R.B.; Posch, V.; Duettmann, W.; Hoenigl, K.; Wolfler, A.; et al. Performance of galactomannan, beta-D-glucan, aspergillus lateral-flow device, conventional culture, and pcr tests with bronchoalveolar lavage fluid for diagnosis of invasive pulmonary aspergillosis. *J. Clin. Microbiol.* **2014**, *52*, 2039–2045. [CrossRef] [PubMed]

44. White, P.L.; Parr, C.; Thornton, C.; Barnes, R.A. Evaluation of real-time pcr, galactomannan enzyme-linked immunosorbent assay (elisa), and a novel lateral-flow device for diagnosis of invasive aspergillosis. *J. Clin. Microbiol.* **2013**, *51*, 1510–1516. [CrossRef] [PubMed]

45. Willinger, B.; Lackner, M.; Lass-Florl, C.; Prattes, J.; Posch, V.; Selitsch, B.; Eschertzhuber, S.; Honigl, K.; Koidl, C.; Sereinigg, M.; et al. Bronchoalveolar lavage lateral-flow device test for invasive pulmonary aspergillosis in solid organ transplant patients: A semiprospective multicenter study. *Transplantation* **2014**, *98*, 898–902. [CrossRef] [PubMed]

46. Hoenigl, M.; Eigl, S.; Heldt, S.; Duettmann, W.; Thornton, C.; Prattes, J. Clinical evaluation of the newly formatted lateral-flow device for invasive pulmonary aspergillosis. *Mycoses* **2018**, *61*, 40–43. [CrossRef]

47. Jenks, J.D.; Mehta, S.R.; Taplitz, R.; Aslam, S.; Reed, S.L.; Hoenigl, M. Point-of care diagnosis of invasive aspergillosis in in non-neutropenic patients: Aspergillus galactomannan lateral flow assay versus aspergillus-specific lateral flow device test in bronchoalveolar lavage. *Mycoses* **2018**. [CrossRef]

48. Clancy, C.J.; Nguyen, M.H. T2 magnetic resonance for the diagnosis of bloodstream infections: Charting a path forward. *J. Antimicrob. Chemother.* **2018**, *73*, iv2–iv5. [CrossRef]

49. Clancy, C.J.; Nguyen, M.H. Non-culture diagnostics for invasive candidiasis: Promise and unintended consequences. *J. Fungi (Basel)* **2018**, *4*, 27. [CrossRef]

50. Zacharioudakis, I.M.; Zervou, F.N.; Mylonakis, E. T2 magnetic resonance assay: Overview of available data and clinical implications. *J. Fungi (Basel)* **2018**, *4*, 45. [CrossRef]

51. Clancy, C.J.; Pappas, P.G.; Vazquez, J.; Judson, M.A.; Kontoyiannis, D.P.; Thompson, G.R., 3rd; Garey, K.W.; Reboli, A.; Greenberg, R.N.; Apewokin, S.; et al. Detecting infections rapidly and easily for candidemia trial, part 2 (direct2): A prospective, multicenter study of the t2candida panel. *Clin. Infect. Dis.* **2018**, *66*, 1678–1686. [CrossRef]

52. Mylonakis, E.; Clancy, C.J.; Ostrosky-Zeichner, L.; Garey, K.W.; Alangaden, G.J.; Vazquez, J.A.; Groeger, J.S.; Judson, M.A.; Vinagre, Y.M.; Heard, S.O.; et al. T2 magnetic resonance assay for the rapid diagnosis of candidemia in whole blood: A clinical trial. *Clin. Infect. Dis.* **2015**, *60*, 892–899. [CrossRef] [PubMed]

53. Spreadbury, C.; Holden, D.; Aufauvre-Brown, A.; Bainbridge, B.; Cohen, J. Detection of aspergillus fumigatus by polymerase chain reaction. *J. Clin. Microbiol.* **1993**, *31*, 615–621. [PubMed]

54. Ascioglu, S.; Rex, J.H.; de Pauw, B.; Bennett, J.E.; Bille, J.; Crokaert, F.; Denning, D.W.; Donnelly, J.P.; Edwards, J.E.; Erjavec, Z.; et al. Defining opportunistic invasive fungal infections in immunocompromised patients with cancer and hematopoietic stem cell transplants: An international consensus. *Clin. Infect. Dis.* **2002**, *34*, 7–14. [CrossRef] [PubMed]

55. De Pauw, B.; Walsh, T.J.; Donnelly, J.P.; Stevens, D.A.; Edwards, J.E.; Calandra, T.; Pappas, P.G.; Maertens, J.; Lortholary, O.; Kauffman, C.A.; et al. Revised definitions of invasive fungal disease from the european organization for research and treatment of cancer/invasive fungal infections cooperative group and the national institute of allergy and infectious diseases mycoses study group (eortc/msg) consensus group. *Clin. Infect. Dis.* **2008**, *46*, 1813–1821.

56. White, P.L.; Bretagne, S.; Klingspor, L.; Melchers, W.J.; McCulloch, E.; Schulz, B.; Finnstrom, N.; Mengoli, C.; Barnes, R.A.; Donnelly, J.P.; et al. Aspergillus pcr: One step closer to standardization. *J. Clin. Microbiol.* **2010**, *48*, 1231–1240. [CrossRef]

57. Mengoli, C.; Cruciani, M.; Barnes, R.A.; Loeffler, J.; Donnelly, J.P. Use of pcr for diagnosis of invasive aspergillosis: Systematic review and meta-analysis. *Lancet Infect. Dis.* **2009**, *9*, 89–96. [CrossRef]

58. White, P.L.; Mengoli, C.; Bretagne, S.; Cuenca-Estrella, M.; Finnstrom, N.; Klingspor, L.; Melchers, W.J.; McCulloch, E.; Barnes, R.A.; Donnelly, J.P.; et al. Evaluation of aspergillus pcr protocols for testing serum specimens. *J. Clin. Microbiol.* **2011**, *49*, 3842–3848. [CrossRef]

59. White, P.L.; Barnes, R.A.; Springer, J.; Klingspor, L.; Cuenca-Estrella, M.; Morton, C.O.; Lagrou, K.; Bretagne, S.; Melchers, W.J.; Mengoli, C.; et al. Clinical performance of aspergillus pcr for testing serum and plasma: A study by the european aspergillus pcr initiative. *J. Clin. Microbiol.* **2015**, *53*, 2832–2837. [CrossRef]

60. Cruciani, M.; Mengoli, C.; Loeffler, J.; Donnelly, P.; Barnes, R.; Jones, B.L.; Klingspor, L.; Morton, O.; Maertens, J. Polymerase chain reaction blood tests for the diagnosis of invasive aspergillosis in immunocompromised people. *Cochrane Database Syst. Rev.* **2015**, CD009551. [CrossRef]

61. Rogers, T.R.; Morton, C.O.; Springer, J.; Conneally, E.; Heinz, W.; Kenny, C.; Frost, S.; Einsele, H.; Loeffler, J. Combined real-time pcr and galactomannan surveillance improves diagnosis of invasive aspergillosis in high risk patients with haematological malignancies. *Br. J. Haematol.* **2013**, *161*, 517–524. [CrossRef] [PubMed]

62. Arvanitis, M.; Anagnostou, T.; Mylonakis, E. Galactomannan and polymerase chain reaction-based screening for invasive aspergillosis among high-risk hematology patients: A diagnostic meta-analysis. *Clin. Infect. Dis.* **2015**, *61*, 1263–1272. [CrossRef] [PubMed]
63. Aguado, J.M.; Vazquez, L.; Fernandez-Ruiz, M.; Villaescusa, T.; Ruiz-Camps, I.; Barba, P.; Silva, J.T.; Batlle, M.; Solano, C.; Gallardo, D.; et al. Serum galactomannan versus a combination of galactomannan and polymerase chain reaction-based aspergillus DNA detection for early therapy of invasive aspergillosis in high-risk hematological patients: A randomized controlled trial. *Clin. Infect. Dis.* **2015**, *60*, 405–414. [CrossRef] [PubMed]
64. Barnes, R.A.; White, P.L.; Morton, C.O.; Rogers, T.R.; Cruciani, M.; Loeffler, J.; Donnelly, J.P. Diagnosis of aspergillosis by pcr: Clinical considerations and technical tips. *Med. Mycol.* **2018**, *56*, 60–72. [CrossRef] [PubMed]

Journal of
Fungi

MDPI

Review

A Moldy Application of MALDI: MALDI-ToF Mass Spectrometry for Fungal Identification

Robin Patel [1,2]

[1] Department of Laboratory Medicine and Pathology, Division of Clinical Microbiology, Mayo Clinic, Rochester, MN 55905, USA; patel.robin@mayo.edu; Tel.: +1-507-538-0579; Fax: +1-507-284-4272
[2] Department of Medicine, Mayo Clinic, Division of Infectious Diseases, Rochester, MN 55905, USA

Received: 15 November 2018; Accepted: 25 December 2018; Published: 3 January 2019

Abstract: As a result of its being inexpensive, easy to perform, fast and accurate, matrix-assisted laser desorption ionization time-of-flight mass spectrometry (MALDI-ToF MS) is quickly becoming the standard means of bacterial identification from cultures in clinical microbiology laboratories. Its adoption for routine identification of yeasts and even dimorphic and filamentous fungi in cultures, while slower, is now being realized, with many of the same benefits as have been recognized on the bacterial side. In this review, the use of MALDI-ToF MS for identification of yeasts, and dimorphic and filamentous fungi grown in culture will be reviewed, with strengths and limitations addressed.

Keywords: MALDI-ToF MS; yeast; fungus

1. Background and Introduction

The concept of using mass spectrometry for bacterial identification was suggested by Catherine Fenselau and John Anhalt in 1975 [1], but at the time, intact proteins were not analyzable due to fragmentation during the mass spectrometry (MS) process, with mass spectrometric analysis of intact proteins only becoming possible a decade later. In 1985, Koichi Tanaka described a "soft desorption ionization" technique allowing mass spectrometry of biological macromolecules achieved using ultrafine metal powder and glycerol; for his discovery, he was awarded the Nobel Prize in Chemistry [2]. About the same time, Franz Hillenkamp and Michael Karas reported a soft desorption ionization using an organic compound matrix [3]; it was their approach for which the designation "matrix-assisted laser desorption ionization" or MALDI, was coined and on which subsequent clinical microbiology applications were based. Tied with time-of-flight or ToF analysis, this advancement made it possible to perform mass spectrometry on intact bacterial and, ultimately, fungal cells. However, it took until advances in informatics allowed the connection of microbial MALDI-ToF MS databases to automated computer-based analytics for MALDI-ToF MS to ultimately become usable for routine identification of bacteria and, eventually, fungi in clinical laboratories. In due course, these developments led to the commercialization and, ultimately, regulatory approval of MALDI-ToF MS systems for clinical microbiology laboratories. Although initial applications of MALDI-ToF MS for rapid, inexpensive identification of microorganisms in culture focused on bacteria, it was quickly realized that this new technology could be equally applied to yeasts and, with some caveats, dimorphic and filamentous fungi.

MALDI designates matrix that assists in desorption and ionization of highly abundant bacterial or fungal proteins through laser energy [4]. Like bacteria, fungi may be tested either by "direct transfer" to a MALDI-ToF MS target plate, with or without the addition of an on-plate formic acid treatment (to lyse cells, also referred to as "on-plate extraction" and "extended direct transfer"), or following a more formal (and time-consuming) off-plate protein extraction step. The former is most commonly used for yeasts and the latter for filamentous fungi. For direct transfer testing, whole cells from colonies are simply moved to the target plate using a loop, plastic or wooden stick, or pipette tip, to a "spot" on a MALDI-ToF MS target plate (a reusable or disposable plate with multiple test spots) (Figure 1). For on-plate formic acid treatment, a formic acid solution is incorporated, either by adding the formic acid solution prior to colony transfer or by overlaying the transferred colony with formic acid solution, followed by drying. Then, the microbial mass, either alone or after formic acid treatment, is overlain with matrix; following drying of the matrix, the target plate is moved into a mass spectrometer (Figure 2). After this point, the rest is automated vis-à-vis the described clinical microbiology applications. The matrix (e.g., α-cyano-4-hydroxycinnamic acid dissolved in 50% acetonitrile and 2.5% trifluoroacetic acid), which is used for bacteria and fungi alike, isolates microbial molecules from one another, protecting them from breaking up and allowing their desorption by laser energy; a majority of the energy is absorbed by the matrix, changing it to an ionized state. As a result of random impacts occurring in the gas phase, charge is moved from matrix to microbial molecules; ionized microbial molecules are then accelerated through a positively charged electrostatic field into a ToF, tube, which is under vacuum. In the tube, ions travel to an ion detector, with smaller analytes migrating fastest, followed by increasingly larger analytes; a mass spectrum is thereby generated, signifying the quantity of ions of a specified mass hitting the detector over time. The resultant mass spectrum represents the most abundant proteins, mainly ribosomal proteins, though with this application the specific proteins generating the mass spectrum are not separately identified. The overall mass spectrum is used as a signature profile of individual fungi (or bacteria), with peaks specific to groups, complexes, genera and/or species, depending on relatedness of the test organisms to other closely related ones. The mass spectrum of an individual isolate is compared to a database or library of reference spectra, producing a list of the most closely interrelated fungi (or bacteria) with numeric rankings (assessed as percentages or scores, depending on the system). As with any identification system, it is critical to have a comprehensive and well-curated database; this has been a notable limitation of historical fungal databases, especially those for dimorphic and filamentous fungi. Depending on relatedness of the test organism to the top match (and allowing for the next best matches), the organism is then identified at the group, complex, genus, species or subspecies-level. Usually, organisms are either appropriately identified or yield a low match, indicating that identification has not been attained; the latter suggests that the species being tested is not in the database, or that there is heterogeneity in individual species or genera, but may occur due to an insufficient amount of biomass being tested or poor technical preparation (in which case repeat testing, or testing after incubation for further growth, may be helpful). A Clinical and Laboratory Standards Institute guideline on MALDI-ToF MS was published in 2017 [5].

In the past, fungal identification has been a perplexing, multi-step process, tailored by organism-type. Clinical microbiology students were pedantically educated to interpret colony and microscopic morphology of fungi on solid media as a preface to choosing appropriate further testing, such as biochemical tests or sequencing. With MALDI-ToF MS, cultured yeasts may be correctly identified in minutes without a priori knowledge of organism-type; since it doesn't matter whether a bacterium or yeast is being tested, the decision-making procedure characteristically surrounding differentiation of bacteria or yeasts growing on solid media prior to selecting further testing is obviated. Filamentous fungi can also be identified, though usually their processing prior to MALDI-ToF MS analysis typically takes longer than yeasts. MALDI-ToF MS is enabling implementation of total laboratory automation in clinical microbiology laboratories, allowing automated specimen processing, plating, incubation, plate reading using digital imaging, and spotting to MALDI-ToF MS plates. Early growth detection by digital imaging, paired with MALDI-ToF MS may result in earlier detection of fungi than conventional techniques [6]. MALDI-ToF MS is also changing the educational needs of clinical microbiology laboratory management staff, medical technologists, as well as medical students, fellows and residents. For the curriculum of those who won't practice laboratory medicine, conventional biochemical-based identification is being deemphasized.

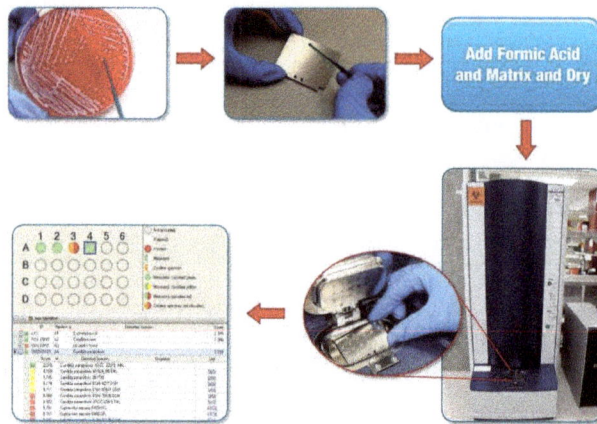

Figure 1. Process of matrix-assisted laser desorption ionization time-of-flight mass spectrometry (MALDI-ToF MS) for yeast identification [7,8]. A colony is picked from a culture plate to a spot on a MALDI-ToF MS target plate (a disposable or reusable plate with a number of spots, each of which may be used to test different colonies). For yeast applications, cells are typically treated with formic acid on the target plate, followed by drying. The spot is overlain with 1–2 µL of matrix and dried. The plate is placed in the ionization chamber of the mass spectrometer (Figure 2). A mass spectrum is produced and compared against a library of mass spectra by the software, resulting in identification of the yeast (*Candida parapsilosis* in position A4 in the example). Used with permission of the Mayo Foundation for Medical Education and Research. All rights reserved.

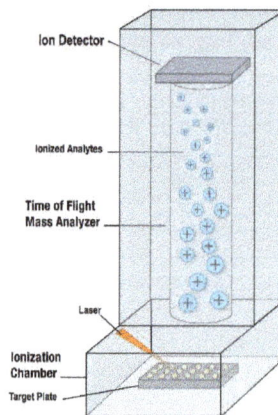

Figure 2. Mass spectrometer used for MALDI-ToF MS [7,8]. The MALDI-ToF MS plate is placed into the chamber of the instrument. Each spot to be analyzed is shot by a laser, resulting in desorption and ionization of bacterial or fungal and matrix molecules from the target plate. The cloud of ionized molecules is accelerated into the time-of-flight mass analyzer, toward a detector. Lighter molecules travel quicker, followed by progressively heavier ones. A mass spectrum is produced; it denotes the number of ions hitting the detector with time. Separation is by mass-to-charge ratio; because charge is typically single for this application, separation is by molecular weight. Used with permission of the Mayo Foundation for Medical Education and Research. All rights reserved.

MALDI-ToF MS instruments used in clinical microbiology laboratories are typically specific for clinical microbiology applications, though other testing may be performed on them and alternative instruments may be used for clinical microbiology purposes. For purposes of efficiency and biosafety, however, such instruments are typically located in (or near) clinical microbiology laboratories themselves, rather than in a centralized mass spectrometry core facility. Commercial MALDI-ToF MS systems for clinical microbiology laboratories are available from bioMérieux, Inc. (Durham, NC, USA) and Bruker Daltonics, Inc. (Billerica, MA, USA). (Other systems, such as Andromas (Paris, France), Clin-TOF (China) Quan TOF (China), Autof ms 1000 (China), and Microtyper MS (China) will not be discussed). In 2010, bioMérieux acquired a microbial database called Spectral Archiving and Microbial Identification System (SARAMIS) marketed by AnagnosTec (Zossen, Germany) and used with Shimadzu's AXIMA Assurance mass spectrometer (Shimadzu, Columbia, MD, USA), and transformed the label to VITEK MS research use only (RUO); bioMérieux then established a new database, software, and algorithms called VITEK MS IVD. bioMérieux's FDA-approved/cleared platform, available since 2013, is named Vitek MS. A RUO version, VITEK MS Plus is available, incorporating the VITEK MS and SARAMIS databases. Bruker began developing a system for identification of cultured microorganisms circa 2005, the so-called Biotyper system, obtaining FDA-approval/clearance shortly after bioMérieux in 2013, with a system referred to as the MALDI Biotyper CA System. Like bioMérieux, Bruker offers a more extensive RUO database. Bruker also has a specific RUO Filamentous Fungi Library. Bruker's mass spectrometer used for clinical microbiology testing is a desktop system, whereas bioMérieux's is a larger instrument that sits on the floor.

Yeasts and filamentous fungi claimed by the FDA cleared/approved versions of at least one commercial MALDI-ToF MS system are shown in Tables 1 and 2, respectively. Both companies' systems claim an extensive portfolio of yeasts commonly encountered in clinical practice, though there are some nomenclature differences and inclusion differences. In some cases, one system may use the teleomorph with the other using the anamorph name; for example, *Cyberlindnera jadinii* is officially claimed by the MALDI Biotyper CA system, whereas *Candida utilis* is officially claimed by the Vitek MS system. The MALDI Biotyper CA system claims *Trichosporon mucoides* group (which per the company's package insert includes *Trichosporon mucoides* and *Trichosporon dermatis*), whereas the Vitek MS system claims *T. mucoides*. Reporting of *Cryptococcus neoformans* and *Cryptococcus gattii* varies between the two systems (Table 1). Species uniquely claimed by the MALDI Biotyper CA system include *Candida boidinii*, *Candida duobushaemulonii*, *Candida metapsilosis*, *Candida orthopsilosis*, *Candida pararugosa*, *Candida valida*, and *Geotrichum candidum*, with *Candida rugosa* being uniquely claimed by the Vitek MS system. The Bruker and bioMérieux systems are different not just in databases, but also in database matching and relatedness reporting strategies. In most comparative studies, performance of the two has been similar, though not identical, assuming that the specific species being studied are represented in both databases [9–11]. Since there have been iterative and rapid growths and curations in both companies' databases over time, in reviewing the published literature, it is important to note not just the company whose system was studied, but also the specimen preparation method and the library version applied, alongside the cutoffs used for identification at the species-, genus-, group- or complex-level. The organism testing sets studied (i.e., supplemented with unusual organisms or not), and reference (i.e., comparator) identification procedures should also be considered. With both systems, users have the option to develop their own database(s), which can enhance performance; this, however, makes generalization to other users challenging. In addition, user-developed databases must be validated to meet regulatory requirements for clinical use. User-developed databases can be used in conjunction with commercial databases; alternatively or additionally, multiple databases from the same company can be used together. Success rates may be compromised if spectra in a particular library were not created from isolates prepared in the same way as they are being tested (e.g., on-plate formic acid preparation versus off-plate protein extraction) [12].

Table 1. Reportable yeasts for the FDA-approved/cleared Vitek MS and MALDI Biotyper CA systems as of October 2018 [13]. Those entries marked with "V" are FDA-approved/cleared for the Vitek MS system only and those marked with "B" are FDA-approved/cleared for the MALDI Biotyper CA system only. Those with marked with neither a "V" nor a "B" are FDA-approved/cleared on both systems.

Candida albicans	Candida krusei	Candida tropicalis	Kodamaea/Pichia ohmeri ***
Candida boidinii B	Candida lambica	Candida utilis/Cyberlindnera jadinii *	Malassezia furfur
Candida dubliniensis	Candida lipolytica	Candida valida B	Malassezia pachydermatis
Candida duobushaemulonii B	Candida lusitaniae	Candida zeylanoides	Rhodotorula mucilaginosa
Candida famata	Candida metapsilosis B	Cryptococcus gattii B	Saccharomyces cerevisiae
Candida glabrata	Candida norvegensis	Cryptococcus neoformans V	Trichosporon asahii
Candida guilliermondii	Candida orthopsilosis B	Cryptococcus neoformans var grubii B	Trichosporon inkin
Candida haemulonii	Candida parapsilosis	Cryptococcus neoformans var neoformans B	Trichosporon mucoides V
Candida inconspicua	Candida pararugosa B	Geotrichum candidum B	
Candida intermedia	Candida pelliculosa B	Geotrichum capitatum/Saprochaete capitate **	Trichosporon mucoides group B
Candida kefyr	Candida rugosa V	Kloeckera apiculata	

* Cyberlindnera jadinii (teleomorph) is approved/cleared on the MALDI Biotyper CA system, whereas Candida utilis (anamorph) is approved/cleared on the Vitek MS system.
** Geotrichum capitatum is approved/cleared on the MALDI Biotyper CA system, whereas Saprochaete capitate is approved/cleared on the Vitek MS system **; *** Kodamaea ohmeri is approved/cleared on the Vitek MS system whereas Pichia ohmeri is approved/cleared on the MALDI Biotyper CA system.

Table 2. Reportable filamentous and dimorphic fungi for the FDA-approved/cleared Vitek MS system as of October 2018 [13].

Acremonium sclerotigenum	Blastomyces dermatitidis	Histoplasma capsulatum	Rhizopus arrhizus complex
Alternaria alternata	Cladophialophora bantiana	Lecythophora hoffmannii	Rhizopus microsporus complex
Aspergillus brasiliensis	Coccidioides immitis/posadasii	Lichtheimia corymbifera	Sarocladium kiliense
Aspergillus calidoustus	Curvularia hawaiiensis	Microsporum audouinii	Scedosporium apiospermum
Aspergillus flavus/oryzae	Curvularia spicifera	Microsporum canis	Scedosporium prolificans
Aspergillus fumigatus	Epidermophyton floccosum	Microsporum gypseum	Sporothrix schenckii complex
Aspergillus lentulus	Exophiala dermatitidis	Mucor racemosus complex	Trichophyton interdigitale
Aspergillus nidulans	Exophiala xenobiotica	Paecilomyces variotii complex	Trichophyton rubrum
Aspergillus niger complex	Exserohilum rostratum	Penicillium chrysogenum	Trichophyton tonsurans
Aspergillus sydowii	Fusarium oxysporum complex	Pseudallescheria boydii	Trichophyton verrucosum
Aspergillus terreus complex	Fusarium proliferatum	Purpureocillium lilacinum	Trichophyton violaceum
Aspergillus versicolor	Fusarium solani complex	Rasamsonia argillacea complex	

MALDI-ToF MS turnaround time is five or fewer minutes per isolate for direct target plate methods; the turnaround time is longer with off-plate protein extraction. Compared to standard methods, yeast and bacterial identification is achieved an average of 1.45 days faster [14], and since only a slight amount of organism is required, testing can be completed on small amounts of growth on primary culture plates without subculture. MALDI-ToF MS has a low reagent cost [14], being less expensive than biochemical- or sequencing-based identification. One study showed that a projected 87% of bacterial and yeast isolates may be identified on the first day using MALDI-ToF MS (versus 9% historically) [14]. Using MALDI-ToF MS, DNA sequencing expenses can be avoided, waste disposal reduced [15,16], and quality control and technologist labor/training for retired tests/replaced tests avoided.

2. Yeasts, with a Focus on *Candida* and *Cryptococcus* Species

MALDI-ToF MS has rapidly become a standard method for yeast identification, out-performing some historical phenotypic systems, and differentiating *Candida albicans* from *Candida dubliniensis*; *C. pararugosa* from *Candida rugosa*; *Candida krusei*, *Candida norvegensis*, and *Candida inconspicua* from one another; *C. orthopsilosis*, *C. metapsilosis* and *C. parapsilosis* from one another [17]; and *C. gattii* from *C. neoformans*, dependent on spectral database representation [18,19]. MALDI-ToF MS may outdo other identification systems for esoteric species, such as *C. famata* and *C. auris*, for example [12,17]. Although older studies used off-plate extraction for yeasts, on-plate extraction with formic acid is now favored for its simplicity; on-plate formic acid preparation yields higher identification rates than does direct transfer alone [20,21].

Dhiman et al. evaluated the Bruker system for identification of 138 common and 103 unusual yeast isolates, reporting 96% and 85% accurate species-level identification, respectively [22]. Westblade et al. assessed the Vitek MS v2.0 system for identification of 852 yeast isolates, including *Candida* species, *C. neoformans*, and other clinically relevant yeasts, using on-plate formic acid preparation, in a multicenter study, reporting 97% and 86% identification to the genus- and species-level, respectively [20]. Won et al. assessed the accuracy of yeast bloodstream isolate identification using the Vitek MS system; correct identification, misidentification and no identification were achieved in 96%, 1% and 3% of cases, respectively [23]. Mancini et al. compared the Bruker and Vitek MS systems for identification of yeasts; correct species-level identifications were comparable using the commercial databases (90% and 84%, respectively), with 100% identified using the Bruker system and a user-developed database [24]. More misidentifications were reported with the Vitek MS system compared to the Bruker system. Rosenvinge et al. studied the Bruker system with 200 yeast isolates, reporting 88% species-level identification (species cutoff of $\geq$1.700) using on-plate formic acid testing [25]. Lacroix et al. demonstrated that the Bruker system with protein extraction and using the manufacturer's species-level cutoff identified 97% of 1383 regularly isolated *Candida* isolates [26]. Pence et al. compared the VITEK MS (IVD Knowledgebase v.2.0) and Biotyper (software v3.1) for identification of 117 yeast isolates, showing correct identification of 95% and 83% of isolates, respectively, using on-plate formic acid testing [27]. Jamal et al. evaluated the Bruker and VITEK MS systems for identification of 188 clinically significant yeast isolates [28], reporting accurate identification of 93% of isolates with both. Three isolates were not identified by VITEK MS, while nine *C. orthopsilosis* were incorrectly identified as *C. parapsilosis*, which was not unexpected since *C. orthopsilosis* was not included in the database studied. Eleven isolates were not identified or misidentified by the Bruker system and although another 14 were identified correctly, their score was <1.700. Hamprecht et al. compared the VITEK MS (V2.0 knowledge base) and the Biotyper (v3.0 software, v3.0.10.0 database, using a species-level cutoff $\geq$2.000) systems for identification of 210 yeasts using on-plate formic acid testing, showing identification of 96% and 91%, respectively [29]. De Carolis et al. made an in-house database using spectra from 156 reference and clinical yeast isolates generated with a sample preparation procedure using suspension of a colony in 10% formic acid, and using 1 μL of the lysate for analysis [30]. Using their library and processing method, and the Bruker system (software v3.0) with a species-level cutoff of ≥ 2.000, they identified 96% of 4232 routinely isolated yeasts. Fatania et al. evaluated the Bruker system with 200 clinically significant yeasts, representing 19 species and five genera, showing agreement between MALDI-ToF MS and conventional methods for 91% [31]. Wang et al. evaluated 2683 yeast isolates comprising 41 species from the National China Hospital Invasive Fungal Surveillance Net program, reporting that the Bruker Biotyper MS system exhibited greater accuracy than the Vitek MS system for all isolates (99% and 95%, respectively) and for *Candida* and related species (99% and 96%, respectively) [32]. Fraser et al. evaluated MALDI-ToF MS using the Bruker system for identification of 6343 clinical isolates of yeasts representing 71 species using a user-developed simplified rapid extraction method, reporting correct identification of 94% of isolates, with a further 6% identified after full extraction [33]. Lee et al. compared the Bruker and VITEK MS systems for identification of 309 clinical isolates of four common *Candida* species, *C. neoformans*, as well as 37 uncommon yeast species, using on-plate formic acid preparation [34]. If "no identification" was obtained, isolates were retested using on-plate formic acid preparation and, for the Bruker system, tube-based extraction. Both systems accurately identified all 158 isolates of the common *Candida* species with initial analysis. The Bruker system correctly identified 9%, 30%, and 100% of 23 *C. neoformans* isolates after initial on-plate formic acid preparation, repeat on-plate formic acid preparation, and tube-based extraction, respectively; VITEK MS identified all *C. neoformans* isolates after initial on-plate formic acid preparation. Both systems had comparable identification rates for 37 uncommon yeast species following initial on-plate formic acid preparation (Bruker, 74%; VITEK MS, 73%) and repeat on-plate formic acid preparation (Bruker, 82%; VITEK MS, 73%). Marucco et al. compared identification of *Candida* species obtained by BD Phoenix (Becton Dickinson, Franklin Lakes, NJ, USA) and the Bruker system using 192 isolates from the strain collection of the Mycology Network of the Autonomous City of Buenos Aires, Argentina, reporting an observed concordance of 95%, with 5% of isolates not correctly identified by the BD Phoenix system [35]. Wilson et al. reported results of a multicenter assessment of the Bruker MALDI Biotyper CA system for identification of clinically significant bacteria and yeasts, including 815 yeast isolates evaluated using three processing methods [36]. The percentage identified and the percentage identified with a high level of confidence were 98% and 88%, respectively, with the extended direct transfer method being superior to the direct transfer method (74% and 49% success, respectively) [36]. Turhan et al. assessed the Bruker system with 117 yeasts, including 115 candidemia-associated *Candida* species, reporting 98% and 87% identification to the genus- and species-level, respectively [37]. Porte et al. compared the two commercial MALDI-ToF MS systems in a routine laboratory in Chile, in a study that included 47 yeasts; the bioMérieux system yielded higher rates of yeast identification to species-level than did the Bruker system (46 and 37 respectively) [38].

2.1. Malassezia Species

Malassezia furfur and Malassezia pachydermatis are included in both FDA-approved/cleared databases (Table 1). Denis et al. developed and evaluated a MALDI-ToF MS database for identifying *Malassezia* species using the Bruker system [39]. Forty-five isolates of *M. furfur*, *Malassezia slooffiae*, *Malassezia sympodialis*, *M. pachydermatis*, *Malassezia restricta* and *Malassezia globosa* were used to create a database, with 40 different isolates used to test the database; all isolates were identified with scores of >2.000.

2.2. Trichosporon Species

Trichosporon inkin and Trichosporon asahii are included in both FDA-approved/cleared databases, with *T. mucoides* additionally claimed by the Vitek MS database and *T. mucoides* group claimed by the MALDI Biotyper CA system (Table 1). de Almeida et al. subjected 16 *Trichosporon* species isolates to MALDI-ToF MS using the Bruker system, evaluating several extraction methods [40]. Overall, incubation for 30 min with 70% formic acid yielded spectra with the highest scores; among the six libraries studied, a library made of 18 strains plus seven clinical isolates yielded the best results, correctly identifying 99% of 68 clinical isolates.

3. Filamentous Fungi

Filamentous fungi demonstrate variable phenotypes as a result of which protein spectra may vary; heterogeneity can be affected by growth conditions and the zone of fungal mycelium examined. Nevertheless, filamentous fungi can be identified using MALDI-ToF MS [41]. The FDA-approved/ cleared Vitek MS system claims 47 filamentous fungi, either species or complexes, including dimorphic pathogens, alongside dermatophytes. As mentioned above, Bruker has an RUO Filamentous Fungi Library. Sample preparation has varied from study-to-study, with sample preparation for molds recommended by companies having changed over time [41]; the FDA-approved/cleared Vitek MS system uses off-plate protein extraction.

McMullen et al. evaluated the Vitek MS using the Vitek MS Knowledge Base, v3.0 for identification of 319 mold isolates, representative of 43 genera, reporting 67% correct identification; when a modified SARAMIS database was used to supplement the v3.0 Knowledge Base, 77% were identified [42]. Rychert et al. reported correct species-level identification of 301/324 clinical isolates of various *Aspergillus* species tests as part of an FDA trial of the Vitek MS v3.0 system; species evaluated included *Aspergillus brasiliensis*, *Aspergillus calidoustus*, *Aspergillus flavus/oryzae*, *Aspergillus fumigatus*, *Aspergillus lentulus*, *Aspergillus nidulans*, *Aspergillus niger* complex, *Aspergillus sydowii*, *Aspergillus terreus* complex, and *Aspergillus versicolor* [43]. Rychert et al. also reported correct species identification of 205/325 clinical isolates of dematiaceous fungi in the same study, including *Alternaria alternata*, *Curvularia hawaiiensis*, *Curvularia spicifera*, *Exserohilum rostratum*, *Exophiala dermatitidis*, *Exophiala xenobiotica*, *Scedosporium boydii*, *Scedosporium apiospermum*, *Scedosporium prolificans* and *Cladophialophora bantiana* [43]. Finally, Rychert et al. reported correct species-level identification of 298/315 clinical isolates of "other potential pathogens", including *Fusarium oxysporum* complex, *Fusarium proliferatum*, *Fusarium solani* complex, *Paecilomyces variotii*, *Penicillium chrysogenum*, *Rasamsonia argillacea*, *Acremonium sclerotigenum*, *Lecythophora hoffmannii*, *Sarocladium kiliense* and *Purpureocillium lilacinum* [43].

De Carolis et al. established their own library of *Fusarium* species, *Aspergillus* species, and Mucorales using the Biotyper system and identified 97% of 94 isolates to the species-level [44]. Gautier et al. used an in-house database to assess the level to which MALDI-ToF MS performed using the Bruker platform enhanced identification; implementation of MALDI-ToF MS resulted in marked enhancement in mold identification at the species-level (to 98%) [45]. Lau et al. used a special extraction technique with a user-developed library representing 294 isolates of 76 genera and 152 species and the Bruker system, to test 421 mold isolates, achieving correct species- and genus-level identifications of 89% and 93% of isolates, respectively [46]. Zvezdanova et al. recently assessed the Bruker system with the Filamentous Fungi Library 1.0 for clinical mold identification using direct target plate testing and simplified processing consisting of mechanical lysis of molds preparatory to protein extraction [47]. They reported accurate species-level identification of 25/34 *Fusarium* species and all 10 *Mucor circinelloides* isolates tested. In addition, 1/21 *Pseudallescheria/Scedosporium* and 7/34 *Fusarium* species isolates were correctly identified to the genus level. The remaining 60 isolates were not identified using the commercial database. They then constructed an in-house database with 63 isolates, which allowed identification of 91% and 100% identification to the species- and genus-levels, respectively.

Normand et al. reported decision criteria for MALDI-ToF MS identification of molds and dermatophytes using the Bruker system [48]. They employed user-developed and Bruker databases as well as 422 isolates of 126 species to evaluate a number of thresholds and one to four spots. They found optimal results with a decision algorithm in which only the uppermost score of four spots was applied, with a 1.700 score threshold. Testing the complete panel enabled identification of 87% and 35% of isolates with the user-developed and Bruker databases, respectively. Applying the same rules to isolates with species represented by at least three strains in the database allowed identification of 92% and 47% of isolates with the user-developed and Bruker databases, respectively. Huang et al. described their findings using the Bruker system and 374 clinical filamentous fungal isolates with correct species and genus identification realized in 99% and 100% of isolates, respectively [49]. Riat et al. used the Bruker

Filamentous Fungi Library 1.0, reporting that an identification score of >1.700 was obtained for 92% of 48 mold isolates studied [50]. Using the Bruker system and a user-developed database, Masih et al. identified 95% of *Aspergillus* species [51]. Park et al. evaluated the Bruker's Filamentous Fungi Library 1.0 with 345 clinical *Aspergillus* isolates; compared with findings of internal transcribed spacer (ITS) sequencing, rates of accurate identification at the species-complex level were 95% and 99%, with cutoff values of 2.000 and 1.700, respectively [52]. Compared with β-tubulin gene sequencing, rates of accurate identification to the species-level were 96% (cutoff 2.000) and 100% (cutoff 1.700) for 303 *Aspergillus* isolates of five common species, but only 5% (cutoff 1.700) and 0% (cutoff 2.000) for 42 *Aspergillus* isolates of six rare species. Schulthess et al. evaluated Bruker's Filamentous Fungi Library 1.0, first studying 83 phenotypically- and molecularly-characterized, non-dermatophyte, non-dematiaceous molds from a clinical isolate collection [53]. Using manufacturer-recommended interpretative criteria, genus and species identification frequencies were 78% and 54%, respectively. Decreasing the species cutoff to 1.700 increased species identification to 71%, without impacting misidentification. In a follow-on prospective study, 200 successive clinical mold isolates were assessed; genus and species identification rates were 84% and 79%, respectively, with a species cutoff of 1.700. Sleiman et al. developed a database for identification of *Aspergillus*, *Fusarium* and *Scedosporium* species [54]. Using 117 isolates, species-level identification was enhanced when the user-developed database was used in conjunction with the Bruker Filamentous Fungi Library compared with the Bruker database alone (*Aspergillus* species, 93% versus 69%; *Fusarium* species, 84% versus 42%; and *Scedosporium* species, 94% versus 18%, respectively). Becker et al. employed a user-developed library and the Bruker system to evaluate 390 clinical isolates, reporting correct identification of 86% of isolates to the species-level using a cutoff of 1.700 [55]. Vidal-Acuña et al. created their own library using 42 clinical *Aspergillus* isolates and 11 strains, cultured in liquid medium, including 23 different species [56]. One hundred and ninety isolates cultured on solid media (179 clinical isolates identified by sequencing and the 11 strains) were studied, with species- and genus-level identifications of 87 and 100%, respectively. They then prospectively challenged their library with 200 *Aspergillus* clinical isolates grown on solid media; species identification was obtained in 96%. Stein et al. evaluated the Bruker system with clinical isolates and reference strains of molds using the Bruker mold, National Institutes of Health, and Mass Spectrometry Identification (MSI) online libraries, comparing results to morphological and molecular identification methods [57]. All libraries studied showed better accuracy in genus identification (≥95%) compared to conventional methods (86%), with 73% of isolates identified to the species-level. The MSI library showed the highest rate of species-level identification (72%) compared to National Institutes of Health (20%) and Bruker (14%) libraries. More than 20% of molds were unidentified to the species-level by all libraries studied, a finding attributed to library limitations and/or poor spectra. Triest et al. evaluated the Bruker system with a user-developed database for identification of 289 *Fusarium* isolates encompassing 40 species from the Belgian Coordinated Collections of Microorganisms/Institute of Hygiene and Epidemiology Mycology culture collection, observing no incorrect species complex identifications [58]. 83% of identifications were accurate to the species-level.

Rychert et al. reported correct species-level identification of clinical isolates of 24/30 *Mucor racemosus* complex, 22/28 *Rhizopus arrhizus* complex, 26/29 *Rhizopus microsporus* complex and 29/31 *Lichtheimia corymbifera*, as part of an FDA trial of the Vitek MS v3.0 system [43]. Dolatabadi et al. utilized the Bruker system with a user-developed database for identification of *R. arrhizus* and its varieties, *delemar* and *arrhizus*, as well as *R. microspores* [59]. Chen et al. assessed the Bruker system with 50 clinically encountered mold isolates, including *Talaromyces marneffei*, *Rhizopus* species, *Paecilomyces* species, *Fusarium solani*, and *Pseudallescheria boydii* [60]. The correct identification rate of *T. marneffei* (score ≥2.000) was 86% based on their user-developed library. Although all seven *P. variotii* isolates, two of the four *P. lilacinus*, four of the six *F. solani*, and two of the three isolates of *Rhizopus* species, and the *P. boydii* isolate had concordant identifications between MALDI-ToF MS and sequencing analysis, scores were all <1.700 [60]. Shao et al. studied 111 isolates of Mucorales belonging to six genera from the Research Center for Medical Mycology of Peking University, initially using the Bruker

Filamentous Fungi library (v1.0), showing 50% and 67% identification to species- and genus-levels, respectively [61]. They then created an in-house library, the Beijing Medical University database, using [11] strains of *Mucor hiemalis*, *Mucor racemosus*, *Mucor irregularis*, *Cunninghamella phaeospora*, *Cunninghamella bertholletiae*, and *Cunninghamella echinulate*. Using the Beijing Medical University and Bruker databases together, all 111 isolates were identified, 81% and 100% to the species- and genus-levels, respectively.

Singh et al. analyzed 72 melanized clinical fungal isolates from patients in 19 Indian medical centers using the Bruker system and a user-developed database, reporting 100% identification [62]. Paul et al. created an in-house database of 59 melanized fungi using a modified protein extraction protocol, and tested 117 clinical isolates using the database [63]. Whereas using the Bruker database only 29 (25%) molds were identified, all were accurately identified accurately by supplementing the Bruker database with the in-house library.

Dermatophytes

With appropriate databases, dermatophytes may be identified using MALDI-ToF MS [64–66]. *Microsporum audouinii*, *Microsporum canis*, *Microsporum gypseum*, *Epidermophyton floccosum*, *Trichophyton rubrum*, *Trichophyton interdigitale*, *Trichophyton tonsurans*, *Trichophyton verrucosum*, and *Trichophyton violaceum* are included in the FDA-approved/cleared Vitek MS system, with no dermatophytes in the FDA-approved/cleared Bruker system. Rychert et al. reported correct species-level identification of clinical isolates of 30/33 *M. audouinii*, 30/31 *M. canis*, 32/35 *M. gypseum*, 30/31 *E. floccosum*, 31/31 *T. rubrum*, 29/30 *T. interdigitale*, 30/33 *T. tonsurans*, 18/31 *T. verrucosum*, and 13/34 *T. violaceum*, as part of an FDA trial of the Vitek MS v3.0 system [43].

Packeu et al. evaluated the Bruker system with a user-developed library for the identification of 176 clinical dermatophyte isolates [67]. MALDI-ToF MS yielded accurate identifications of 97 and 90% of isolates with lowered scores and application of the user-supplemented database, respectively, versus 52% and 14% correct identifications with the unmodified library and recommended scores at the genus- and species-levels, respectively. Calderaro et al. determined the ability of a user-developed database with the Bruker system to identify 64 clinical isolates; all were correctly identified (score of >2.000 for 47 isolates, and 1.700 to 2.000 for the other 17 isolates) [68]. An on-plate procedure after 3 days of incubation produced 40% accurate identification; prolonging incubation time and using an extraction procedure both yielded 100% accurate identification. Karabicak et al. evaluated the Bruker system using a user-developed database with 126 dermatophytes, including 115 clinical isolates and [9] strains; using a combination of the user-developed database and lowered cutoff scores, genus and species identifications were achieved for 97% and 90% of the isolates [69]. L'Ollivier et al. appraised ten studies published between 2008 and 2015 showing accuracy of MALDI-ToF MS-based identification of dermatophytes to vary between 14 and 100% [70]; they ascribed inconsistencies, in part, to processing variability. Use of a tube-based extraction step and a manufacturer database augmented with user-developed spectra were helpful for accurate species identification. Da Cunha et al. assessed whether the direct transfer method can be used with dermatophytes [71]. They built their own library using the Bruker system and evaluated its performance with a panel of mass spectra produced with molecularly-identified isolates and, compared MALDI-ToF MS to morphology-based identification. Although dermatophyte identification using the Bruker library was poor, their database yielded 97% concordance between ITS sequencing and MALDI-ToF MS with 276 isolates. The direct transfer method using unpolished target plates permitted the correct identification of 85% of the clinical dermatophyte isolates.

4. Dimorphic Fungi

The Vitek MS database includes *Blastomyces dermatitidis*, *Coccidioides immitis/posadasii*, *Histoplasma capsulatum*, and *Sporothrix schenckii* complex (Table 2); Rychert et al. evaluated 40, 38, 32 and 31 of these, respectively, as part of an FDA trial of the Vitek MS v3.0 system, reporting 100% identification [43].

Lau et al. assessed the Bruker system for identification of 39 isolates of *T. marneffei* [72]. Using the Filamentous Fungi Library 1.0, MALDI-ToF MS did not identify the isolates; when the database was expanded by including spectra from 21 *T. marneffei* isolates, all isolates in the mold or yeast phase were identified to the species-level. De Almeida et al. showed that the Bruker system with a user-developed database, could identify *Paracoccidioides brasiliensis* and *Paracoccidioides lutzii* [73]. Valero et al. established their own *H. capsulatum* Bruker database using six strains [74]. Then, 30 *H. capsulatum* isolates from the Collection of the Spanish National Centre for Microbiology were studied and correctly identified, 87% with scores above 1.700. The created database was able to identify both growth phases of the fungus, with the most reliable results for the mycelial phase.

5. Limitations

MALDI-ToF MS has limitations. Unlike publicly available sequence databases, such as GenBank, commercial MALDI-ToF MS databases are typically exclusive to companies. Although low identification rates for some organisms may be enhanced by user addition of mass spectral entries of underrepresented species or strains (to cover intraspecies variability), or even re-addition of reference strain spectra to the library, especially those created using parallel growth conditions and preparation methods, doing so may be beyond the know-how of some laboratories. Because of low scores/percentages, repeat testing of isolates may be required [14]. Growth on some media may yield low scores/percentages [75], and small or mucoid colonies may fail. Using experimental capsule size manipulation, it was demonstrated that capsule size of *C. neoformans* and *C. gattii* can compromise identification by the Bruker system [76]. Refined interpretive criteria may be needed to discriminate closely related species and distinguish them from the next best taxon match. For some species of fungi, genus- or species-specific (including lowered) cutoffs may be needed. Mistakes that may occur include testing mixed colonies, spreading amongst spots, spotting into incorrect target plate positions, not properly cleaning re-usable target plates, and wrongly entering results into laboratory information systems. There is a learning curve to depositing ideal biomass onto target plates [77]. Although results are normally reproducible, sources of variability include the technologist, mass spectrometer and especially laser age, matrix and solvent make-up, biological variability, and culture conditions [48]. Instrument (e.g., laser) and software failure may happen. As a result of the simplicity of MALDI-ToF MS, technologists may lose or never develop fine-tuned abilities to visually identify fungi, macroscopically and microscopically.

6. Conclusions

In summary, MALDI-ToF MS has become a routine method for the identification of yeasts and is also being applied to filamentous and dimorphic fungi. Although databases are slowly becoming more complete with regards to clinically-relevant fungi, due to evolving nomenclature and constant description of new species/genera, systematic and continuing library updates will be needed to deliver quality fungal identification into the anticipatable future.

This manuscript has its limitation as the appreciation of MALDI-ToF MS to fungi is rapidly evolving such that some of the cited studies, even if recently published, may rapidly be antiquated.

This paper is based in part on [4,7,8,13,78].

Funding: This manuscript received no external funding.

Conflicts of Interest: Patel reports grants from CD Diagnostics, BioFire, Curetis, Merck, Contrafect, Hutchison Bioflim Medical Solutions, Accelerate Diagnostics, Allergan, EnBiotix, Contrafect and The Medicines Company. Patel is or has been a consultant to Curetis, Specific Technologies, Selux Dx, Genmark Diagnostics, PathoQuest, Heraeus Medical, and Qvella; monies are paid to Mayo Clinic. In addition, Patel has a patent on *Bordetella pertussis/parapertussis* PCR issued, a patent on a device method for sonication with royalties paid by Samsung to Mayo Clinic, and a patent on an anti-biofilm substance issued. Patel received travel reimbursmnet from ASM and IDSA and an editor's stipend from ASM and IDSA, as well as honoraria from the NBME, Up-to-Date and the Infectious Diseases Board Review Course.

References

1. Anhalt, J.; Fenselau, C. Identification of bacteria using mass spectrometry. *Anal. Chem.* **1975**, *47*, 219–225. [CrossRef]
2. Tanaka, K. The origin of macromolecule ionization by laser irradiation (Nobel Lecture). *Angew. Chem.* **2003**, *42*, 3860–3870. [CrossRef] [PubMed]
3. Karas, M.; Hillenkamp, F. Laser desorption ionization of proteins with molecular masses exceeding 10,000 Daltons. *Anal. Chem.* **1988**, *60*, 2299–2301. [CrossRef] [PubMed]
4. Patel, R. MALDI-TOF mass spectrometry: Transformative proteomics for clinical microbiology. *Clin. Chem.* **2013**, *59*, 340–342. [CrossRef] [PubMed]
5. Clinical and Laboratory Standards Institute. *Methods for the Identification of Cultured Microorganisms Using Matrix-Assisted Laser Desorption/Ionization Time-of-Flight Mass Spectrometry*, 1st ed.; Clinical and Laboratory Standards Institute: Wayne, PA, USA, 2017; Volume M58.
6. Mutters, N.T.; Hodiamont, C.J.; de Jong, M.D.; Overmeijer, H.P.; van den Boogaard, M.; Visser, C.E. Performance of Kiestra total laboratory automation combined with MS in clinical microbiology practice. *Ann. Lab. Med.* **2014**, *34*, 111–117. [CrossRef] [PubMed]
7. Patel, R. Matrix-assisted laser desorption ionization-time of flight mass spectrometry in clinical microbiology. *Clin. Infect. Dis.* **2013**, *57*, 564–572. [CrossRef]
8. Patel, R. MALDI-TOF MS for the diagnosis of infectious diseases. *Clin. Chem.* **2015**, *61*, 100–111. [CrossRef]
9. McElvania TeKippe, E.; Burnham, C.A. Evaluation of the Bruker Biotyper and VITEK MS MALDI-TOF MS systems for the identification of unusual and/or difficult-to-identify microorganisms isolated from clinical specimens. *Eur. J. Clin. Microbiol. Infect. Dis.* **2014**, *33*, 2163–2171. [CrossRef]
10. Bilecen, K.; Yaman, G.; Ciftci, U.; Laleli, Y.R. Performances and reliability of Bruker Microflex LT and VITEK MS MALDI-TOF mass spectrometry systems for the identification of clinical microorganisms. *BioMed Res. Int.* **2015**, *2015*, 516410. [CrossRef]
11. Levesque, S.; Dufresne, P.J.; Soualhine, H.; Domingo, M.C.; Bekal, S.; Lefebvre, B.; Tremblay, C. A side by side comparison of Bruker Biotyper and VITEK MS: Utility of MALDI-TOF MS technology for microorganism identification in a public health reference laboratory. *PLoS ONE* **2015**, *10*, e0144878. [CrossRef]
12. Bao, J.R.; Master, R.N.; Azad, K.N.; Schwab, D.A.; Clark, R.B.; Jones, R.S.; Moore, E.C.; Shier, K.L. Rapid, accurate identification of *Candida auris* by using a novel matrix-assisted laser desorption ionization-time of flight mass spectrometry (MALDI-TOF MS) database (library). *J. Clin. Microbiol.* **2018**, *56*. [CrossRef] [PubMed]
13. Carroll, K.; Patel, R. Systems for identification of bacteria and fungi. In *Manual of Clinical Microbiology*, 12th ed.; ASM Press: Washington, DC, USA, 2018; Volume 1, in press.
14. Tan, K.E.; Ellis, B.C.; Lee, R.; Stamper, P.D.; Zhang, S.X.; Carroll, K.C. Prospective evaluation of a matrix-assisted laser desorption ionization-time of flight mass spectrometry system in a hospital clinical microbiology laboratory for identification of bacteria and yeasts: A bench-by-bench study for assessing the impact on time to identification and cost-effectiveness. *J. Clin. Microbiol.* **2012**, *50*, 3301–3308. [CrossRef] [PubMed]
15. Ge, M.C.; Kuo, A.J.; Liu, K.L.; Wen, Y.H.; Chia, J.H.; Chang, P.Y.; Lee, M.H.; Wu, T.L.; Chang, S.C.; Lu, J.J. Routine identification of microorganisms by matrix-assisted laser desorption ionization time-of-flight mass spectrometry: Success rate, economic analysis, and clinical outcome. *J. Microbiol. Immunol. Infect.* **2016**. [CrossRef] [PubMed]
16. Sparbier, K.; Weller, U.; Boogen, C.; Kostrzewa, M. Rapid detection of *Salmonella* sp. by means of a combination of selective enrichment broth and MALDI-TOF MS. *Eur. J. Clin. Microbiol. Infect. Dis.* **2012**, *31*, 767–773. [CrossRef] [PubMed]
17. Castanheira, M.; Woosley, L.N.; Diekema, D.J.; Jones, R.N.; Pfaller, M.A. *Candida guilliermondii* and other species of candida misidentified as *Candida famata*: Assessment by vitek 2, DNA sequencing analysis, and matrix-assisted laser desorption ionization-time of flight mass spectrometry in two global antifungal surveillance programs. *J. Clin. Microbiol.* **2013**, *51*, 117–124. [CrossRef] [PubMed]
18. Firacative, C.; Trilles, L.; Meyer, W. MALDI-TOF MS enables the rapid identification of the major molecular types within the *Cryptococcus neoformans*/*C. gattii* species complex. *PLoS ONE* **2012**, *7*, e37566. [CrossRef]

19. Sendid, B.; Ducoroy, P.; Francois, N.; Lucchi, G.; Spinali, S.; Vagner, O.; Damiens, S.; Bonnin, A.; Poulain, D.; Dalle, F. Evaluation of MALDI-TOF mass spectrometry for the identification of medically-important yeasts in the clinical laboratories of Dijon and Lille hospitals. *Med. Mycol.* **2013**, *51*, 25–32. [CrossRef]

20. Westblade, L.F.; Jennemann, R.; Branda, J.A.; Bythrow, M.; Ferraro, M.J.; Garner, O.B.; Ginocchio, C.C.; Lewinski, M.A.; Manji, R.; Mochon, A.B.; et al. Multicenter study evaluating the Vitek MS system for identification of medically important yeasts. *J. Clin. Microbiol.* **2013**, *51*, 2267–2272. [CrossRef]

21. Theel, E.S.; Schmitt, B.H.; Hall, L.; Cunningham, S.A.; Walchak, R.C.; Patel, R.; Wengenack, N.L. Formic acid-based direct, on-plate testing of yeast and *Corynebacterium* species by Bruker Biotyper matrix-assisted laser desorption ionization-time of flight mass spectrometry. *J. Clin. Microbiol.* **2012**, *50*, 3093–3095. [CrossRef]

22. Dhiman, N.; Hall, L.; Wohlfiel, S.L.; Buckwalter, S.P.; Wengenack, N.L. Performance and cost analysis of matrix-assisted laser desorption ionization-time of flight mass spectrometry for routine identification of yeast. *J. Clin. Microbiol.* **2011**, *49*, 1614–1616. [CrossRef]

23. Won, E.J.; Shin, J.H.; Lee, K.; Kim, M.N.; Lee, H.S.; Park, Y.J.; Joo, M.Y.; Kim, S.H.; Shin, M.G.; Suh, S.P.; et al. Accuracy of species-level identification of yeast isolates from blood cultures from 10 university hospitals in South Korea by use of the matrix-assisted laser desorption ionization-time of flight mass spectrometry-based Vitek MS system. *J. Clin. Microbiol.* **2013**, *51*, 3063–3065. [CrossRef]

24. Mancini, N.; De Carolis, E.; Infurnari, L.; Vella, A.; Clementi, N.; Vaccaro, L.; Ruggeri, A.; Posteraro, B.; Burioni, R.; Clementi, M.; et al. Comparative evaluation of the Bruker Biotyper and Vitek MS matrix-assisted laser desorption ionization-time of flight (MALDI-TOF) mass spectrometry systems for identification of yeasts of medical importance. *J. Clin. Microbiol.* **2013**, *51*, 2453–2457. [CrossRef] [PubMed]

25. Rosenvinge, F.S.; Dzajic, E.; Knudsen, E.; Malig, S.; Andersen, L.B.; Lovig, A.; Arendrup, M.C.; Jensen, T.G.; Gahrn-Hansen, B.; Kemp, M. Performance of matrix-assisted laser desorption-time of flight mass spectrometry for identification of clinical yeast isolates. *Mycoses* **2013**, *56*, 229–235. [CrossRef]

26. Lacroix, C.; Gicquel, A.; Sendid, B.; Meyer, J.; Accoceberry, I.; Francois, N.; Morio, F.; Desoubeaux, G.; Chandenier, J.; Kauffmann-Lacroix, C.; et al. Evaluation of two matrix-assisted laser desorption ionization-time of flight mass spectrometry (MALDI-TOF MS) systems for the identification of *Candida* species. *Clin. Microbiol. Infect.* **2014**, *20*, 153–158. [CrossRef]

27. Pence, M.A.; McElvania Tekippe, E.; Wallace, M.A.; Burnham, C.A. Comparison and optimization of two MALDI-TOF MS platforms for the identification of medically relevant yeast species. *Eur. J. Clin. Microbiol. Infect. Dis.* **2014**. [CrossRef] [PubMed]

28. Jamal, W.Y.; Ahmad, S.; Khan, Z.U.; Rotimi, V.O. Comparative evaluation of two matrix-assisted laser desorption/ionization time-of-flight mass spectrometry (MALDI-TOF MS) systems for the identification of clinically significant yeasts. *Int. J. Infect. Dis.* **2014**, *26*, 167–170. [CrossRef] [PubMed]

29. Hamprecht, A.; Christ, S.; Oestreicher, T.; Plum, G.; Kempf, V.A.; Gottig, S. Performance of two MALDI-TOF MS systems for the identification of yeasts isolated from bloodstream infections and cerebrospinal fluids using a time-saving direct transfer protocol. *Med. Microbiol. Immunol.* **2014**, *203*, 93–99. [CrossRef] [PubMed]

30. De Carolis, E.; Vella, A.; Vaccaro, L.; Torelli, R.; Posteraro, P.; Ricciardi, W.; Sanguinetti, M.; Posteraro, B. Development and validation of an in-house database for matrix-assisted laser desorption ionization-time of flight mass spectrometry-based yeast identification using a fast protein extraction procedure. *J. Clin. Microbiol.* **2014**, *52*, 1453–1458. [CrossRef] [PubMed]

31. Fatania, N.; Fraser, M.; Savage, M.; Hart, J.; Abdolrasouli, A. Comparative evaluation of matrix-assisted laser desorption ionisation-time of flight mass spectrometry and conventional phenotypic-based methods for identification of clinically important yeasts in a UK-based medical microbiology laboratory. *J. Clin. Pathol.* **2015**, *68*, 1040–1042. [CrossRef]

32. Wang, H.; Fan, Y.Y.; Kudinha, T.; Xu, Z.P.; Xiao, M.; Zhang, L.; Fan, X.; Kong, F.; Xu, Y.C. A comprehensive evaluation of the Bruker Biotyper MS and Vitek MS matrix-assisted laser desorption ionization-time of flight mass spectrometry systems for identification of yeasts, part of the national China hospital invasive fungal surveillance net (CHIF-NET) study, 2012 to 2013. *J. Clin. Microbiol.* **2016**, *54*, 1376–1380. [CrossRef]

33. Fraser, M.; Brown, Z.; Houldsworth, M.; Borman, A.M.; Johnson, E.M. Rapid identification of 6328 isolates of pathogenic yeasts using MALDI-ToF MS and a simplified, rapid extraction procedure that is compatible with the Bruker Biotyper platform and database. *Med. Mycol.* **2016**, *54*, 80–88. [CrossRef] [PubMed]

34. Lee, H.S.; Shin, J.H.; Choi, M.J.; Won, E.J.; Kee, S.J.; Kim, S.H.; Shin, M.G.; Suh, S.P. Comparison of the Bruker Biotyper and VITEK MS matrix-assisted laser desorption/ionization time-of-flight mass spectrometry systems using a formic acid extraction method to identify common and uncommon yeast isolates. *Ann. Lab. Med.* **2017**, *37*, 223–230. [CrossRef] [PubMed]

35. Marucco, A.P.; Minervini, P.; Snitman, G.V.; Sorge, A.; Guelfand, L.I.; Moral, L.L.; Integrantes de la Red de Micologia CABA. Comparison of the identification results of *Candida* species obtained by BD Phoenix and Maldi-TOF (Bruker Microflex LT Biotyper 3.1). *Rev. Argent. Microbiol.* **2018**. [CrossRef] [PubMed]

36. Wilson, D.A.; Young, S.; Timm, K.; Novak-Weekley, S.; Marlowe, E.M.; Madisen, N.; Lillie, J.L.; Ledeboer, N.A.; Smith, R.; Hyke, J.; et al. Multicenter evaluation of the Bruker MALDI Biotyper CA system for the identification of clinically important bacteria and yeasts. *Am. J. Clin. Pathol.* **2017**, *147*, 623–631. [CrossRef] [PubMed]

37. Turhan, O.; Ozhak-Baysan, B.; Zaragoza, O.; Er, H.; Saritas, Z.E.; Ongut, G.; Ogunc, D.; Colak, D.; Cuenca-Estrella, M. Evaluation of MALDI-TOF-MS for the identification of yeast isolates causing bloodstream infection. *Clin. Lab.* **2017**, *63*, 699–703. [CrossRef] [PubMed]

38. Porte, L.; Garcia, P.; Braun, S.; Ulloa, M.T.; Lafourcade, M.; Montana, A.; Miranda, C.; Acosta-Jamett, G.; Weitzel, T. Head-to-head comparison of Microflex LT and Vitek MS systems for routine identification of microorganisms by MALDI-TOF mass spectrometry in Chile. *PLoS ONE* **2017**, *12*, e0177929. [CrossRef]

39. Denis, J.; Machouart, M.; Morio, F.; Sabou, M.; Kauffmann-LaCroix, C.; Contet-Audonneau, N.; Candolfi, E.; Letscher-Bru, V. Performance of matrix-assisted laser desorption ionization-time of flight mass spectrometry for identifying clinical *Malassezia* isolates. *J. Clin. Microbiol.* **2017**, *55*, 90–96. [CrossRef]

40. de Almeida, J.N.; Figueiredo, D.S.; Toubas, D.; Del Negro, G.M.; Motta, A.L.; Rossi, F.; Guitard, J.; Morio, F.; Bailly, E.; Angoulvant, A.; et al. Usefulness of matrix-assisted laser desorption ionisation-time-of-flight mass spectrometry for identifying clinical *Trichosporon* isolates. *Clin. Microbiol. Infect.* **2014**, *20*, 784–790. [CrossRef]

41. Sanguinetti, M.; Posteraro, B. Identification of molds by matrix-assisted laser desorption ionization-time of flight mass spectrometry. *J. Clin. Microbiol.* **2017**, *55*, 369–379. [CrossRef]

42. McMullen, A.R.; Wallace, M.A.; Pincus, D.H.; Wilkey, K.; Burnham, C.A. Evaluation of the Vitek MS matrix-assisted laser desorption ionization-time of flight mass spectrometry system for identification of clinically relevant filamentous fungi. *J. Clin. Microbiol.* **2016**, *54*, 2068–2073. [CrossRef]

43. Rychert, J.; Slechta, E.S.; Barker, A.P.; Miranda, E.; Babady, N.E.; Tang, Y.W.; Gibas, C.; Wiederhold, N.; Sutton, D.; Hanson, K.E. Multicenter Evaluation of the Vitek MS v3.0 System for the Identification of Filamentous Fungi. *J. Clin. Microbiol.* **2018**, *56*, e01353-17. [CrossRef] [PubMed]

44. De Carolis, E.; Posteraro, B.; Lass-Florl, C.; Vella, A.; Florio, A.R.; Torelli, R.; Girmenia, C.; Colozza, C.; Tortorano, A.M.; Sanguinetti, M.; et al. Species identification of *Aspergillus*, *Fusarium* and Mucorales with direct surface analysis by matrix-assisted laser desorption ionization time-of-flight mass spectrometry. *Clin. Microbiol. Infect.* **2012**, *18*, 475–484. [CrossRef] [PubMed]

45. Gautier, M.; Ranque, S.; Normand, A.C.; Becker, P.; Packeu, A.; Cassagne, C.; L'Ollivier, C.; Hendrickx, M.; Piarroux, R. Matrix-assisted laser desorption ionization time-of-flight mass spectrometry: Revolutionizing clinical laboratory diagnosis of mould infections. *Clin. Microbiol. Infect.* **2014**, *20*, 1366–1371. [CrossRef] [PubMed]

46. Lau, A.F.; Drake, S.K.; Calhoun, L.B.; Henderson, C.M.; Zelazny, A.M. Development of a clinically comprehensive database and a simple procedure for identification of molds from solid media by matrix-assisted laser desorption ionization-time of flight mass spectrometry. *J. Clin. Microbiol.* **2013**, *51*, 828–834. [CrossRef]

47. Zvezdanova, M.E.; Escribano, P.; Ruiz, A.; Martinez-Jimenez, M.C.; Pelaez, T.; Collazos, A.; Guinea, J.; Bouza, E.; Rodriguez-Sanchez, B. Increased species-assignment of filamentous fungi using MALDI-TOF MS coupled with a simplified sample processing and an in-house library. *Med. Mycol.* **2018**. [CrossRef]

48. Normand, A.C.; Cassagne, C.; Gautier, M.; Becker, P.; Ranque, S.; Hendrickx, M.; Piarroux, R. Decision criteria for MALDI-TOF MS-based identification of filamentous fungi using commercial and in-house reference databases. *BMC Microbiol.* **2017**, *17*, 25. [CrossRef]

49. Huang, Y.; Zhang, M.; Zhu, M.; Wang, M.; Sun, Y.; Gu, H.; Cao, J.; Li, X.; Zhang, S.; Wang, J.; et al. Comparison of two matrix-assisted laser desorption ionization-time of flight mass spectrometry systems for the identification of clinical filamentous fungi. *World J. Microbiol. Biotechnol.* **2017**, *33*, 142. [CrossRef]

50. Riat, A.; Hinrikson, H.; Barras, V.; Fernandez, J.; Schrenzel, J. Confident identification of filamentous fungi by matrix-assisted laser desorption/ionization time-of-flight mass spectrometry without subculture-based sample preparation. *Int. J. Infect. Dis.* **2015**, *35*, 43–45. [CrossRef]
51. Masih, A.; Singh, P.K.; Kathuria, S.; Agarwal, K.; Meis, J.F.; Chowdhary, A. Identification by molecular methods and matrix-assisted laser desorption ionization-time of flight mass spectrometry and antifungal susceptibility profiles of clinically significant rare *Aspergillus* species in a referral chest hospital in Delhi, India. *J. Clin. Microbiol.* **2016**, *54*, 2354–2364. [CrossRef]
52. Park, J.H.; Shin, J.H.; Choi, M.J.; Choi, J.U.; Park, Y.J.; Jang, S.J.; Won, E.J.; Kim, S.H.; Kee, S.J.; Shin, M.G.; et al. Evaluation of matrix-assisted laser desorption/ionization time-of-fight mass spectrometry for identification of 345 clinical isolates of *Aspergillus* species from 11 Korean hospitals: Comparison with molecular identification. *Diagn. Microbiol. Infect. Dis.* **2017**, *87*, 28–31. [CrossRef]
53. Schulthess, B.; Ledermann, R.; Mouttet, F.; Zbinden, A.; Bloemberg, G.V.; Bottger, E.C.; Hombach, M. Use of the Bruker MALDI Biotyper for identification of molds in the clinical mycology laboratory. *J. Clin. Microbiol.* **2014**, *52*, 2797–2803. [CrossRef]
54. Sleiman, S.; Halliday, C.L.; Chapman, B.; Brown, M.; Nitschke, J.; Lau, A.F.; Chen, S.C. Performance of matrix-assisted laser desorption ionization-time of flight mass spectrometry for identification of *Aspergillus*, Scedosporium, and *Fusarium* spp. in the Australian clinical setting. *J. Clin. Microbiol.* **2016**, *54*, 2182–2186. [CrossRef]
55. Becker, P.T.; de Bel, A.; Martiny, D.; Ranque, S.; Piarroux, R.; Cassagne, C.; Detandt, M.; Hendrickx, M. Identification of filamentous fungi isolates by MALDI-TOF mass spectrometry: Clinical evaluation of an extended reference spectra library. *Med. Mycol.* **2014**, *52*, 826–834. [CrossRef]
56. Vidal-Acuna, M.R.; Ruiz-Perez de Pipaon, M.; Torres-Sanchez, M.J.; Aznar, J. Identification of clinical isolates of *Aspergillus*, including cryptic species, by matrix assisted laser desorption ionization time-of-flight mass spectrometry (MALDI-TOF MS). *Med. Mycol.* **2018**, *56*, 838–846. [CrossRef] [PubMed]
57. Stein, M.; Tran, V.; Nichol, K.A.; Lagace-Wiens, P.; Pieroni, P.; Adam, H.J.; Turenne, C.; Walkty, A.J.; Normand, A.C.; Hendrickx, M.; et al. Evaluation of three MALDI-TOF mass spectrometry libraries for the identification of filamentous fungi in three clinical microbiology laboratories in Manitoba, Canada. *Mycoses* **2018**, *61*, 743–753. [CrossRef]
58. Triest, D.; Stubbe, D.; De Cremer, K.; Pierard, D.; Normand, A.C.; Piarroux, R.; Detandt, M.; Hendrickx, M. Use of matrix-assisted laser desorption ionization-time of flight mass spectrometry for identification of molds of the *Fusarium* genus. *J. Clin. Microbiol.* **2015**, *53*, 465–476. [CrossRef] [PubMed]
59. Dolatabadi, S.; Kolecka, A.; Versteeg, M.; de Hoog, S.G.; Boekhout, T. Differentiation of clinically relevant Mucorales *Rhizopus microsporus* and *R. arrhizus* by matrix-assisted laser desorption ionization time-of-flight mass spectrometry (MALDI-TOF MS). *J. Med. Microbiol.* **2015**, *64*, 694–701. [CrossRef]
60. Chen, Y.S.; Liu, Y.H.; Teng, S.H.; Liao, C.H.; Hung, C.C.; Sheng, W.H.; Teng, L.J.; Hsueh, P.R. Evaluation of the matrix-assisted laser desorption/ionization time-of-flight mass spectrometry Bruker Biotyper for identification of *Penicillium marneffei*, *Paecilomyces* species, *Fusarium solani*, *Rhizopus* species, and *Pseudallescheria Boydii*. *Front. Microbiol.* **2015**, *6*, 679. [CrossRef] [PubMed]
61. Shao, J.; Wan, Z.; Li, R.; Yu, J. Species identification and delineation of pathogenic Mucorales by matrix-assisted laser desorption ionization-time of flight mass spectrometry. *J. Clin. Microbiol.* **2018**, *56*. [CrossRef]
62. Singh, A.; Singh, P.K.; Kumar, A.; Chander, J.; Khanna, G.; Roy, P.; Meis, J.F.; Chowdhary, A. Molecular and matrix-assisted laser desorption ionization-time of flight mass spectrometry-based characterization of clinically significant melanized fungi in India. *J. Clin. Microbiol.* **2017**, *55*, 1090–1103. [CrossRef]
63. Paul, S.; Singh, P.; Sharma, S.; Prasad, G.S.; Rudramurthy, S.M.; Chakrabarti, A.; Ghosh, A.K. MALDI-TOF MS-based identification of melanized fungi is faster and reliable after the expansion of in-house database. *Proteom. Clin. Appl.* **2018**. [CrossRef] [PubMed]
64. Theel, E.S.; Hall, L.; Mandrekar, J.; Wengenack, N.L. Dermatophyte identification using matrix-assisted laser desorption ionization-time of flight mass spectrometry. *J. Clin. Microbiol.* **2011**, *49*, 4067–4071. [CrossRef] [PubMed]
65. Nenoff, P.; Erhard, M.; Simon, J.C.; Muylowa, G.K.; Herrmann, J.; Rataj, W.; Graser, Y. MALDI-TOF mass spectrometry—A rapid method for the identification of dermatophyte species. *Med. Mycol.* **2013**, *51*, 17–24. [CrossRef] [PubMed]

66. de Respinis, S.; Tonolla, M.; Pranghofer, S.; Petrini, L.; Petrini, O.; Bosshard, P.P. Identification of dermatophytes by matrix-assisted laser desorption/ionization time-of-flight mass spectrometry. *Med. Mycol.* **2013**, *51*, 514–521. [CrossRef] [PubMed]
67. Packeu, A.; De Bel, A.; l'Ollivier, C.; Ranque, S.; Detandt, M.; Hendrickx, M. Fast and accurate identification of dermatophytes by matrix-assisted laser desorption ionization-time of flight mass spectrometry: Validation in the clinical laboratory. *J. Clin. Microbiol.* **2014**, *52*, 3440–3443. [CrossRef] [PubMed]
68. Calderaro, A.; Motta, F.; Montecchini, S.; Gorrini, C.; Piccolo, G.; Piergianni, M.; Buttrini, M.; Medici, M.C.; Arcangeletti, M.C.; Chezzi, C.; et al. Identification of dermatophyte species after implementation of the in-house MALDI-TOF MS database. *Int. J. Mol. Sci.* **2014**, *15*, 16012–16024. [CrossRef]
69. Karabicak, N.; Karatuna, O.; Ilkit, M.; Akyar, I. Evaluation of the Bruker matrix-assisted laser desorption-ionization time-of-flight mass spectrometry (MALDI-TOF MS) system for the identification of clinically important dermatophyte species. *Mycopathologia* **2015**, *180*, 165–171. [CrossRef] [PubMed]
70. L'Ollivier, C.; Ranque, S. MALDI-TOF-based dermatophyte identification. *Mycopathologia* **2017**, *182*, 183–192. [CrossRef] [PubMed]
71. da Cunha, K.C.; Riat, A.; Normand, A.C.; Bosshard, P.P.; de Almeida, M.T.G.; Piarroux, R.; Schrenzel, J.; Fontao, L. Fast identification of dermatophytes by MALDI-TOF/MS using direct transfer of fungal cells on ground steel target plates. *Mycoses* **2018**. [CrossRef] [PubMed]
72. Lau, S.K.; Lam, C.S.; Ngan, A.H.; Chow, W.N.; Wu, A.K.; Tsang, D.N.; Tse, C.W.; Que, T.L.; Tang, B.S.; Woo, P.C. Matrix-assisted laser desorption ionization time-of-flight mass spectrometry for rapid identification of mold and yeast cultures of *Penicillium Marneffei*. *BMC Microbiol.* **2016**, *16*, 36. [CrossRef] [PubMed]
73. de Almeida, J.N.; Del Negro, G.M.; Grenfell, R.C.; Vidal, M.S.; Thomaz, D.Y.; de Figueiredo, D.S.; Bagagli, E.; Juliano, L.; Benard, G. Matrix-assisted laser desorption ionization-time of flight mass spectrometry for differentiation of the dimorphic fungal species *Paracoccidioides brasiliensis* and *Paracoccidioides Lutzii*. *J. Clin. Microbiol.* **2015**, *53*, 1383–1386. [CrossRef] [PubMed]
74. Valero, C.; Buitrago, M.J.; Gago, S.; Quiles-Melero, I.; Garcia-Rodriguez, J. A matrix-assisted laser desorption/ionization time of flight mass spectrometry reference database for the identification of *Histoplasma Capsulatum*. *Med. Mycol.* **2018**, *56*, 307–314. [CrossRef] [PubMed]
75. Anderson, N.W.; Buchan, B.W.; Riebe, K.M.; Parsons, L.N.; Gnacinski, S.; Ledeboer, N.A. Effects of solid-medium type on routine identification of bacterial isolates by use of matrix-assisted laser desorption ionization-time of flight mass spectrometry. *J. Clin. Microbiol.* **2012**, *50*, 1008–1013. [CrossRef] [PubMed]
76. Thomaz, D.Y.; Grenfell, R.C.; Vidal, M.S.; Giudice, M.C.; Del Negro, G.M.; Juliano, L.; Benard, G.; de Almeida Junior, J.N. Does the capsule interfere with performance of matrix-assisted laser desorption ionization-time of flight mass spectrometry for identification of *Cryptococcus neoformans* and *Cryptococcus gattii*? *J. Clin. Microbiol.* **2016**, *54*, 474–477. [CrossRef] [PubMed]
77. Harris, P.; Winney, I.; Ashhurst-Smith, C.; O'Brien, M.; Graves, S. Comparison of Vitek MS (MALDI-TOF) to standard routine identification methods: An advance but no panacea. *Pathology* **2012**, *44*, 583–585. [CrossRef] [PubMed]
78. Heaton, P.; Patel, R. Mass spectrometry applications in infectious disease and pathogens identification. In *Principles and Applications of Clinical Mass Spectrometry, Small Molecules, Peptides, and Pathogens*; Elsevier: Amsterdam, The Netherlands, 2018; pp. 93–114.

Journal of
Fungi

MDPI

Review

Antifungal Resistance: Specific Focus on Multidrug Resistance in *Candida auris* and Secondary Azole Resistance in *Aspergillus fumigatus*

Sevtap Arikan-Akdagli [1,*], Mahmoud Ghannoum [2] and Jacques F. Meis [3,4]

[1] Department of Medical Microbiology, Mycology Laboratory, Hacettepe University Medical School, TR-06100 Ankara, Turkey
[2] Center for Medical Mycology, Department of Dermatology, University Hospitals Cleveland Medical Center, Case Western Reserve University, Cleveland, OH 44106, USA; mag3@case.edu
[3] Department of Medical Microbiology and Infectious Diseases, Canisius Wilhelmina Hospital (CWZ), 6532 Nijmegen, The Netherlands; jacques.meis@gmail.com
[4] Centre of Expertise in Mycology Radboudumc/CWZ, 6532 Nijmegen, The Netherlands
* Correspondence: sarikanakdagli@gmail.com

Received: 15 November 2018; Accepted: 3 December 2018; Published: 5 December 2018

Abstract: Antifungal resistance is a topic of concern, particularly for specific fungal species and drugs. Among these are the multidrug-resistant *Candida auris* and azole-resistant *Aspergillus fumigatus*. While the knowledge on molecular mechanisms of resistance is now accumulating, further data are also available for the clinical implications and the extent of correlation of in vitro resistance to clinical outcomes. This review article summarizes the epidemiology of *C. auris* infections, animal models focusing on the activity of novel antifungal compounds in *C. auris* infections, virulence factors, and the mechanisms of antifungal resistance for this multi-resistant *Candida* species. Regarding *A. fumigatus*, the significance of azoles in the treatment of *A. fumigatus* infections, reference methods available for the detection of resistance in vitro, molecular mechanisms of secondary azole resistance, routes of acquisition, and clinical implications of in vitro resistance are covered to provide guidance for the current status of azole resistance in *A. fumigatus*.

Keywords: *Candida auris*; *Aspergillus fumigatus*; antifungal resistance; multidrug resistance; mechanisms of antifungal resistance

1. Candida auris

1.1. Epidemiology and Risk Factors for Candida auris Infection

Nosocomial infections with resistant *Candida* species are increasing and candidemia is becoming a public health concern in Europe, the Americas, and Asia. This is associated with increasing numbers of immunocompromised individuals, the rampant empirical use of broad-spectrum antibiotics and fluconazole, and the widespread use of implanted medical devices. Invasive non-*albicans* candidiasis was mainly reported, until recently, due to *C. glabrata*, *C. parapsilosis*, *C. tropicalis*, and *Pichia kudriavzevii* (*C. krusei*). *C. parapsilosis* is common among newborns, while *C. glabrata* is more prevalent among older adults and patients with cancer. *C. tropicalis*, on the other hand, is more commonly seen in patients with leukemia and neutropenia. *C. parapsilosis*, a skin colonizer, is a common pathogen in intravascular catheter-related infections. *Pichia kudriavzevii*, in turn, is found more often among patients with leukemia and associated neutropenia, who receive fluconazole prophylaxis [1]. A new species, *C. auris*, associated with resistance to several antifungal drugs and difficulty in identification, has been observed to be emerging in the last decade. This yeast was first described in East Asia in 2009, after being isolated from a Japanese patient with otitis externa and three Korean patients with candidemia [2,3]. These observations did not attract much attention from the medical community at the time until clonal outbreaks were observed in several Indian hospitals [4,5]. Shortly after these seminal publications, reports followed from Kuwait, South Africa, and Venezuela [6–10]. *C. auris*, which was never heard of prior to the first publication in 2009, became an emerging global health threat with outbreaks occurring in many health facilities. It is highly likely that *C. auris* was an underreported infection in the first years after 2009 due to difficulties in identification [11–16]. At present, infected and colonized patients have been identified in Australia [17], Austria [18], Belgium [19], Canada [20], China [21–23], Colombia [24,25], Egypt (unpublished), France [19], Germany [19], India [4,26–29], Iran (unpublished), Israel [30,31], Kenya [15], Kuwait [7,9,10], Korea [3,32], Malaysia [33], the Netherlands [15], Norway [19], Oman [34,35], Pakistan [36], Panama [37], Russia [15], Saudi Arabia [38], Singapore [39], South Africa [8,40], Spain [41], Switzerland [42], Thailand [15], United Kingdom [43–46], United States [36,47], United Arab Emirates [48], and Venezuela [6]. More than 4000 cases of infection and colonization, the majority from India and South Africa, have been recorded to date, but it is highly likely that we are observing only the tip of the iceberg.

C. auris is a novel *Candida* species in the *Candida haemulonii* species complex, which causes a wide range of infections, especially in debilitated patients residing in intensive care units (ICUs). A large 18-month prospective study in Indian ICUs recorded 1400 candidemia cases; *C. auris* was identified as the fifth most common cause found in 19 out of 27 ICUs, with a prevalence of 5.3% [27]. In some tertiary care Indian hospitals, *C. auris* is the second most common cause of candidemia after *C. tropicalis* [49]. A tertiary medical center in South America reported *C. auris* as the sixth most common cause of nosocomial bloodstream infections between March 2012 and July 2013 [6]. The mode of spread within the hospital setting is through person to person transmission and via contaminated surfaces and/or equipment. During outbreaks, *C. auris* can contaminate the room of colonized or infected patients [50]. It is therefore of utmost importance to quickly identify contaminated surfaces and screen specimens of patients. Real-time detection and identification of *C. auris* is the target of several molecular kits [51–59]. The survival of *C. auris* for weeks, even months, within the hospital confirms the importance of infection prevention programs [60–62]. Transmission from patient to patient has been documented to lead to skin colonization by *C. auris* and increased risk for candidemia. The hospital environment represents a reservoir that contributes to the nosocomial transmission of *C. auris* similar to that seen with multi-resistant bacterial pathogens [63,64]. Risk factors for infection with *C. auris* are related to immunosuppression, hospitalization in intensive care units over prolonged periods, use of central venous and urinary catheters, and empirical use of antibiotics or antifungals. Adults are mainly affected, but in an outbreak situation in Venezuela 13/18 cases were pediatric patients [6]. As observed in many other studies, all isolates were initially mis-identified as *Candida haemulonii*, a commonly reported mistake [65–67]. Sequencing of the internal transcribed spacer (ITS) region and MALDI-TOF analysis were necessary to identify isolates of the outbreak involved as *C. auris*. The predisposing risk factors for *C. auris* infection are similar to other opportunistic *Candida* species [1]; that is, immunocompromised patients (diabetes mellitus, malignancy, chronic renal disease, neutropenia, HIV), concomitant bacteremia, broad spectrum antibacterial or antifungal therapy within 90 days, surgery within 90 days, presence of central venous catheters or urinary catheters, ICU stay, and parenteral nutrition (PN) administration confer an increased risk of acquiring *C. auris*. A case-control study in an Indian center was conducted to determine specific risk factors predisposing to *C. auris* candidemia [29]. Patients with *C. auris* (*n* = 74) and non-*auris* (*n* = 1087) fungemia cases were analyzed. Multivariate analysis showed that patients with respiratory diseases, vascular surgery, and prolonged exposure to fluconazole were more likely to develop ICU-onset *C. auris* fungemia. In describing the epidemiology of *C. auris* infections, the Center for Disease Control (CDC) used whole genome sequencing of 54 isolates collected from India, Pakistan, South Africa, Japan, and Venezuela [36]. Four distinct geographical clades were observed, suggesting emergence at the same time on three continents. Similar geographic clustering was observed with Amplified Fragment Length

Polymorphism(AFLP) and proteomic analysis of *C. auris* isolates from three different continents—Asia, Africa, and Latin America [68]. A recent Whole Genome Sequencing (WGS) and single-nucleotide polymorphism (SNP) analysis of *C. auris* strains isolated in the USA showed multiple introductions of *C. auris* isolates belonging to the four clades, and spread among healthcare facilities [47]. Most *C. auris* strains (>60–90%) are resistant to fluconazole, 10–30% exhibit a high minimum inhibitory concentration (MIC) for amphotericin B, and <5% can be considered resistant to echinocandins [28,69]. Given the recent unprecedented worldwide spread and multidrug resistance, *C. auris* is included in the world's 10 most feared fungi [70].

1.2. Virulence Factors of C. auris

To determine the virulence properties of *C. auris* relative to *C. albicans*, a set of clinical strains were investigated regarding the ability to germinate, adhere, and produce extracellular enzymes [71]. *C. auris* strains failed to germinate but, in contrast and as expected, *C. albicans* germinated profusely. Similarly, *C. auris* exhibited a significantly reduced ability to adhere to silicon elastomer disks relative to *C. albicans*. Moreover, the *C. auris* isolates produced phospholipase and proteinase in a strain-dependent manner (37.5% of the *C. auris* strains possessed phospholipase activity, while 64% evaluated secreted proteinase activity). The last virulence factor evaluated was the ability of *C. auris* to form biofilms. Our data showed that the formed biofilms were mainly composed of yeast cells, while biofilms formed by *C. albicans* had a heterogeneous architecture of biofilms comprised of yeast and hyphae morphology embedded within the extracellular matrix. Furthermore, *C. auris* biofilms had a limited amount of extracellular matrix relative to *C. albicans* and its biofilm thickness was significantly less than the biofilms formed by *C. albicans*. Taken together, these data show that *C. auris* is relatively less pathogenic than *C. albicans*.

1.3. C. auris Animal Models and Activity of Experimental Antifungals

To gain insight into the in vivo virulence of *C. auris*, an immunosuppressed murine model was developed [72,73]. Once the model was established it was used to evaluate the efficacy of two experimental antifungals (rezafungin and APX001A). The data showed that rezafungin had a significantly reduced CFUs/g kidneys fungal burden compared with vehicle- or amphotericin B-treated groups. Furthermore, treatment with rezafungin resulted in a significantly lower CFUs/g tissue fungal burden compared to micafungin-treated animals [72].

Evaluation of the efficacy of APX001 using the optimized immunocompromised mouse model showed that treatment with this experimental drug resulted in a significant increase in animal survival (between 80 and 100% survival in the three treatment groups). In contrast, treatment with anidulafungin led to only a 50% survival rate. In addition, APX001 treatment led to a significant reduction in CFUs/g of kidneys, lung, and brain tissue compared to the vehicle-treated group [73]. In an immunocompetent murine model, virulence was also highest for *C. albicans*, closely followed by *C. auris*, *C. glabrata*, and *C. haemulonii*, respectively [74].

1.4. Resistance of C. auris

Besides being antifungal-resistant, *C. auris* is thermotolerant, grows well up to 42 °C, and is salt-tolerant (up to 10%). These characteristics can be used to design selective media for the detection of *C. auris* for screening purposes which have been used successfully in outbreak investigations [75]. Concerning resistance to antifungal agents, *C. auris* has demonstrated extensive resistance to azoles and amphotericin B [24,28]. The ATP-binding cassette (ABC) transporter activity was significantly higher in *C. auris* than in *C. glabrata* [31]. Several genes show encoding of ABC transporters and the important families of *C. auris* major facilitator superfamily (MFS) genes [76]. An Indian study with a large number of isolates showed that 41% of *C. auris* from India showed resistance to two antifungal classes and 4% to three antifungal agents [28]. Molecular mechanisms responsible for antifungal resistance point to efflux pumps and mutations in the lanosterol 14-alpha-demethylase (*ERG11*) gene to explain the high rate of resistance to fluconazole [28,31]. The latter study demonstrated that 90% of *C. auris* isolates were resistant to fluconazole (MICs 32 to ≥64 mg/L). *ERG11* sequences of resistant *C. auris* exhibited substitutions of the Y132 and K143 amino acids in 77% of the fluconazole-resistant strains. No substitutions at these positions were observed in isolates with low fluconazole MICs (1–2 mg/L), suggesting that these substitutions confer the fluconazole resistance phenotype similar to that described for *C. albicans* [77].

Another study, in a murine model, showed that micafungin was superior compared to fluconazole or amphotericin B, with greater fungicidal activity [78]. These findings make echinocandins the drugs of choice to treat *C. auris* infections and clinical trials are on their way to explore the therapeutic potential of new drugs.

The combination of antifungals such as voriconazole and echinocandins has been shown to be promising in vitro against resistant *C. auris* [79]. Although some studies show variable susceptibility of *C. auris* to the echinocandin class [80], the good news is that there are new drugs in development with excellent activity against *C. auris* [71–73,81–85]. SCY-078, a novel orally bioavailable 1,3-β-D-glucan synthesis inhibitor, has been shown to exhibit both in vitro and in vivo activity against *C. auris*, including some echinocandin-resistant isolates. VT-1598 is another new azole drug with broad activity including *C. auris* isolates (MIC range 0.03–8 mg/L) [86,87].

The cleaning and terminal disinfection of rooms where *C. auris*-colonized patients have been problematic [61,88]. Moore et al. [89] showed that chlorine-based disinfectants and iodine-based skin antiseptics were effective against *C. auris* and reduced environmental contamination and skin colonization. Chlorhexidine-based products may also be effective. Abdolrasouli and collaborators [90] demonstrated that *C. auris* isolates were inhibited by chlorhexidine gluconate at 0.125–1.5% and by iodinated povidone at a concentration of 0.07–1.25%.

2. *Aspergillus fumigatus*

2.1. Azole Resistance in A. fumigatus

Aspergillus remains significant as one of the causative agents of invasive infections in immunocompromised individuals and frequently constitutes the most common mold genus isolated in this setting. While voriconazole is the primary drug of choice in the treatment of invasive aspergillosis, the emergence of azole resistance in *Aspergillus* has been a concern since the first report of secondary resistance of *A. fumigatus* to itraconazole in 1997 [91,92]. Antifungal drugs which exert activity against *Aspergillus* spp. are amphotericin B, triazoles, and echinocandins. Furthermore, triazoles are of particular significance due to the availability of oral formulations. Based on this, triazoles constitute significant therapeutic options for patients with chronic pulmonary aspergillosis and allergic bronchopulmonary aspergillosis who require long-term therapy [93] and azole resistance in *A. fumigatus* is thus a concern in this respect as well.

Secondary azole resistance in *A. fumigatus* has been reported from many countries and centers in six continents at extensively varying rates. Similar to those for strains isolated from clinical samples, resistance rates detected for environmental strains are also diverse [94–101]. The ISHAM/ECMM *Aspergillus* Resistance Surveillance Working Group aims to facilitate surveillance studies to determine resistance epidemiology in countries where data are currently lacking and provide further insight in terms of clinical implications [102].

While secondary azole resistance in *A. fumigatus* draws remarkable attention, the awareness and knowledge on primary antifungal resistance in *Aspergillus* strains are also increasing. Among the species which are relatively common causes of invasive infections and exhibit primary resistance or reduced susceptibility to one or more antifungal drugs are *Aspergillus lentulus* (resistance to azoles and amphotericin B and varied susceptibility to caspofungin), *Aspergillus flavus* (reduced susceptibility to amphotericin B and varied susceptibility to caspofungin), *Aspergillus alliaceus* (reduced susceptibility to amphotericin B and caspofungin), and *Aspergillus terreus* (resistance to amphotericin B) [96].

2.2. Detection of Antifungal Resistance In Vitro by Reference Methods

Reference CLSI [103] and EUCAST [104,105] microdilution susceptibility testing methods are available for testing antifungal drugs against *Aspergillus* and recommended for routine use [106,107]. A disk diffusion method of CLSI for testing non-dermatophyte molds and thus applying also to *Aspergillus* is also available [108]. Epidemiological cut-off values have been determined for the interpretation of the results obtained by the CLSI method [109–112], while both epidemiological cut-off values and clinical breakpoints are available for interpreting EUCAST minimum inhibitory concentration values (MIC, mg/L) for some drugs and species [113]. The official reading method for amphotericin B and azole MICs against *Aspergillus* is visual reading for both CLSI and EUCAST methodologies. A spectrophotometric reading alternative for EUCAST amphotericin B and azole MICs at 5% growth cut-off (vs. complete inhibition of growth visually) proved to be a reliable alternative [105].

An agar screening method for the detection of secondary azole resistance in *A. fumigatus* strains has also been validated recently by a multicenter study undertaken by EUCAST [114]. This method uses (in-house or commercially available) 4-well agar plates containing itraconazole (4 mg/L), voriconazole (2 mg/L), and posaconazole (0.5 mg/L); the fourth well serves as the growth control well without any antifungal drug. The ranges of 80–100% and 97–100% were obtained, respectively, for interobserver agreement rate and overall sensitivity. The inter-plate (in-house vs. commercial) agreement rate was high. Similarly, the sensitivity for simulated mixed samples of wild-type and mutant strains and the overall specificity rates also proved to be acceptably high (83–100% and 95–100%, respectively). Based on these data, the assay was validated and is now available as a reference method as documented in EUCAST E.DEF 10.1 [115]. It is an easy and reliable method recommended to be used for routine laboratory work-up, to be followed by reference MIC testing for confirmation in case of the detection of a resistant strain [106].

2.3. Molecular Mechanisms Involved in Secondary Azole Resistance and the Resulting Azole Susceptibility Profiles

Point mutations in the *cyp51A* gene associated with amino acid changes of M220, G54, G138, G448S, as well as L98H are the most common mechanisms of secondary azole resistance in *A. fumigatus*. Extra copies of the *cyp51A* gene (e.g., tandem repeats of a 34- or 46-bp sequence in the promoter of the *cyp51A* gene) may also accompany specific amino acid changes. The typical examples of this combined pattern are TR34/L98H and TR46/Y121F/T289A [116]. A tandem repeat of 53 bp without any accompanying amino acid change has also been described [117,118]. In addition, non-*cyp51* mutations and increased expression of efflux pumps may play a role in the development of secondary azole resistance. On the other hand, the mechanism remains unknown for a number of isolates. The expected azole susceptibility profiles in relation to the associated amino acid changes and/or tandem repeats are summarized in Table 1 [95,116].

Table 1. Expected azole susceptibility profiles with respect to the detected resistance mechanism(s).

Associated Amino Acid Change/Tandem Repeat	Resistance	Reduced Susceptibility	Variable Susceptibility Profile
G54	ITC, POS		
G138	ITC, POS		
G448S	VRC	ITC, POS	
M220	ITC	VRC	POS
TR34/L98H *	ITC, VRC, POS, ISV		
TR46/Y121F/T289A	VRC		ITC, POS
TR53	ITC, VRC	POS	

ISV: isavuconazole; ITC: itraconazole; POS: posaconazole; VRC: voriconazole; *: Isolates with TR34/L98H/S297T/F495I changes may have lower minimum inhibitory concentrations (MICs) of voriconazole in the wild-type range. The S297T mutation might be a compensatory mutation in these cases [119,120].

2.4. Acquisition of Secondary Azole Resistance

There are two mechanisms that play a role in the development of secondary azole resistance in *A. fumigatus*. First is the (long-term) azole therapy in an individual patient with chronic pulmonary lung disease mostly in existence of a pulmonary cavity, and second is the direct acquisition of a resistant strain from the environment. The latter develops due to the use of azole fungicides (penconazole, difenoconazole, tetraconazole, and tebuconazole) in the environment in agriculture for plant protection [98,121]. The molecular mechanisms leading to resistance also differ in general for these two routes of acquisition. In the patient-acquired route, M220, G54, and G138 changes are more common while TR34/L98H and TR46/Y121F/T289A patterns are mostly (but not always) observed following environmental acquisition [95,116].

2.5. Clinical Implications and Current Recommendations for Treatment of Aspergillosis due to Azole-Resistant A. fumigatus

While high azole MICs [122,123] or the existence of *cyp51A* mutations [124] were found to be correlated with clinical failure in some studies, other investigators were not able to detect any correlation between MICs and survival rates [125]. This may also emphasize the influence of host factors as well as several others on clinical outcomes in invasive fungal infections observed in immunocompromised individuals. The low rates of resistance, i.e., the low number of infections due to azole-resistant strains included in the analysis, may also render it more difficult to detect any possibly existing in vitro–in vivo correlation. "Strong" or "Moderate" recommendations for the treatment of documented azole-resistant aspergillosis, as included in the recently published ESCMID-ECMM-ERS Guideline [106], are liposomal amphotericin B monotherapy (Strength of Recommendation (SoR) and Quality of Evidence (QoE): AIIu) and voriconazole and anidulafungin combination (BIII), respectively. Other options with a "Marginal" level of recommendation (CIII for all noted alternatives) include amphotericin B lipid complex monotherapy, posaconazole and caspofungin combination, caspofungin or micafungin monotherapy. The expert opinion, on the other hand, recommends a modification in primary therapeutic choice of voriconazole in case of local environmental resistance rates of >10%. Voriconazole and echinocandin combination or liposomal amphotericin B monotherapy is recommended for initial therapy under these settings [126].

3. Concluding Remark

The emerging field of molecular mechanisms of antifungal resistance has been an underestimated area of global public health concern, but significant progress has been made lately in *A. fumigatus* and *C. auris*, although research challenges remain formidable.

Author Contributions: S.A.-A., M.G. and J.F.M. prepared the manuscript.

Funding: This research received no external funding.

Conflicts of Interest: JFM reports grants from F2G, Merck and Pulmozyme, consultancy fees from Scynexis and speaker fees from Merck, United Medical, Gilead Sciences and TEVA, outside the submitted work. MAG declares receiving grants, and/or acting as a consultant for the following companies: Scynexis, Amplyx, Cidara, and F2G. SAA reports investigator-initiated research grant from Pfizer and lecture honoraria from Astellas, Gilead, Merck, and Pfizer, outside the submitted work.

References

1. Pappas, P.G.; Lionakis, M.S.; Arendrup, M.C.; Ostrosky-Zeichner, L.; Kullberg, B.J. Invasive candidiasis. *Nat. Rev. Dis. Primers* **2018**, *4*, 18026. [CrossRef] [PubMed]
2. Satoh, K.; Makimura, K.; Hasumi, Y.; Nishiyama, Y.; Uchida, K.; Yamaguchi, H. *Candida auris* sp. nov., a novel ascomycetous yeast isolated from the external ear canal of an inpatient in a Japanese hospital. *Microbiol. Immunol.* **2009**, *53*, 41–44. [CrossRef] [PubMed]
3. Lee, W.G.; Shin, J.H.; Uh, Y.; Kang, M.G.; Kim, S.H.; Park, K.H.; Jang, H.C. First three reported cases of nosocomial fungemia caused by *Candida auris*. *J. Clin. Microbiol.* **2011**, *49*, 3139–3142. [CrossRef] [PubMed]

4.   Chowdhary, A.; Sharma, C.; Duggal, S.; Agarwal, K.; Prakash, A.; Singh, P.K.; Jain, S.; Kathuria, S.; Randhawa, H.S.; Hagen, F.; et al. New clonal strain of *Candida auris*, Delhi, India. *Emerg. Infect. Dis.* **2013**, *19*, 1670–1673. [CrossRef] [PubMed]

5.   Chowdhary, A.; Sharma, C.; Meis, J.F. *Candida auris*: A rapidly emerging cause of hospital-acquired multidrug-resistant fungal infections globally. *PLoS Pathog.* **2017**, *13*, e1006290. [CrossRef] [PubMed]

6.   Calvo, B.; Melo, A.S.; Perozo-Mena, A.; Hernandez, M.; Francisco, E.C.; Hagen, F.; Meis, J.F.; Colombo, A.L. First report of *Candida auris* in America: Clinical and microbiological aspects of 18 episodes of candidemia. *J. Infect.* **2016**, *73*, 369–374. [CrossRef] [PubMed]

7.   Emara, M.; Ahmad, S.; Khan, Z.; Joseph, L.; Al-Obaid, I.; Purohit, P.; Bafna, R. *Candida auris* candidemia in Kuwait, 2014. *Emerg. Infect. Dis.* **2015**, *21*, 1091–1092. [CrossRef]

8.   Magobo, R.E.; Corcoran, C.; Seetharam, S.; Govender, N.P. *Candida auris*-associated candidemia, South Africa. *Emerg. Infect. Dis.* **2014**, *20*, 1250–1251. [CrossRef]

9.   Khan, Z.; Ahmad, S.; Al-Sweih, N.; Joseph, L.; Alfouzan, W.; Asadzadeh, M. Increasing prevalence, molecular characterization and antifungal drug susceptibility of serial *Candida auris* isolates in Kuwait. *PLoS ONE* **2018**, *13*, e0195743. [CrossRef]

10.   Khan, Z.; Ahmad, S.; Benwan, K.; Purohit, P.; Al-Obaid, I.; Bafna, R.; Emara, M.; Mokaddas, E.; Abdullah, A.A.; Al-Obaid, K.; et al. Invasive *Candida auris* infections in Kuwait hospitals: epidemiology, antifungal treatment and outcome. *Infection* **2018**, *46*, 641–650. [CrossRef]

11.   Bidaud, A.L.; Chowdhary, A.; Dannaoui, E. *Candida auris*: An emerging drug resistant yeast-A mini-review. *J. Mycol Med.* **2018**, *28*, 568–573. [CrossRef] [PubMed]

12.   Clancy, C.J.; Nguyen, M.H. Emergence of *Candida auris*: An international call to arms. *Clin. Infect. Dis.* **2017**, *64*, 141–143. [CrossRef] [PubMed]

13.   Forsberg, K.; Woodworth, K.; Walters, M.; Berkow, E.L.; Jackson, B.; Chiller, T.; Vallabhaneni, S. *Candida auris*: The recent emergence of a multidrug-resistant fungal pathogen. *Med. Mycol.* **2018**. [CrossRef] [PubMed]

14.   Jeffery-Smith, A.; Taori, S.K.; Schelenz, S.; Jeffery, K.; Johnson, E.M.; Borman, A.; Candida auris Incident Management, T.; Manuel, R.; Brown, C.S. *Candida auris*: a Review of the literature. *Clin. Microbiol. Rev.* **2018**, *31*, e00029-17. [CrossRef] [PubMed]

15.   Saris, K.; Meis, J.F.; Voss, A. *Candida auris*. *Curr Opin Infect. Dis.* **2018**, *31*, 334–340. [CrossRef] [PubMed]

16.   Tsay, S.; Kallen, A.; Jackson, B.R.; Chiller, T.M.; Vallabhaneni, S. Approach to the investigation and management of patients with *Candida auris*, an emerging multidrug-resistant yeast. *Clin. Infect. Dis.* **2018**, *66*, 306–311. [CrossRef]

17.   Heath, C.H.; Dyer, J.R.; Pang, S.; Coombs, G.W.; Gardam, D.J. *Candida auris* sternal osteomyelitis diagnosis in man from Kenya visiting Australia, 2015. *Emerg. Infect. Dis.* **2019**, *25*. (in press).

18.   Pekard-Amenitsch, S.; Schriebl, A.; Posawetz, W.; Willinger, B.; Kolli, B.; Buzina, W. Isolation of *Candida auris* from Ear of Otherwise Healthy Patient, Austria, 2018. *Emerg. Infect. Dis.* **2018**, *24*, 1596–1597. [CrossRef]

19.   Kohlenberg, A.; Struelens, M.J.; Monnet, D.L.; Plachouras, D.; The Candida Auris Survey Collaborative, G. *Candida auris*: epidemiological situation, laboratory capacity and preparedness in European Union and European Economic Area countries, 2013 to 2017. *Euro Surveill* **2018**, *23*. [CrossRef]

20.   Schwartz, I.S.; Hammond, G.W. First reported case of multidrug-resistant *Candida auris* in Canada. *Can. Commun Dis. Rep.* **2017**, *43*, 150–153. [CrossRef]

21.   Chen, Y.; Zhao, J.; Han, L.; Qi, L.; Fan, W.; Liu, J.; Wang, Z.; Xia, X.; Chen, J.; Zhang, L. Emergency of fungemia cases caused by fluconazole-resistant *Candida auris* in Beijing, China. *J. Infect.* **2018**. [CrossRef]

22.   Tian, S.; Rong, C.; Nian, H.; Li, F.; Chu, Y.; Cheng, S.; Shang, H. First cases and risk factors of super yeast *Candida auris* infection or colonization from Shenyang, China. *Emerg. Microbes Infect.* **2018**, *7*, 128. [CrossRef] [PubMed]

23.   Wang, X.; Bing, J.; Zheng, Q.; Zhang, F.; Liu, J.; Yue, H.; Tao, L.; Du, H.; Wang, Y.; Wang, H.; et al. The first isolate of *Candida auris* in China: clinical and biological aspects. *Emerg. Microbes Infect.* **2018**, *7*, 93. [CrossRef] [PubMed]

24.   Escandon, P.; Chow, N.A.; Caceres, D.H.; Gade, L.; Berkow, E.L.; Armstrong, P.; Rivera, S.; Misas, E.; Duarte, C.; Moulton-Meissner, H.; et al. Molecular epidemiology of *Candida auris* in Colombia reveals a highly-related, country-wide colonization with regional patterns in Amphotericin B resistance. *Clin. Infect. Dis.* **2018**. [CrossRef] [PubMed]

25. Morales-Lopez, S.E.; Parra-Giraldo, C.M.; Ceballos-Garzon, A.; Martinez, H.P.; Rodriguez, G.J.; Alvarez-Moreno, C.A.; Rodriguez, J.Y. Invasive infections with multidrug-resistant yeast *Candida auris*, Colombia. *Emerg. Infect. Dis.* **2017**, *23*, 162–164. [CrossRef] [PubMed]

26. Biswal, M.; Rudramurthy, S.M.; Jain, N.; Shamanth, A.S.; Sharma, D.; Jain, K.; Yaddanapudi, L.N.; Chakrabarti, A. Controlling a possible outbreak of *Candida auris* infection: lessons learnt from multiple interventions. *J. Hosp. Infect.* **2017**, *97*, 363–370. [CrossRef] [PubMed]

27. Chakrabarti, A.; Sood, P.; Rudramurthy, S.M.; Chen, S.; Kaur, H.; Capoor, M.; Chhina, D.; Rao, R.; Eshwara, V.K.; Xess, I.; et al. Incidence, characteristics and outcome of ICU-acquired candidemia in India. *Intensive Care Med.* **2015**, *41*, 285–295. [CrossRef] [PubMed]

28. Chowdhary, A.; Prakash, A.; Sharma, C.; Kordalewska, M.; Kumar, A.; Sarma, S.; Tarai, B.; Singh, A.; Upadhyaya, G.; Upadhyay, S.; et al. A multicentre study of antifungal susceptibility patterns among 350 *Candida auris* isolates (2009–17) in India: role of the ERG11 and FKS1 genes in azole and echinocandin resistance. *J. Antimicrob. Chemother.* **2018**, *73*, 891–899. [CrossRef]

29. Rudramurthy, S.M.; Chakrabarti, A.; Paul, R.A.; Sood, P.; Kaur, H.; Capoor, M.R.; Kindo, A.J.; Marak, R.S.K.; Arora, A.; Sardana, R.; et al. *Candida auris* candidaemia in Indian ICUs: analysis of risk factors. *J. Antimicrob. Chemother.* **2017**, *72*, 1794–1801. [CrossRef]

30. Belkin, A.; Gazit, Z.; Keller, N.; Ben-Ami, R.; Wieder-Finesod, A.; Novikov, A.; Rahav, G.; Brosh-Nissimov, T. *Candida auris* infection leading to nosocomial transmission, Israel, 2017. *Emerg. Infect. Dis.* **2018**, *24*, 801–804. [CrossRef]

31. Ben-Ami, R.; Berman, J.; Novikov, A.; Bash, E.; Shachor-Meyouhas, Y.; Zakin, S.; Maor, Y.; Tarabia, J.; Schechner, V.; Adler, A.; et al. Multidrug-resistant *Candida haemulonii* and *Candida auris*, Tel Aviv, Israel. *Emerg. Infect. Dis.* **2017**, *23*. [CrossRef]

32. Choi, H.I.; An, J.; Hwang, J.J.; Moon, S.Y.; Son, J.S. Otomastoiditis caused by *Candida auris*: Case report and literature review. *Mycoses* **2017**, *60*, 488–492. [CrossRef] [PubMed]

33. Mohd Tap, R.; Lim, T.C.; Kamarudin, N.A.; Ginsapu, S.J.; Abd Razak, M.F.; Ahmad, N.; Amran, F. A fatal case of *Candida auris* and *Candida tropicalis* candidemia in neutropenic patient. *Mycopathologia* **2018**, *183*, 559–564. [CrossRef]

34. Al-Siyabi, T.; Al Busaidi, I.; Balkhair, A.; Al-Muharrmi, Z.; Al-Salti, M.; Al'Adawi, B. First report of *Candida auris* in Oman: Clinical and microbiological description of five candidemia cases. *J. Infect.* **2017**, *75*, 373–376. [CrossRef] [PubMed]

35. Mohsin, J.; Hagen, F.; Al-Balushi, Z.A.M.; de Hoog, G.S.; Chowdhary, A.; Meis, J.F.; Al-Hatmi, A.M.S. The first cases of *Candida auris* candidaemia in Oman. *Mycoses* **2017**, *60*, 569–575. [CrossRef] [PubMed]

36. Lockhart, S.R.; Etienne, K.A.; Vallabhaneni, S.; Farooqi, J.; Chowdhary, A.; Govender, N.P.; Colombo, A.L.; Calvo, B.; Cuomo, C.A.; Desjardins, C.A.; et al. Simultaneous emergence of multidrug-resistant *Candida auris* on three continents confirmed by whole-genome sequencing and epidemiological analyses. *Clin. Infect. Dis.* **2017**, *64*, 134–140. [CrossRef] [PubMed]

37. Araúz, A.B.; Caceres, D.H.; Santiago, E.; Armstrong, P.; Arosemena, S.; Ramos, C.; Espinosa-Bode, A.; Borace, J.; Hayer, L.; Cedeno, I.; et al. Isolation of *Candida auris* from nine patients in Central America: Importance of accurate diagnosis and susceptibility testing. *Mycoses* **2018**, *61*, 44–47. [CrossRef] [PubMed]

38. Abdalhamid, B.; Almaghrabi, R.; Althawadi, S.; Omrani, A. First report of *Candida auris* infections from Saudi Arabia. *J. Infect. Public Health* **2018**, *11*, 598–599. [CrossRef] [PubMed]

39. Tan, Y.E.; Tan, A.L. Arrival of *Candida auris* fungus in Singapore: Report of the first three cases. *Ann. Acad Med. Singapore* **2018**, *47*, 260–262.

40. Govender, N.P.; Magobo, R.E.; Mpembe, R.; Mhlanga, M.; Matlapeng, P.; Corcoran, C.; Govind, C.; Lowman, W.; Senekal, M.; Thomas, J. *Candida auris* in South Africa, 2012–2016. *Emerg. Infect. Dis.* **2018**, *24*, 2036–2040. [CrossRef]

41. Ruiz-Gaitan, A.; Moret, A.M.; Tasias-Pitarch, M.; Aleixandre-Lopez, A.I.; Martinez-Morel, H.; Calabuig, E.; Salavert-Lleti, M.; Ramirez, P.; Lopez-Hontangas, J.L.; Hagen, F.; et al. An outbreak due to *Candida auris* with prolonged colonisation and candidaemia in a tertiary care European hospital. *Mycoses* **2018**, *61*, 498–505. [CrossRef]

42. Riat, A.; Neofytos, D.; Coste, A.; Harbarth, S.; Bizzini, A.; Grandbastien, B.; Pugin, J.; Lamoth, F. First case of *Candida auris* in Switzerland: discussion about preventive strategies. *Swiss Med. Wkly.* **2018**, *148*, w14622. [PubMed]

43.   Borman, A.M.; Szekely, A.; Johnson, E.M. Comparative pathogenicity of United Kingdom isolates of the emerging pathogen *Candida auris* and other key pathogenic *Candida* species. *mSphere* **2016**, *1*. [CrossRef] [PubMed]

44.   Schelenz, S.; Hagen, F.; Rhodes, J.L.; Abdolrasouli, A.; Chowdhary, A.; Hall, A.; Ryan, L.; Shackleton, J.; Trimlett, R.; Meis, J.F.; et al. First hospital outbreak of the globally emerging *Candida auris* in a European hospital. *Antimicrob. Resist. Infect. Control.* **2016**, *5*, 35. [CrossRef] [PubMed]

45.   Eyre, D.W.; Sheppard, A.E.; Madder, H.; Moir, I.; Moroney, R.; Quan, T.P.; Griffiths, D.; George, S.; Butcher, L.; Morgan, M.; et al. A *Candida auris* outbreak and its Control in an intensive care setting. *N. Engl. J. Med.* **2018**, *379*, 1322–1331. [CrossRef] [PubMed]

46.   Rhodes, J.; Abdolrasouli, A.; Farrer, R.A.; Cuomo, C.A.; Aanensen, D.M.; Armstrong-James, D.; Fisher, M.C.; Schelenz, S. Genomic epidemiology of the UK outbreak of the emerging human fungal pathogen *Candida auris*. *Emerg. Microbes Infect.* **2018**, *7*, 43. [CrossRef] [PubMed]

47.   Chow, N.A.; Gade, L.; Tsay, S.V.; Forsberg, K.; Greenko, J.A.; Southwick, K.L.; Barrett, P.M.; Kerins, J.L.; Lockhart, S.R.; Chiller, T.M.; et al. Multiple introductions and subsequent transmission of multidrug-resistant *Candida auris* in the USA: a molecular epidemiological survey. *Lancet Infect. Dis.* **2018**, *18*, 1377–1384. [CrossRef]

48.   Alatoom, A.; Sartawi, M.; Lawlor, K.; AbdelWareth, L.; Thomsen, J.; Nusair, A.; Mirza, I. Persistent candidemia despite appropriate fungal therapy: First case of *Candida auris* from the United Arab Emirates. *Int. J. Infect. Dis.* **2018**, *70*, 36–37. [CrossRef] [PubMed]

49.   Mathur, P.; Hasan, F.; Singh, P.K.; Malhotra, R.; Walia, K.; Chowdhary, A. Five-year profile of candidaemia at an Indian trauma centre: High rates of *Candida auris* blood stream infections. *Mycoses* **2018**. [CrossRef] [PubMed]

50.   Piedrahita, C.T.; Cadnum, J.L.; Jencson, A.L.; Shaikh, A.A.; Ghannoum, M.A.; Donskey, C.J. Environmental surfaces in healthcare facilities are a potential source for transmission of *Candida auris* and other *Candida* species. *Infect. Control. Hosp. Epidemiol.* **2017**, *38*, 1107–1109. [CrossRef] [PubMed]

51.   Yamamoto, M.; Alshahni, M.M.; Tamura, T.; Satoh, K.; Iguchi, S.; Kikuchi, K.; Mimaki, M.; Makimura, K. Rapid detection of *Candida auris* based on loop-mediated isothermal amplification (LAMP). *J. Clin. Microbiol.* **2018**, *56*. [CrossRef]

52.   Kordalewska, M.; Zhao, Y.; Lockhart, S.R.; Chowdhary, A.; Berrio, I.; Perlin, D.S. Rapid and accurate molecular identification of the emerging multidrug-resistant pathogen *Candida auris*. *J. Clin. Microbiol.* **2017**, *55*, 2445–2452. [CrossRef] [PubMed]

53.   Leach, L.; Zhu, Y.; Chaturvedi, S. Development and validation of a real-time PCR assay for rapid detection of *Candida auris* from surveillance samples. *J. Clin. Microbiol.* **2018**, *56*. [CrossRef] [PubMed]

54.   Martinez-Murcia, A.; Navarro, A.; Bru, G.; Chowdhary, A.; Hagen, F.; Meis, J.F. Internal validation of GPS() MONODOSE CanAur dtec-qPCR kit following the UNE/EN ISO/IEC 17025:2005 for detection of the emerging yeast *Candida auris*. *Mycoses* **2018**, *61*, 877–884. [CrossRef] [PubMed]

55.   Ruiz-Gaitan, A.C.; Fernandez-Pereira, J.; Valentin, E.; Tormo-Mas, M.A.; Eraso, E.; Peman, J.; de Groot, P.W.J. Molecular identification of *Candida auris* by PCR amplification of species-specific GPI protein-encoding genes. *Int. J. Med. Microbiol.* **2018**, *308*, 812–818. [CrossRef] [PubMed]

56.   Sexton, D.J.; Kordalewska, M.; Bentz, M.L.; Welsh, R.M.; Perlin, D.S.; Litvintseva, A.P. Direct detection of emergent fungal pathogen *Candida auris* in clinical skin swabs by SYBR Green qPCR assay. *J. Clin. Microbiol.* **2018**. [CrossRef] [PubMed].

57.   Sexton, D.J.; Bentz, M.L.; Welsh, R.M.; Litvintseva, A.P. Evaluation of a new T2 Magnetic Resonance assay for rapid detection of emergent fungal pathogen *Candida auris* on clinical skin swab samples. *Mycoses* **2018**, *61*, 786–790. [CrossRef] [PubMed]

58.   Theill, L.; Dudiuk, C.; Morales-Lopez, S.; Berrio, I.; Rodriguez, J.Y.; Marin, A.; Gamarra, S.; Garcia-Effron, G. Single-tube classical PCR for *Candida auris* and *Candida haemulonii* identification. *Rev. Iberoam Micol.* **2018**, *35*, 110–112. [CrossRef] [PubMed]

59.   Hou, X.; Lee, A.; Jimenez-Ortigosa, C.; Kordalewska, M.; Perlin, D.S.; Zhao, Y. Rapid detection of ERG11-associated azole resistance and FKS-associated echinocandin resistance in *Candida auris*. *Antimicrob. Agents Chemother.* **2018**. [CrossRef]

60. Cadnum, J.L.; Shaikh, A.A.; Piedrahita, C.T.; Jencson, A.L.; Larkin, E.L.; Ghannoum, M.A.; Donskey, C.J. Relative resistance of the emerging fungal pathogen *Candida auris* and other *Candida* species to killing by ultraviolet light. *Infect. Control. Hosp. Epidemiol* **2018**, *39*, 94–96. [CrossRef]

61. Kean, R.; Sherry, L.; Townsend, E.; McKloud, E.; Short, B.; Akinbobola, A.; Mackay, W.G.; Williams, C.; Jones, B.L.; Ramage, G. Surface disinfection challenges for *Candida auris*: an in-vitro study. *J. Hosp. Infect.* **2018**, *98*, 433–436. [CrossRef]

62. Welsh, R.M.; Bentz, M.L.; Shams, A.; Houston, H.; Lyons, A.; Rose, L.J.; Litvintseva, A.P. Survival, persistence, and isolation of the emerging multidrug-resistant pathogenic yeast *Candida auris* on a plastic health care surface. *J. Clin. Microbiol.* **2017**, *55*, 2996–3005. [CrossRef] [PubMed]

63. Chowdhary, A.; Voss, A.; Meis, J.F. Multidrug-resistant *Candida auris*: 'new kid on the block' in hospital-associated infections? *J. Hosp. Infect.* **2016**, *94*, 209–212. [CrossRef] [PubMed]

64. Meis, J.F.; Chowdhary, A. *Candida auris*: a global fungal public health threat. *Lancet Infect. Dis.* **2018**, *18*, 1298–1299. [CrossRef]

65. Kathuria, S.; Singh, P.K.; Sharma, C.; Prakash, A.; Masih, A.; Kumar, A.; Meis, J.F.; Chowdhary, A. Multidrug-resistant *Candida auris* misidentified as *Candida haemulonii*: Characterization by matrix-assisted laser desorption ionization-time of flight mass spectrometry and DNA sequencing and its antifungal susceptibility profile variability by Vitek 2, CLSI broth microdilution, and Etest method. *J. Clin. Microbiol.* **2015**, *53*, 1823–1830. [PubMed]

66. Kumar, A.; Prakash, A.; Singh, A.; Kumar, H.; Hagen, F.; Meis, J.F.; Chowdhary, A. *Candida haemulonii* species complex: an emerging species in India and its genetic diversity assessed with multilocus sequence and amplified fragment-length polymorphism analyses. *Emerg. Microbes Infect.* **2016**, *5*, e49. [CrossRef] [PubMed]

67. Mizusawa, M.; Miller, H.; Green, R.; Lee, R.; Durante, M.; Perkins, R.; Hewitt, C.; Simner, P.J.; Carroll, K.C.; Hayden, R.T.; et al. Can multidrug-resistant *Candida auris* be reliably identified in clinical microbiology laboratories? *J. Clin. Microbiol.* **2017**, *55*, 638–640. [CrossRef] [PubMed]

68. Prakash, A.; Sharma, C.; Singh, A.; Kumar Singh, P.; Kumar, A.; Hagen, F.; Govender, N.P.; Colombo, A.L.; Meis, J.F.; Chowdhary, A. Evidence of genotypic diversity among *Candida auris* isolates by multilocus sequence typing, matrix-assisted laser desorption ionization time-of-flight mass spectrometry and amplified fragment length polymorphism. *Clin. Microbiol. Infect.* **2016**, *22*, e271–e279. [CrossRef]

69. Arendrup, M.C.; Prakash, A.; Meletiadis, J.; Sharma, C.; Chowdhary, A. Comparison of EUCAST and CLSI reference microdilution MICs of eight antifungal compounds for *Candida auris* and associated tentative epidemiological cutoff values. *Antimicrob. Agents Chemother.* **2017**, *61*. [CrossRef]

70. Hyde, K.D.; Al-Hatmi, A.M.S.; Andersen, B.; Boekhout, T.; Buzina, W.; Dawson, T.L.; Eastwood, D.C.; Jones, E.B.G.; de Hoog, S.; Kang, Y.; et al. The world's ten most feared fungi. *Fungal Diversity* **2018**. [CrossRef]

71. Larkin, E.; Hager, C.; Chandra, J.; Mukherjee, P.K.; Retuerto, M.; Salem, I.; Long, L.; Isham, N.; Kovanda, L.; Borroto-Esoda, K.; et al. The emerging pathogen *Candida auris*: Growth phenotype, virulence factors, activity of antifungals, and effect of SCY-078, a novel glucan synthesis inhibitor, on growth morphology and biofilm formation. *Antimicrob. Agents Chemother.* **2017**, *61*. [CrossRef]

72. Hager, C.L.; Larkin, E.L.; Long, L.A.; Ghannoum, M.A. Evaluation of the efficacy of rezafungin, a novel echinocandin, in the treatment of disseminated *Candida auris* infection using an immunocompromised mouse model. *J. Antimicrob. Chemother.* **2018**, *73*, 2085–2088. [CrossRef]

73. Hager, C.L.; Larkin, E.L.; Long, L.; Zohra Abidi, F.; Shaw, K.J.; Ghannoum, M.A. In vitro and in vivo evaluation of the antifungal activity of APX001A/APX001 against *Candida auris*. *Antimicrob. Agents Chemother.* **2018**, *62*. [CrossRef] [PubMed]

74. Fakhim, H.; Vaezi, A.; Dannaoui, E.; Chowdhary, A.; Nasiry, D.; Faeli, L.; Meis, J.F.; Badali, H. Comparative virulence of *Candida auris* with *Candida haemulonii*, *Candida glabrata* and *Candida albicans* in a murine model. *Mycoses* **2018**, *61*, 377–382. [CrossRef] [PubMed]

75. Adams, E.; Quinn, M.; Tsay, S.; Poirot, E.; Chaturvedi, S.; Southwick, K.; Greenko, J.; Fernandez, R.; Kallen, A.; Vallabhaneni, S.; et al. *Candida auris* in Healthcare Facilities, New York, USA, 2013–2017. *Emerg. Infect. Dis.* **2018**, *24*, 1816–1824. [CrossRef] [PubMed]

76. Chatterjee, S.; Alampalli, S.V.; Nageshan, R.K.; Chettiar, S.T.; Joshi, S.; Tatu, U.S. Draft genome of a commonly misdiagnosed multidrug resistant pathogen *Candida auris*. *BMC Genomics* **2015**, *16*, 686. [CrossRef] [PubMed]

77. Healey, K.R.; Kordalewska, M.; Jimenez Ortigosa, C.; Singh, A.; Berrio, I.; Chowdhary, A.; Perlin, D.S. Limited ERG11 mutations identified in isolates of *Candida auris* directly contribute to reduced azole susceptibility. *Antimicrob. Agents Chemother.* **2018**, *62*, e01427-18. [CrossRef] [PubMed]

78. Lepak, A.J.; Zhao, M.; Berkow, E.L.; Lockhart, S.R.; Andes, D.R. Pharmacodynamic optimization for treatment of invasive *Candida auris* infection. *Antimicrob. Agents Chemother.* **2017**, *61*. [CrossRef] [PubMed]

79. Fakhim, H.; Chowdhary, A.; Prakash, A.; Vaezi, A.; Dannaoui, E.; Meis, J.F.; Badali, H. In vitro interactions of echinocandins with triazoles against multidrug-resistant *Candida auris*. *Antimicrob. Agents Chemother.* **2017**, *61*. [CrossRef]

80. Kordalewska, M.; Lee, A.; Park, S.; Berrio, I.; Chowdhary, A.; Zhao, Y.; Perlin, D.S. Understanding echinocandin resistance in the emerging pathogen *Candida auris*. *Antimicrob. Agents Chemother.* **2018**, *62*. [CrossRef]

81. Arendrup, M.C.; Chowdhary, A.; Astvad, K.M.T.; Jorgensen, K.M. APX001A in vitro activity against contemporary blood isolates and *Candida auris* determined by the EUCAST reference method. *Antimicrob. Agents Chemother.* **2018**, *62*. [CrossRef]

82. Berkow, E.L.; Angulo, D.; Lockhart, S.R. In vitro activity of a novel glucan synthase inhibitor, SCY-078, against clinical isolates of *Candida auris*. *Antimicrob. Agents Chemother.* **2017**, *61*. [CrossRef] [PubMed]

83. Berkow, E.L.; Lockhart, S.R. Activity of novel antifungal compound APX001A against a large collection of *Candida auris*. *J. Antimicrob. Chemother.* **2018**, *73*, 3060–3062. [CrossRef] [PubMed]

84. Berkow, E.L.; Lockhart, S.R. Activity of CD101, a long-acting echinocandin, against clinical isolates of *Candida auris*. *Diagn Microbiol. Infect. Dis.* **2018**, *90*, 196–197. [CrossRef] [PubMed]

85. Lepak, A.J.; Zhao, M.; Andes, D.R. Pharmacodynamic evaluation of rezafungin (CD101) against *Candida auris* in the neutropenic mouse invasive candidiasis model. *Antimicrob. Agents Chemother.* **2018**, *62*. [CrossRef] [PubMed]

86. Wiederhold, N.P.; Patterson, H.P.; Tran, B.H.; Yates, C.M.; Schotzinger, R.J.; Garvey, E.P. Fungal-specific Cyp51 inhibitor VT-1598 demonstrates in vitro activity against Candida and Cryptococcus species, endemic fungi, including Coccidioides species, Aspergillus species and Rhizopus arrhizus. *J. Antimicrob. Chemother.* **2018**, *73*, 404–408. [CrossRef] [PubMed]

87. Break, T.J.; Desai, J.V.; Healey, K.R.; Natarajan, M.; Ferre, E.M.N.; Henderson, C.; Zelazny, A.; Siebenlist, U.; Yates, C.M.; Cohen, O.J.; et al. VT-1598 inhibits the in vitro growth of mucosal *Candida* strains and protects against fluconazole-susceptible and -resistant oral candidiasis in IL-17 signalling-deficient mice. *J. Antimicrob. Chemother.* **2018**, *73*, 2089–2094. [CrossRef] [PubMed]

88. Ku, T.S.N.; Walraven, C.J.; Lee, S.A. *Candida auris*: Disinfectants and implications for infection control. *Front. Microbiol.* **2018**, *9*, 726. [CrossRef]

89. Moore, G.; Schelenz, S.; Borman, A.M.; Johnson, E.M.; Brown, C.S. Yeasticidal activity of chemical disinfectants and antiseptics against *Candida auris*. *J. Hosp. Infect.* **2017**, *97*, 371–375. [CrossRef] [PubMed]

90. Abdolrasouli, A.; Armstrong-James, D.; Ryan, L.; Schelenz, S. In vitro efficacy of disinfectants utilised for skin decolonisation and environmental decontamination during a hospital outbreak with *Candida auris*. *Mycoses* **2017**, *60*, 758–763. [CrossRef]

91. Denning, D.W.; Venkateswarlu, K.; Oakley, K.L.; Anderson, M.J.; Manning, N.J.; Stevens, D.A.; Warnock, D.W.; Kelly, S.L. Itraconazole resistance in *Aspergillus fumigatus*. *Antimicrob. Agents Chemother.* **1997**, *41*, 1364–1368. [CrossRef]

92. Verweij, P.E.; Chowdhary, A.; Melchers, W.J.; Meis, J.F. Azole resistance in *Aspergillus fumigatus*: Can we retain the clinical use of mold-active antifungal azoles? *Clin. Infect. Dis.* **2016**, *62*, 362–368. [CrossRef] [PubMed]

93. Denning, D.W.; Perlin, D.S. Azole resistance in *Aspergillus*: a growing public health menace. *Future Microbiol.* **2011**, *6*, 1229–1232. [CrossRef] [PubMed]

94. van der Linden, J.W.; Arendrup, M.C.; Warris, A.; Lagrou, K.; Pelloux, H.; Hauser, P.M.; Chryssanthou, E.; Mellado, E.; Kidd, S.E.; Tortorano, A.M.; et al. Prospective multicenter international surveillance of azole resistance in *Aspergillus fumigatus*. *Emerg. Infect. Dis.* **2015**, *21*, 1041–1044. [CrossRef] [PubMed]

95. Meis, J.F.; Chowdhary, A.; Rhodes, J.L.; Fisher, M.C.; Verweij, P.E. Clinical implications of globally emerging azole resistance in *Aspergillus fumigatus*. *Philos Trans. R Soc. Lond B Biol. Sci.* **2016**, *371*. [CrossRef] [PubMed]

96. Goncalves, S.S.; Souza, A.C.; Chowdhary, A.; Meis, J.F.; Colombo, A.L. Epidemiology and molecular mechanisms of antifungal resistance in *Candida* and *Aspergillus*. *Mycoses* **2016**. [CrossRef]

97. Mortensen, K.L.; Mellado, E.; Lass-Florl, C.; Rodriguez-Tudela, J.L.; Johansen, H.K.; Arendrup, M.C. Environmental study of azole-resistant *Aspergillus fumigatus* and other aspergilli in Austria, Denmark, and Spain. *Antimicrob. Agents Chemother.* **2010**, *54*, 4545–4549. [CrossRef] [PubMed]

98. Snelders, E.; Huis In 't Veld, R.A.; Rijs, A.J.; Kema, G.H.; Melchers, W.J.; Verweij, P.E. Possible environmental origin of resistance of *Aspergillus fumigatus* to medical triazoles. *Appl. Environ. Microbiol.* **2009**, *75*, 4053–4057. [CrossRef]

99. Badali, H.; Vaezi, A.; Haghani, I.; Yazdanparast, S.A.; Hedayati, M.T.; Mousavi, B.; Ansari, S.; Hagen, F.; Meis, J.F.; Chowdhary, A. Environmental study of azole-resistant *Aspergillus fumigatus* with TR34/L98H mutations in the *cyp51A* gene in Iran. *Mycoses* **2013**, *56*, 659–663. [CrossRef]

100. Ozmerdiven, G.E.; Ak, S.; Ener, B.; Agca, H.; Cilo, B.D.; Tunca, B.; Akalin, H. First determination of azole resistance in *Aspergillus fumigatus* strains carrying the TR34/L98H mutations in Turkey. *J. Infect. Chemother.* **2015**, *21*, 581–586. [CrossRef]

101. Koehler, P.; Hamprecht, A.; Bader, O.; Bekeredjian-Ding, I.; Buchheidt, D.; Doelken, G.; Elias, J.; Haase, G.; Hahn-Ast, C.; Karthaus, M.; et al. Epidemiology of invasive aspergillosis and azole resistance in patients with acute leukaemia: the SEPIA Study. *Int J. Antimicrob. Agents* **2017**, *49*, 218–223. [CrossRef]

102. Resendiz Sharpe, A.; Lagrou, K.; Meis, J.F.; Chowdhary, A.; Lockhart, S.R.; Verweij, P.E.; ISHAM ECMM Aspergillus Resistance Surveillance Working Group. Triazole resistance surveillance in *Aspergillus fumigatus*. *Med. Mycol.* **2018**, *56*, 83–92. [CrossRef]

103. CLSI. *CLSI Document M38-A2. Reference Method for Broth Dilution Antifungal Susceptibility Testing of Filamentous Fungi; Approved Standard-Second Edition*; Clinical and Laboratory Standards Institute: Wayne, PA, USA, 2008.

104. Arendrup, M.C.; Meletiadis, J.; Mouton, J.W.; Lagrou, K.; Hamal, P.; Guinea, J.; Subcommittee on AFST of the ESCMID for EUCAST. EUCAST Definitive Document E.DEF 9.3.1 Method for the Determination of Broth Dilution Minimum Inhibitory Concentrations of Antifungal Agents for Conidia Forming Moulds. Available online: http://www.eucast.org/ast_of_fungi/methodsinantifungalsusceptibilitytesting/susceptibility_testing_of_moulds/ (accessed on 1 November 2018).

105. Meletiadis, J.; Leth Mortensen, K.; Verweij, P.E.; Mouton, J.W.; Arendrup, M.C. Spectrophotometric reading of EUCAST antifungal susceptibility testing of *Aspergillus fumigatus*. *Clin. Microbiol. Infect.* **2017**, *23*, 98–103. [CrossRef] [PubMed]

106. Ullmann, A.J.; Aguado, J.M.; Arikan-Akdagli, S.; Denning, D.W.; Groll, A.H.; Lagrou, K.; Lass-Flörl, C.; Lewis, R.E.; Munoz, P.; Verweij, P.E.; et al. Diagnosis and management of *Aspergillus* diseases: executive summary of the 2017 ESCMID-ECMM-ERS guideline. *Clin. Microbiol. Infect.* **2018**, *24*, e1–e38. [CrossRef] [PubMed]

107. Patterson, T.F.; Thompson, G.R., 3rd; Denning, D.W.; Fishman, J.A.; Hadley, S.; Herbrecht, R.; Kontoyiannis, D.P.; Marr, K.A.; Morrison, V.A.; Nguyen, M.H.; et al. Practice guidelines for the diagnosis and management of aspergillosis: 2016 Update by the Infectious Diseases Society of America. *Clin. Infect. Dis.* **2016**, *63*, e1–e60. [CrossRef] [PubMed]

108. CLSI. *CLSI Document M51-A. Method for Antifungal Disk Diffusion Susceptibility Testing of Nondermatophyte Filamentous Fungi; Approved Guideline*; Clinical and Laboratory Standards Institute: Wayne, PA, USA, 2010.

109. Espinel-Ingroff, A.; Chowdhary, A.; Gonzalez, G.M.; Lass-Florl, C.; Martin-Mazuelos, E.; Meis, J.; Pelaez, T.; Pfaller, M.A.; Turnidge, J. Multicenter study of isavuconazole MIC distributions and epidemiological cutoff values for *Aspergillus* spp. for the CLSI M38-A2 broth microdilution method. *Antimicrob. Agents Chemother.* **2013**, *57*, 3823–3828. [CrossRef] [PubMed]

110. Espinel-Ingroff, A.; Cuenca-Estrella, M.; Fothergill, A.; Fuller, J.; Ghannoum, M.; Johnson, E.; Pelaez, T.; Pfaller, M.A.; Turnidge, J. Wild-type MIC distributions and epidemiological cutoff values for amphotericin B and *Aspergillus* spp. for the CLSI broth microdilution method (M38-A2 document). *Antimicrob. Agents Chemother.* **2011**, *55*, 5150–5154. [CrossRef]

111. Espinel-Ingroff, A.; Diekema, D.J.; Fothergill, A.; Johnson, E.; Pelaez, T.; Pfaller, M.A.; Rinaldi, M.G.; Canton, E.; Turnidge, J. Wild-type MIC distributions and epidemiological cutoff values for the triazoles and six *Aspergillus* spp. for the CLSI broth microdilution method (M38-A2 document). *J. Clin. Microbiol.* **2010**, *48*, 3251–3257. [CrossRef]

112. CLSI. *CLSI Document M59. Epidemiological Cutoff Values for Antifungal Susceptibility Testing*, 1st ed.; Clinical and Laboratory Standards Institute: Wayne, PA, USA, 2016.

113. EUCAST. Available online: http://www.eucast.org/ (accessed on 1 November 2018).

114. Arendrup, M.C.; Verweij, P.E.; Mouton, J.W.; Lagrou, K.; Meletiadis, J. Multicentre validation of 4-well azole agar plates as a screening method for detection of clinically relevant azole-resistant *Aspergillus fumigatus*. *J. Antimicrob. Chemother.* **2017**, *72*, 3325–3333. [CrossRef]

115. Guinea, J.; Verweij, P.E.; Meletiadis, J.; Mouton, J.W.; Barchiesi, F.; Arendrup, M.C.; Subcommittee on Antifungal Susceptibility Testing of the ESCMID European Committee for Antimicrobial Susceptibility Testing (EUCAST), EUCAST-AFST; Arendrup, M.C.; Arikan-Akdagli, S.; Barchiesi, F.; et al. How to: EUCAST recommendations on the screening procedure E.Def 10.1 for the detection of azole resistance in *Aspergillus fumigatus* isolates using four well azole-containing agar plates. *Clin. Microbiol. Infect.* **2018**. [CrossRef]

116. Chowdhary, A.; Sharma, C.; Meis, J.F. Azole-resistant aspergillosis: Epidemiology, molecular mechanisms, and treatment. *J. Infect. Dis.* **2017**, *216*, S436–S444. [CrossRef]

117. Alvarez-Moreno, C.; Lavergne, R.A.; Hagen, F.; Morio, F.; Meis, J.F.; Le Pape, P. Azole-resistant *Aspergillus fumigatus* harboring TR34/L98H, TR46/Y121F/T289A and TR53 mutations related to flower fields in Colombia. *Sci Rep.* **2017**, *7*, 45631. [CrossRef] [PubMed]

118. Hodiamont, C.J.; Dolman, K.M.; Ten Berge, I.J.; Melchers, W.J.; Verweij, P.E.; Pajkrt, D. Multiple-azole-resistant *Aspergillus fumigatus* osteomyelitis in a patient with chronic granulomatous disease successfully treated with long-term oral posaconazole and surgery. *Med. Mycol.* **2009**, *47*, 217–220. [CrossRef] [PubMed]

119. Abdolrasouli, A.; Rhodes, J.; Beale, M.A.; Hagen, F.; Rogers, T.R.; Chowdhary, A.; Meis, J.F.; Armstrong-James, D.; Fisher, M.C. Genomic context of azole resistance mutations in *Aspergillus fumigatus* determined using whole-genome sequencing. *MBio* **2015**, *6*, e00536. [CrossRef]

120. Deng, S.; Zhang, L.; Ji, Y.; Verweij, P.E.; Tsui, K.M.; Hagen, F.; Houbraken, J.; Meis, J.F.; Abliz, P.; Wang, X.; et al. Triazole phenotypes and genotypic characterization of clinical *Aspergillus fumigatus* isolates in China. *Emerg. Microbes Infect.* **2017**, *6*, e109. [CrossRef] [PubMed]

121. Snelders, E.; van der Lee, H.A.; Kuijpers, J.; Rijs, A.J.; Varga, J.; Samson, R.A.; Mellado, E.; Donders, A.R.; Melchers, W.J.; Verweij, P.E. Emergence of azole resistance in *Aspergillus fumigatus* and spread of a single resistance mechanism. *PLoS Med.* **2008**, *5*, e219. [CrossRef] [PubMed]

122. van der Linden, J.W.; Snelders, E.; Kampinga, G.A.; Rijnders, B.J.; Mattsson, E.; Debets-Ossenkopp, Y.J.; Kuijper, E.J.; Van Tiel, F.H.; Melchers, W.J.; Verweij, P.E. Clinical implications of azole resistance in *Aspergillus fumigatus*, The Netherlands, 2007–2009. *Emerg. Infect. Dis.* **2011**, *17*, 1846–1854. [CrossRef] [PubMed]

123. Lestrade, P.P.; Bentvelsen, R.G.; Schauwvlieghe, A.; Schalekamp, S.; van der Velden, W.; Kuiper, E.J.; van Paassen, J.; van der Hoven, B.; van der Lee, H.A.; Melchers, W.J.G.; et al. Voriconazole resistance and mortality in invasive aspergillosis: A multicenter retrospective cohort study. *Clin. Infect. Dis.* **2018**. [CrossRef]

124. Chong, G.M.; van der Beek, M.T.; von dem Borne, P.A.; Boelens, J.; Steel, E.; Kampinga, G.A.; Span, L.F.; Lagrou, K.; Maertens, J.A.; Dingemans, G.J.; et al. PCR-based detection of *Aspergillus fumigatus* Cyp51A mutations on bronchoalveolar lavage: a multicentre validation of the AsperGenius assay(R) in 201 patients with haematological disease suspected for invasive aspergillosis. *J. Antimicrob. Chemother.* **2016**, *71*, 3528–3535. [CrossRef]

125. Heo, S.T.; Tatara, A.M.; Jimenez-Ortigosa, C.; Jiang, Y.; Lewis, R.E.; Tarrand, J.; Tverdek, F.; Albert, N.D.; Verweij, P.E.; Meis, J.F.; et al. Changes in in vitro susceptibility patterns of *Aspergillus* to triazoles and correlation with aspergillosis outcome in a tertiary care cancer center, 1999–2015. *Clin. Infect. Dis.* **2017**, *65*, 216–225. [CrossRef]

126. Verweij, P.E.; Ananda-Rajah, M.; Andes, D.; Arendrup, M.C.; Bruggemann, R.J.; Chowdhary, A.; Cornely, O.A.; Denning, D.W.; Groll, A.H.; Izumikawa, K.; et al. International expert opinion on the management of infection caused by azole-resistant *Aspergillus fumigatus*. *Drug Resist. Updates* **2015**, *21–22*, 30–40. [CrossRef]

Journal of
Fungi

MDPI

Review

Development and Applications of Prognostic Risk Models in the Management of Invasive Mold Disease

Marta Stanzani [1] and **Russell E. Lewis** [2,*]

1   Institute of Hematology "Lorenzo e Ariosto Seràgnoli", Department of Hematology and Clinical Oncology, S. Orsola-Malpighi Hospital, University of Bologna, 40138 Bologna, Italy; marta.stanzani2@unibo.it
2   Clinic of Infectious Diseases, Department of Medical and Surgical Sciences, S. Orsola-Malpighi Hospital, University of Bologna, 40138 Bologna, Italy
*   Correspondence: russeledward.lewis@unibo.it

Received: 16 November 2018; Accepted: 14 December 2018; Published: 19 December 2018

Abstract: Prognostic models or risk scores are frequently used to aid individualize risk assessment for diseases with multiple, complex risk factors and diagnostic challenges. However, relatively little attention has been paid to the development of risk models for invasive mold diseases encountered in patients with hematological malignancies, despite a large body of epidemiological research. Herein we review recent studies that have described the development of prognostic models for mold disease, summarize our experience with the development and clinical use of one such model (BOSCORE), and discuss the potential impact of prognostic risk scores for individualized therapy, diagnostic and antifungal stewardship, as well as clinical and epidemiological research.

Keywords: prognostic risk model; prediction models; risk score; invasive mold disease; hematological malignancy; risk assessment; antifungal stewardship

1. Introduction

Many decisions involved in the prevention, diagnosis and treatment of invasive mold disease (IMD) depend on an accurate estimate of the patient's future risk for developing the infection [1]. This prediction can be challenging given the multivariate and dynamic nature of risk factors that predispose patients to these infections [2–4]. Surprisingly little attention has been given to the development of prediction models or risk scores to aid individualized IMD risk assessment, even though such models have been developed for invasive candidiasis in non-neutropenic patients and incorporated in clinical trial design [5–8]. The concept of developing and validating clinical prediction models for IMD seems logical given that such models have proven to be useful for diseases in other areas of medicine characterized by multivariate and complex risk factors, as well as diagnostic challenges [9,10]. Individualized risk assessment is also a key component of diagnostic and antifungal stewardship efforts, because many evidence-based interventions such as antifungal prophylaxis have only been proven to be clinically beneficial in select high-risk subpopulations [11].

In this review, we will review key risk factors that predispose patients to IMD during the treatment of hematological malignancies, describe recent attempts towards the development prognostic risk models, and explore how these risk models can be incorporated to improve future clinical and epidemiological research. This review is primarily applicable to the development of models for predicting invasive aspergillosis rather than less common molds given the much higher prevalence of this disease, clinical availability of biomarkers for diagnostic-driven management, and evidence-based recommendations for prophylaxis in select patient groups.

2. Risk Factors for Invasive Mold Disease

Epidemiological studies performed over the last four decades have identified many risk factors that predispose patients with hematological malignancies to developing an IMD [3]. These include host-specific factors of the (i) type of underlying malignancy and status, whether in remission or active; (ii) Type of immunosuppressive chemotherapy and associated conditions such as neutropenia and damage to integument; (iii) The type of hematopoietic stem cell transplantation (HSCT) (autologous versus allogeneic graft), stem cell source, status of the malignancy at the time of transplant, and genetic risk factors for fungal disease in the donor; (iv) immunosuppressive therapy required to manage graft versus host disease (GVHD); (iv) Patient comorbidities and age; (v) Environmental or occupational risk factors associated with fungal spore exposure; and (vi) A prior history of IMD [3,4].

Although many of these risk factors are discussed in current treatment guidelines [12,13], most diagnostic and treatment algorithms are still generally based on only a few host risk factors used in epidemiological definitions for probable or proven IMD or in drug registration trials [14]. As a result, patients are generally separated into heterogenous pools of high-risk versus low-risk at the time of a newly-diagnosed malignancy or following HSCT [15]. This approach often falls short in the "real-life" management of patients who receive multiple lines of chemotherapy with relapsed malignancy, genetic or occupational predisposition for mold infections, a prior history of mold disease, or receive novel biological agents or immunosuppressive chemotherapy. Therefore, a goal for any prognostic model is to provide a practical approach for improving the precision and accuracy of risk prediction for all patients so clinicians can make better informed decisions. Before discussing specific risk models, however, it is important to briefly review the major risk factors that predispose patients with hematological malignancies to developing an IMD.

2.1. Underlying Malignancy and Status

Acute myeloid leukemia (AML) and myelodysplastic syndromes (MDS) have historically been associated with the highest reported rates of IMD, ranging from 5–25% [4,16,17]. The period at greatest risk for IMD is typically during initial induction chemotherapy, which typically results in prolonged (>21 days) and profound (<500 PMN/mm^3) neutropenia [18]. Other factors in AML patients that may be associated with higher rates of IMD include older age, poor prognosis for achieving complete remission (CR) due to unfavorable cytogenetics or relapsed malignancy, and presence of multiple baseline comorbidities or poor performance status [4]. Patients who receive induction chemotherapy in non-HEPA filtered rooms, or an occupational history of heavy spore exposure (e.g., farming, construction) may also be at higher risk. Relapse of IMD is also a concern in any patient undergoing consolidation chemotherapy with a previous history of fungal infection during their induction regimen [19]. Survival rates among patients with AML have improved in recent years despite relatively few changes in frontline chemotherapy protocols due in part to improved supportive care measures and the introduction of more effective prophylaxis options such as posaconazole [20].

Older patients with transformed MDS are at high risk for IMD if they receive AML-like treatment regimens. However, hypomethylating agents such as azacytidine or decitabine are increasingly being used in place of cytosine arabinoside (ARA-C), resulting in lower rates of infection complications, including a probable or proven IMD of 2–5% [21–23].

Adult patients with acute lymphoblastic leukemia (ALL) are generally considered to be at moderate risk for developing IMD, with most case series having reported rates of probable or proven IMD between 2–5% [17]. A recent randomized study comparing prophylaxis with liposomal amphotericin B to placebo in adult patients with ALL reported rates of probable IMD of 7.5% and 9%, respectively [24]. The relatively higher incidence of probable IMD observed in this study may have reflected recent trends in the use of more intensive chemotherapy regimens in adults designed to improve long-term survival, which are associated with higher rates of infection during induction chemotherapy [25]. The risk of IMD may also be increased in patients with relapsed ALL, especially in patients who receive regimens with high doses of dexamethasone [26]. In contrast, patients with

Philadelphia-positive ALL receiving treatment with tyrosine kinase inhibitors (imatinib, dasatinib, nilotinib, and bosutinib) as part of standard or reduced-intensity regimens appear to have a lower risk of developing an IMD (3–5% incidence) [27–29].

Patients with chronic lymphoproliferative disorders such as non-Hodgkin's and Hodgkin's lymphoma, chronic myelogenous leukemia, or multiple myeloma are generally considered to be at low risk for IMD (<2% incidence). However, select subsets of patients with extensively-treated lymphoma who receive intensive chemotherapy regimens of high-dose corticosteroids followed by autologous HSCT often experience prolonged neutropenia and with rates of IMD like patients with AML/MDS undergoing induction chemotherapy [30]. Patients who receive intensive chemotherapy regimens for CNS lymphomas in combination with drugs that target B-lymphocyte pathways, particularly the Bruton's tyrosine kinase inhibitor ibrutinib, may be at especially high risk for cryptococcosis, pneumocystis pneumonia, and aspergillosis involving the central nervous system [31]. In one case series the use of ibrutinib in combination with temozolomide, etoposide, cytarabine and liposomal doxorubicin, rituximab and dexamethasone was associated with rates of invasive aspergillosis of 44% [30–32]. In contrast, the use of ibrutinib monotherapy in patients with chronic lymphocytic leukemia was associated with low rates of invasive fungal disease (0.5–1.6%) that were comparable to treatment with alkylating agents or monoclonal antibodies, which are no longer considered to be the main standard of care [32,33].

Relatively few data are available describing the risk of IMD among patients with myeloproliferative neoplasms including chronic myelogenous leukemia, polycythemia vera, essential thrombocytopenia, and myelofibrosis [4]. Accurate estimates of IMD risk in these populations are confounded by the increasing use of tyrosine kinase inhibitors (imatinib, dasatinib, nilotinib, bosutinib, and ponatinib) for targeting the BCR-ABL oncoprotein in chronic myelogenous leukemia, which leads to disease control in most patients. Similarly, targeted therapy with JAK (Janus Kinase) inhibitors in patients with myelofibrosis appears to be associated with a low risk of IMD [34,35].

Severe aplastic anemia is characterized by a reduction in the production of hematopoietic progenitor cells resulting in severe pancytopenia [36]. Infection is a major cause of death and is directly related to prolonged neutropenia. The risk of fungal infection is further increased by treatments that reduce T-cells (anti-thymocyte globulin) or function (corticosteroids, calcineurin inhibitors), however *Candida* spp. appear to be much more common than invasive molds [37].

2.2. Conditions Associated with Disease Treatment

Autopsy studies performed in the early 1960s identified prolonged neutropenia as a major predisposing risk factor for IMD [38,39]. Gerson and colleagues [40] later demonstrated that neutropenia persisting longer than three weeks was the most important risk factor for development of invasive pulmonary aspergillosis in patients with acute leukemia. During the first two weeks of neutropenia, patients developed signs of invasive risk of invasive pulmonary aspergillosis at a rate of approximately 1% per day that increased to 4.3% per day between the 24th and 36th days of neutropenia. Of the 13 patients who remained neutropenic at 28 days, 7 (54%) had developed signs of invasive pulmonary aspergillosis.

The European Organization for Research and Treatment of Cancer /Invasive Fungal Infections Cooperative Group and the National Institute of Allergy and Infectious Diseases Mycoses Study Group (EORTC/MSG) Consensus Group have defined neutropenia as a host factor for invasive fungal disease when it presents as <500 cells/mm^3 for >10 days [41]. Typically, the median duration of neutropenia prior to the first signs of IPA ranges from 16–25 days, but for some patients with multiple risk factors, disease onset may occur before 10 days [3].

Lymphopenia (<300 cells/mm^3) and monocytopenia (<10 cells/mm^3) are infrequently the sole predisposing risk factors for IMD, but often indicate delayed immune reconstitution after chemotherapy or allogeneic HSCT [4,42]. In an analysis of 1248 allogeneic HSCT recipients, neutropenia (HR 2.2; 1.3–3.6, $p < 0.01$), lymphocytopenia (1.4, 1–2; $p = 0.05$), and monocytopenia

(HR 1.8, 1.7–3.4, $p < 0.01$) were independently associated with increased risk of mold disease within one year of transplantation [43]. Mikulska et al. identified the presence and duration of lymphopenia as independent risk factors for early mortality from invasive aspergillosis in recipients of allogeneic HSCT from alternative donors [42]. Lewis et al. reported that patients with pulmonary mucormycosis who presented with lymphocyte counts <100 cells/mm^3 had a 4-fold higher rate of mortality versus patients without lymphocytopenia [44].

Glucocorticoid therapy at supraphysiologic doses have long been associated with the development of mold infections [45], but the cumulative dose that places a patient at increased risk varies from study to study and often depends on concomitant chemotherapy, type of allogeneic HSCT and severity of GVHD [46]. The EORTC/MSG consensus definitions for invasive fungal disease include corticosteroids as a host factor for mold disease when a patient has received a mean minimum dose of 0.3 mg/kg/day prednisone equivalent for greater than 3 weeks [41]. Notably, the EORTC/MSG criteria do not consider inhaled corticosteroids to be a host factor even though cases of invasive pulmonary aspergillosis have been described in critically-ill patients with chronic-obstructive pulmonary disease [47,48].

In patients who undergo allogeneic HSCT, high cumulative doses of glucocorticoids are the most frequently identified risk factor associated with invasive aspergillosis after engraftment [49–52]. O'Donnell and colleagues reported that use of high-dose prednisone (0.5–1.0 mg/kg per day) for graft-versus-host-disease increased the risk six-fold versus lower dose prednisone regimens (0.25 mg/kg/day) for developing invasive aspergillosis [53]. Similarly, Marr and colleagues reported that glucocorticoid doses of 1.9 mg/kg per day, 1.9–3.0 mg/kg per day, and greater than 3 mg/kg per day were associated with IMD risks of 5%, 10%, and 14%, respectively [49]. Ribaud and co-workers reported that the 60-day risk of death increased from 12% to 80% if allo-HSCT recipients had received a cumulative prednisolone dose of greater than 7 mg/kg in the week preceding diagnosis of IMD [51].

2.3. Hematopoietic Stem Cell Transplantation

Allogeneic HSCT recipients have among the highest reported incidence of IMD in studies ranging between 7 and 15% [4]. The risk period is bimodal, with early risk for disease (first 40 days) associated with prolonged neutropenia prior to stem cell engraftment, and later invasive mold disease (often after day +70 or +100) associated with immunosuppressive therapy for controlling GVHD. Several additional pre- and post-transplant factors have been reported to influence IMD risk. Pre-transplant risk factors include the type of transplant (matched-related donor, umbilical-cord donor, or haploidentical/mismatched donor) [43,54,55] stem cell dose [56], receipt of T-cell depleting agents (e.g., anti-thymocyte globulin or alemtuzumab) or a T-cell depleted graft [43,57], polymorphisms in donor genes important for detection of fungal antigens and antifungal innate immune responses (i.e., Toll-like receptor 4, Dectin-1, and Pentraxin-3) [58–60], and iron overload associated with frequent transfusions [43,61,62].

Post-allogeneic HSCT risk factors for IMD include the rate of immune reconstitution (time to engraftment and recovery from neutropenia, lymphocytopenia, and monocytopenia), recovery of natural-killer cell populations [63], development of acute graft versus host disease and its treatment with immunosuppressive agents (corticosteroids, anti-T cell therapies), and development/reactivation of viral co-infections, particularly cytomegalovirus (CMV) [43].

The incidence of IMD following autologous HSCT is lower, with reported incidence ranging between 3–8% [64–66]. The incidence is influenced by the underlying malignancy, number and types of chemotherapy cycles prior to transplantation, previous history of IMD, as well as antifungal prophylaxis [4].

2.4. Patient Comorbidities

Besides older age, several underlying comorbidities or poor performance status overall are associated with increased susceptibility to IMD. Patients with uncontrolled diabetes mellitus may exhibit impaired neutrophil migration and fungal cell phagocytosis and T-cell dysfunction [67]. In the setting of metabolic acidosis, uncoupling of free iron from carrier proteins in blood enhances fungal growth and is an important risk factor for disseminated mucormycosis [68,69]. Smoking and chronic pulmonary disease were similarly identified as pre-chemotherapy risk factors for the development of IMD during initial remission-induction therapy [16]. Poor nutritional status or cachexia associated with advanced disease, often manifesting as hypoalbuminemia, was associated with 3-fold lower odds of developing IMD following allogeneic HSCT for each gram/deciliter increase in serum albumin [70].

Viral co-infections, particularly cytomegalovirus (CMV) and influenzae are associated with increased risk of invasive pulmonary aspergillosis. CMV viremia and recipient CMV serostatus have been identified as risk factors following allogeneic HSCT for both early and late-onset IMD, which may be enhanced in patients who developed prolonged neutropenia following treatment with ganciclovir [43,71,72]. However, it is still debated whether CMV replication directly impairs antifungal immunity or is a signal of already impaired cellular immunity that allows permissive growth of molds [73,74]. Nevertheless, the clear temporal association of CMV viremia or infection with the development of IMD suggests CMV replication or disease can be a powerful predictive factor for mold disease [71].

The link between severe influenzae and invasive aspergillosis has been increasingly described in both non-immunocompromised and immunocompromised hosts [75–79]. It has been hypothesized that the evolution of more virulent influenza strains, as exemplified by the pandemic H1N1 strain, are associated with more severe lymphopenia and diffuse damage to the respiratory mucosa during infection that predispose hosts to fungal invasion [79,80]. Influenza preceding IMD is associated with high rates of respiratory failure and mortality, highlighting the importance of prompt diagnosis and initiation of antifungal therapy.

2.5. Environmental and Occupational Risk Factors

Repeated exposures to high fungal spore counts associated with farming, construction work, gardening or composting, and perhaps geoclimatic factors increase colonization and persistence of fungal spores in the respiratory tract and place the patient at increased risk for developing IMD during induction chemotherapy [16,81,82]. Outbreaks of invasive mold disease have been repeatedly described following hospital construction and renovation that results in dust contamination and dispersal of fungal spores [83]. Similarly, some studies have documented a relationship between environmental contamination by *Aspergillus* and other fungal species and the incidence of invasive aspergillosis [84]. This risk may be increased if patients are admitted for intensive chemotherapy or transplantation to rooms without positive pressure high efficiency particulate air (HEPA) filtration [85].

3. Risk Models for Invasive Mold Disease

3.1. Neutropenia-Associated Risk Measured by the D-Index

Given the high frequency and importance of neutropenia as a key risk factor for IMD, several investigators have focused on the development of tools that measure both the intensity and duration of neutropenia to predict risk of IMD. Portugal et al. proposed an index (D-index) to improve the assessment of risk for IMD in neutropenic patients. The D-index represents the difference in the area under the curve (AUC) for neutrophil counts over time versus an area resulting from a normal neutrophil count. This difference is geometrically represented as the area over the neutrophil curve (Figure 1) [86]. The investigators also evaluated the prognostic performance of the cumulative D-index score (c-D-Index), which represents the D-index from the start of neutropenia until the date of the first clinical manifestation of IMD. Compared to just measuring the duration of neutropenia, the D-index and c-D-index better discriminated patients ($n = 11$) who developed IMD with area under the receiver operator curve (aROC) values of 0.86 and 0.81 versus patients who did not develop an IMD ($n = 33$). At a cut-off of 6200, the sensitivity and specificity of the D-index was 100% and 58%, respectively and for the c-D-index value of 5800 the sensitivity and specificity were 91% and 58%

respectively. Over an incidence range of 5–15%, the positive predictive value (PPV) of the D-index was relatively low (11–30%) but the negative predictive value was high (97–99%) suggesting that the D-index may be useful for "screening-out" lower risk patients with neutropenia who are unlikely to develop a mold infection [86].

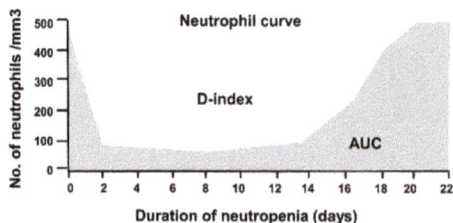

Figure 1. D-index area over the neutrophil curve: AUC areas under the neutrophil curve.

In a follow-up prospective study among 29 patients with acute leukemia undergoing remission-chemotherapy, the investigators utilized the D-index and galactomannan screening to stratify patients as low (<3000), intermediate (3000–5800), and high risk (>5800) for IMD [87]. Although a positive galactomannan result or clinical symptoms triggered a diagnostic workup in similar numbers of patients irrespective of the D-index risk stratification (58–73%), patients in the low risk D-index group were less likely to receive antifungal therapy (17% vs. 54–67%) and no cases of IMD were diagnosed in the low risk group versus 67% and 45% of the patients classified as high and moderate risk, respectively.

3.2. A risk score for Predicting IMD in Lymphoma Patients Receiving Salvage Chemotherapy

Takaoka et al. colleagues retrospectively analyzed 177 consecutive patients who received salvage chemotherapy for active lymphoma (705 courses in total) [88]. The IMD incidence rate was 2.3% (6 probable and 6 possible cases). Multivariate analysis revealed that relapsed refractory disease, receipt of two or more treatment courses, and neutropenia (ANC < 500 cells/mm^3) were independently risk factors associated with the development of IMD. Using these variables, the authors developed a simple weighted risk score: 1 point for refractory therapy, 1 point for two or more treatment lines, 2 points for three or more treatment lines, and 1 point for neutropenia. By applying the score, the authors were able to differentiate a subgroup of lymphoma patients with a higher incidence of IMD by day 80 if the score was above 3 (9% incidence) versus below 2 (0.19% incidence). However, given the retrospective nature of the analysis, it is unclear at what time the score could be applied to predict future IMD. Additionally, the inclusion of EORTC/MSG possible cases of IMD might be questioned as potentially more than half of these cases did not have IMD, especially in the setting of relapsing lymphoma which can produce nodular consolidations in the lung indistinguishable from lymphoma by standard chest CT [89].

3.3. A Risk Score for Predicting IMD Risk Post-Engraftment in Adult Allogeneic HSCT Recipients

Montesinos et al. [90] analyzed risk factors for probable or proven IMD among 404 allogeneic HSCT recipients who engrafted and survived more than 40 days after transplant. The one-year cumulative incidence of IMD in their study cohort was 11%. Five risk factors identified in multivariable analysis (age greater than 40 years, more than one previous HSCT, pre-engraftment neutropenia lasting more than 10 days, extensive and chronic GVHD, and CMV reactivation) were used to construct as risk score for stratifying patients into low risk (0–1 factor, cumulative incidence 0.7%) intermediate risk (2 factors, cumulative incidence 9.9%), and high-risk (3–5 factors, cumulative incidence 24.7%) categories. Although the authors suggest the score could be used to assess risk at patient discharge, two of the five risk factors (extensive chronic GVHD, CMV reactivation) would likely develop later after discharge, therefore patients would need to be reassessed periodically to fully apply the risk score.

3.4. Predicting Invasive Fungal Infection in Pediatric Allogeneic HSCT Recipients

Hol and colleagues retrospectively analyzed pre- and post-transplant predictors of invasive fungal disease among 209 pediatric recipients of allogeneic HSCT with at cumulative incidence of IMD of 12% (mostly molds) [91]. Patients were classified as high or low risk based on pre-transplant risk factors that included: age < 10 years, gender, treatment-related mortality risk predicted by the EBMT risk score, diagnosis type, use of mold-active prophylaxis, prior history of invasive fungal infections, donor type, donor relation and match, conditioning regimen, transplant number, and presence of galactomannan in pre-transplant bronchial alveolar lavage samples. Post-transplant risk factors included duration of neutropenia, aGVHD > grade II, extensive cGVHD, and high-dose corticosteroids (>1 mg/kg/day for at least one week). In multivariate analysis, an EMBT score predicted that treatment related mortality risk >20% was the only factor associated with the occurrence of an IFI, while posttransplant high-dose steroids were the only predictor of invasive fungal infection. After adjustment for pre-transplant treatment related mortality risk and use of corticosteroids, the odds of survival were significantly lower in children who developed IMD versus those who did not (OR 0.30, 95% CI 0.13–0.71, $p = 0.006$).

3.5. Applying Comorbidity Index to Predict IMD after Allogeneic HSCT

Prognostic risk models are frequently used to predict non-relapse mortality among patients undergoing allogeneic HSCT. In 2005, Sorror et al. [92] introduced the hematopoietic stem cell transplantation-comorbidity index (HSCT-CI) assessed prior to transplantation as a means of predicting non-relapse mortality, that was later updated with additional clinical risk factors [93]. Busca et al. recently explored whether the HSCT-CI could have similar utility in predicting the risk for adult patients undergoing allogeneic HSCT for developing fatal IMD [94]. Among 360 retrospectively-analyzed patients who underwent allogeneic HSCT, 8.5% of patients developed EORTC/MSG probable or proven IMD that was significantly higher among patients with and HSCT-CI score of ≥3 (12%) compared to patients with HSCT-CI scores of 0–2 (5%). Pulmonary comorbidities were the most common pre-transplant risk factor associated with the development of IMD. Advanced disease at the time of transplant, acute grade II-IV GVHD, and a comorbidity score ≥ 3 were independent risk factors for non-relapse mortality.

3.6. Development of a Universal IMD Risk Model for Patients with Hematological Malignancies

Many decisions regarding the management of IMD (e.g., galactomannan screening, antifungal prophylaxis computer tomography in symptomatic febrile patients) are made at the time or soon after admission to the hospital. With this idea in mind, we sought to develop a universal prognostic risk model that could be assessed at the time of each hospital admission to predict an individual patient's risk for developing IMD in the future [15]. Seventeen risk factors assessed at the time of patient admission were first retrospectively analyzed using a data registry of 840 patients with hematological malignancies over 1709 admission episodes lasting more than 5 days from 2005–2008. The data registry was maintained by a dedicated data manager and adjudicated periodically by an attending hematologist with expertise in infectious diseases. Although a total of 11/17 analyzed risk factors correlated with IMD in univariate analysis, only four risk factors were retained in the final multivariable model that were predictive of probable or proven IMD within 90 days of admission: (i) active malignancy (not in remission); (ii) PMN < 500 cells/mm^3 > 10 days or projected prolonged with chemotherapy; (iii) severe lymphocytopenia < 50 cells/mm^3, or lymphocyte-impairing therapies such as calcineurin inhibitors; (iv) and prior history of IMD. These variables were then used to construct a weighted risk score (BOSCORE) with a scale of 0–13. A risk score threshold of < 6 differentiated patients with low (<1%) versus higher (>5%) probability thresholds of IMD. The score was then prospectively validated in our institution in 855 patients over 1746 admissions from 2009–2012

(Figure 2). The discrimination and calibration of the score for predicting IMD were similar in the validation cohort of patients despite introduction of routine posaconazole prophylaxis in AML/MDS patients undergoing induction chemotherapy, with an aROC of 0.84 (95% CI 0.79–0.89) and negative predictive value of 0.99 (95% CI 0.98–0.99) at a 5% predicted probability threshold cut-off.

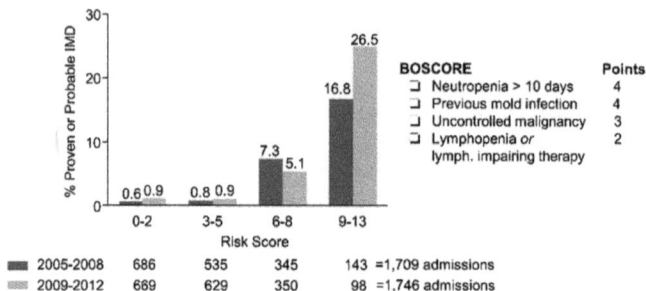

Figure 2. BOSCORE distribution versus observed 90-day cumulative incidence of IMD in the retrospective (2005–2008) development and prospective (2009–2012) validation cohorts.

We recently recalibrated the BOSCORE with additional risk factors to predict the 60-day probability of developing probable or proven IMD using 1944 patients with 4127 admissions from 2007–2016. The overall incidence of probable or proven IMD was 3.3%. Most of the analyzed risk factors were associated with the development of IMD (Figure 3), however only seven risk factors were retained in the final multivariable model: (i) prior history of IMD, (iii) receipt of 0.5 mg/kg prednisone equivalent within 30 days, (iii) uncontrolled malignancy; (iv) receipt of high-risk chemotherapy—e.g., any conditioning for allogeneic HSCT, high-dose ARA-C, fludarabine, and idarubicin (FLAI), or ifosfamide, carboplatin, etoposide (ICE); (v) PMN < 100 cells/mm^3 for >10 days or anticipated prolonged neutropenia; (vi) total lymphocyte count <50 cells/mm^3; and (vii) CMV reactivation (DNA > 1000 IU/mL in serum) or disease.

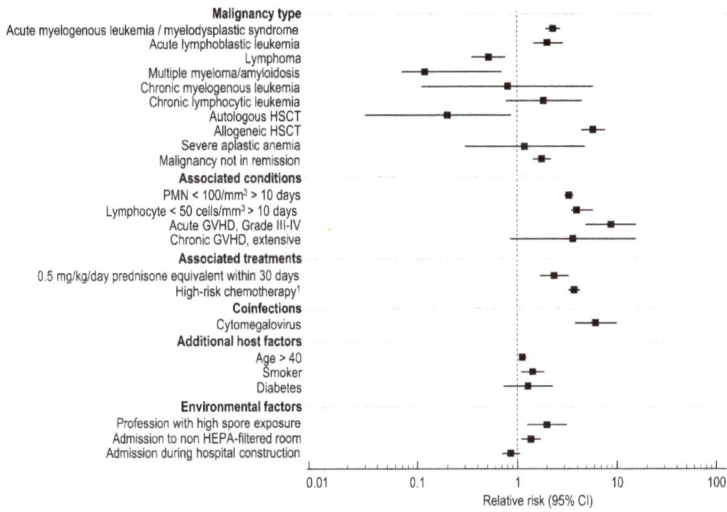

Figure 3. Risk factors in univariate analysis associated with the development of EORTC/MSG defined probable or proven invasive mold disease (*n* = 133) among 1944 adult inpatients (*n* = 4127 admissions) undergoing treatment for a hematological malignancy from 2007–2016 at the Seràgnoli Hematology Institute in Bologna, Italy.

The re-calibrated multivariate risk model displayed good calibration and discrimination in both low and higher risk groups and provided consistent predictions during internal validation with bootstrap resampled populations over an IMD incidence range from 2–9% (Figure 4a). When a 5% threshold was applied to differentiate low versus high-risk patients, the risk model correctly identified 60-day IMD outcomes in 85% of patients (Figure 4b). Among patients who developed IMD but were predicted to be low risk at admission, most (74%) fell just below the provisional 5% threshold (i.e., 3.5%–4.9% risk).

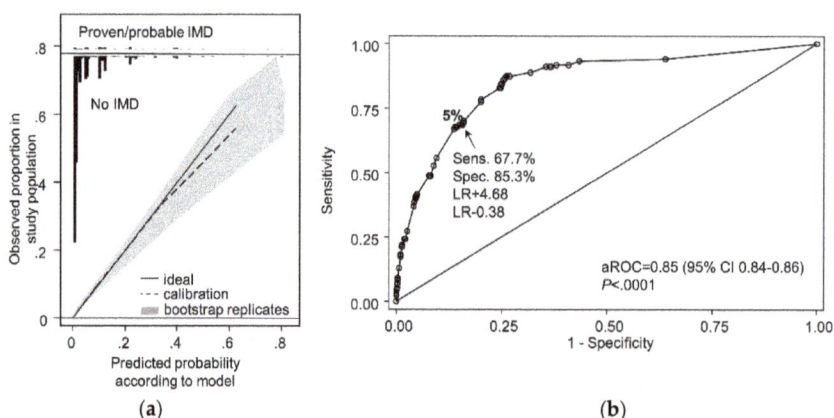

Figure 4. Revised BOSCORE model calibration and discrimination. Panel (**a**) shows the relationship between a perfectly calibrated model (solid line) and the observed incidence of EORTC/MSG proven, or probable mold disease fitted using the Loess smoothing algorithm (dashed line). Black bars on the top frame of the graph show the relative risk distribution of admissions without IMD (downward pointing spikes) or patients with proven or probable IMD (upward pointing spikes). The gray shaded area shows observed versus predicted incidence of proven or probable mold disease in 100 bootstrapped resampled datasets with varying IMD incidence of 2–9%. Panel (**b**) shows the area under the receiver operator curve (aROC) and predicted performance for EORTC/MSG probable or proven IMD at a prediction cut-off of 5%. Sens., sensitivity, spec. specificity, LR, likelihood ratio.

The risk model was subsequently developed as a smartphone application (Figure 5) that requires users to only check boxes of each risk factor present on admission and the 60-day risk for IMD is automatically calculated with links to recommendations institutional diagnostic and treatment algorithms based on the patient's estimated risk.

The true test of any prognostic risk model is its acceptance by practitioners and effects on clinical decision making [95]. The high negative-predictive value of the BOSCORE (0.96–1.0) across a wide range of patient groups with varying incidence of IMD (0.4–8%) is used by clinicians in our institute to screen out the majority (> 67%) of patients admitted who are unlikely to benefit from intensive monitoring of serum galactomannan or posaconazole prophylaxis (Figure 6). Indeed, routine serum testing of galactomannan in serum of patients with a low pretest probability of IMD (i.e., <2%) may result in more frequent false-positive rather than true-positive results resulting in potentially unnecessary invasive diagnostic procedures and antifungal therapy [96]. Patients at greater than 5% risk are considered for more intensive management by a diagnostic-driven algorithm. This algorithm consists of twice weekly serum galactomannan screening plus immediate low-dose CT (within 24–48 h of fever or signs of infection) followed by CT pulmonary angiography ± bronchoscopy with culture or possible biopsy if the patient has evaluable lesions and can tolerate the procedures [97].

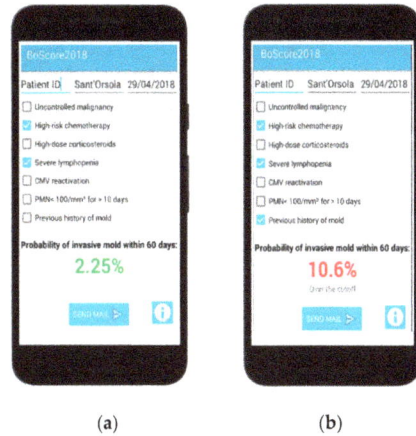

(a)   (b)

Figure 5. Smartphone-assisted prognostic model assessments of lower-risk (**a**) and higher-risk (**b**) nonneutropenic patient. The probability of IMD within 60 days is estimated using the formula: Risk = (0.68 × uncontrolled malignancy) + (0.79 × high-risk chemotherapy) + (0.80 × high-dose corticosteroids) + (0.89 × severe lymphopenia) + (1.14 × CMV reactivation) + (1.52 × prolonged neutropenia) + (1.64 × previous mold disease) − 5.45. To calculate the 60-day probability from the formula, the calculated result is first converted from log odds to odds (eRisk); then odds must be converted to probability using the formulae: Risk/(1 + Risk).

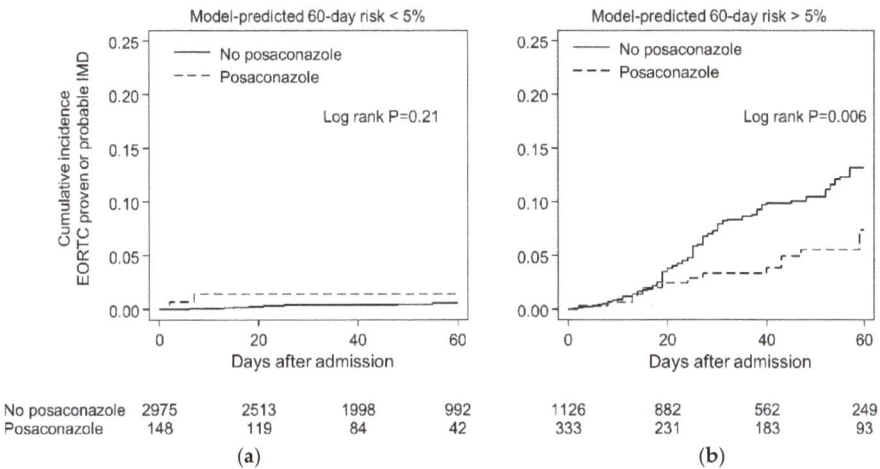

No posaconazole	2975	2513	1998	992	1126	882	562	249
Posaconazole	148	119	84	42	333	231	183	93

(a)   (b)

Figure 6. Impact of posaconazole prophylaxis on the 60-day cumulative incidence of IMD in patients with model predicted risk of IMD <5% (**a**) and >5% (**b**). Note the nonsignificant higher rate of IMD in low risk patients (Panel A) is explained by two patients. One patient had a predicted risk at the risk cut-off (0.5) and prior IMD as the only risk factor (receiving posaconazole secondary prophylaxis). He was diagnosed with breakthrough IMD within 7 days of admission with fever. The second patient receiving posaconazole prophylaxis was admitted for treatment of extensive chronic graft versus host disease (he was not receiving corticosteroids) and only had severe lymphopenia at the time of admission. His predicted risk was 2.5%. During his work-up he had no fever but a serum galactomannan was positive and a CT following admission was consistent with IMD.

Very-high risk patients (>10%) may be considered eligible for posaconazole prophylaxis, even if they do not have AML/MDS or have undergone allogeneic HSCT. For example, we routinely use the risk model to help identify patients with acute leukemia undergoing consolidation chemotherapy who require mold-active antifungal prophylaxis. We have also found that that the risk model identifies small subsets of patients with lymphoma (6%) and myeloma (3.4%) who have equivalent risk for developing IMD as an AML patient undergoing remission/induction chemotherapy. These unique subsets of patients are then targeted for more intensive diagnostic intervention.

4. Future Perspectives

Because accurate risk assessment is fundamental to the effective management of mold infections and stewardship of diagnostic and treatment resources, it is likely that interest in prognostic risk models will increase. However, adherence to several fundamental principles during model development and reporting should be encouraged. First and foremost, there should be a clear explanation of the patient population used to develop the predictive model and the timing of assessment- i.e., when IMD is already present (diagnostic model) or to predict the risk of IMD in the future (prognostic model). Steps involved in the development, analysis and validation of the model should be clearly reported according to standards described in the Transparent Reporting of a Multivariable Prediction Model for Individual Prognosis Or Diagnosis (TRIPOD) Guidelines [98]. Adherence to these guidelines improves the clarity in communication and allows researchers from other institutions to better assess the generalizability and risk of bias of published models [98].

Validation studies are often considered to be the benchmark of whether a developed model is of high quality or clinically useful. The real aim of model validation, however, is to measure the model's predictive performance in either resampled participant data of the development data set (often referred to as internal validation) or in other, independent participant data that were not used for developing the model (often referred to as external validation) [98,99]. In this respect, the key difference is the generalizability of the performance characteristics of the model, which should ideally be confirmed before the model is used in clinical care like any new diagnostic test [95]. Prediction models developed in one institution may not necessarily be equally useful in another setting, as illustrated with recent studies of *Candida* prediction models in non-neutropenic patients [8]. Indeed, it is not uncommon that existing models must be adapted to local circumstances or with new predictors before application in a local setting. This adaption will be impossible if local hospitals do not have surveillance systems in place where these infections are managed. Finally, research on the uptake and effect of prognostic models on clinical decision-making and patients' outcomes (impact studies) are needed to confirm the usefulness of the prognostic model in clinical care.

Outside of daily clinical use, Harrell and colleagues have identified several areas of clinical research where prognostic risk models are applicable [10]. Risk models are useful in the evaluation of new diagnostic technologies, as estimates derived both with and without the new test can be compared to measure the incremental prognostic information provided. White, Parr and Barnes recently used this approach to evaluate how genetic risk factors with early diagnostic markers can improve predictions of which neutropenic patients with hematological malignancies will develop invasive aspergillosis [100]. The investigators found that compared to clinical risk factors alone (allogeneic HSCT, respiratory virus infection), the incorporation of genetic risk factors (DECTIN-1, DC-SIGN mutations) plus a positive PCR result increased the predicted probability of invasive aspergillosis from 7% to 56.7%. This combined prognostic-diagnostic model approach was associated with good discrimination (aROC 0.86) and could substantially reduce the percentage of patients administered preemptive therapy to only 8.4% of the population of patients with febrile neutropenia.

A second means by which a validated prognostic model could benefit clinical research is to enable researchers to better estimate the effect of a single factor (e.g., antifungal prophylaxis) on patients' outcomes in observational data where many uncontrolled confounding factors also affect risk.

Finally, prognostic models can improve the design and enrollment of clinical trials for invasive mold diseases. Both the decision concerning which patients to randomize and the design of the randomization process (for example, stratified randomization using prognostic factors) are improved by more accurate prediction of patient risk before randomization. Accurate prognostic models can be used to test for differential therapeutic benefit or to estimate the clinical benefit for an individual patient in a clinical trial, because low-risk patients are likely to have less absolute benefit. These areas are largely unexplored for prognostic models and clinical studies of IMD.

5. Conclusions

Compared to therapeutic trials and etiological research, prognostic risk models for IMD have received relatively little attention. However, several studies have now shown that development of such models can improve the accuracy of risk prediction, and incorporation of these models into the clinical management of patients has the potential to improve diagnostic and antifungal stewardship, as well as personalized treatment of invasive mold diseases in patients with hematological malignancies.

Author Contributions: M.S. was involved in the writing of the manuscript, study concept and design and collection of data for the BSCORE risk model. R.L. was involved in the writing and editing of the manuscript and statistical analysis of the BOSCORE models.

Funding: Support for the data registry used to construct the BOSCORE was provided in part by AIL foundation and Gilead Sciences, Inc.

Acknowledgments: Nicola Vianelli, Giovanni Vianelli and Nicola Semprini helped develop the BOSCORE smartphone application. Riccardo Ragionieri was responsible for the development and maintenance of the data registry used to construct BOSCORE.

Conflicts of Interest: M.S. have received research funding from Gilead and has received speaking fees from Merck, Pfizer. R.L. has received research funding from Merck and has received speaking fees from Gilead and Cidara.

References

1. Lamoth, F.; Calandra, T. Early diagnosis of invasive mould infections and disease. *J. Antimicrob. Chemother.* **2017**, *72*, i19–i28. [CrossRef] [PubMed]
2. Pagano, L.; Akova, M.; Dimopoulos, G.; Herbrecht, R.; Drgona, L.; Blijlevens, N. Risk assessment and prognostic factors for mould-related diseases in immunocompromised patients. *J. Antimicrob. Chemother.* **2011**, *66*, i5–i14. [CrossRef] [PubMed]
3. Herbrecht, R.; Bories, P.; Moulin, J.-C.; Ledoux, M.-P.; Letscher-Bru, V. Risk stratification for invasive aspergillosis in immunocompromised patients. *Ann. N. Y. Acad. Sci.* **2012**, *1272*, 23–30. [CrossRef]
4. Pagano, L.; Busca, A.; Candoni, A.; Cattaneo, C.; Cesaro, S.; Fanci, R.; Nadali, G.; Potenza, L.; Russo, D.; Tumbarello, M.; et al. Risk stratification for invasive fungal infections in patients with hematological malignancies: SEIFEM recommendations. *Blood Rev.* **2016**, *31*, 17–29. [CrossRef] [PubMed]
5. Ostrosky-Zeichner, L.; Sable, C.; Sobel, J.; Alexander, B.D.; Donowitz, G.; Kan, V.; Kauffman, C.A.; Kett, D.; Larsen, R.A.; Morrison, V.; et al. Multicenter retrospective development and validation of a clinical prediction rule for nosocomial invasive candidiasis in the intensive care setting. *Eur. J. Clin. Microbiol. Infect. Dis.* **2007**, *26*, 271–276. [CrossRef] [PubMed]
6. Ostrosky-Zeichner, L.; Shoham, S.; Vazquez, J.; Reboli, A.; Betts, R.; Barron, M.A.; Schuster, M.; Judson, M.A.; Revankar, S.G.; Caeiro, J.P.; et al. MSG-01: A randomized, double-blind, placebo-controlled trial of caspofungin prophylaxis followed by preemptive therapy for invasive candidiasis in high-risk adults in the critical care setting. *Clin. Infect. Dis.* **2014**, *58*, 1219–1226. [CrossRef] [PubMed]
7. León, C.; Ruiz-Santana, S.; Saavedra, P.; Almirante, B.; Nolla-Salas, J.; Alvarez-Lerma, F.; Garnacho-Montero, J.; León, M.A.; EPCAN Study Group. A bedside scoring system ("Candida score") for early antifungal treatment in nonneutropenic critically ill patients with Candida colonization. *Crit. Care Med.* **2006**, *34*, 730–737. [CrossRef]
8. Guillamet, C.V.; Vazquez, R.; Micek, S.T.; Ursu, O.; Kollef, M. Development and validation of a clinical prediction rule for candidemia in hospitalized patients with severe sepsis and septic shock. *J. Crit. Care* **2015**, *30*, 715–720. [CrossRef]

9. Moons, K.; Royston, P.; Vergouwe, Y.; Grobbee, D.E. Prognosis and prognostic research: What, why, and how? *BMJ* **2009**, *338*, 1317–1320. [CrossRef]

10. Harrell, F.E.; Lee, K.L.; Mark, D.B. Tutorial in biostatistics multivariable prognostic models: Issues in developing models, evaluating assumptions and adequacy, and measuring and reducing errors. *Stat. Med.* **1996**, *15*, 361–387. [CrossRef]

11. Cornely, O.A.; Koehler, P.; Arenz, D.; C Mellinghoff, S. EQUAL Aspergillosis Score 2018: An ECMM score derived from current guidelines to measure QUALity of the clinical management of invasive pulmonary aspergillosis. *Mycoses* **2018**, *61*, 833–836. [CrossRef] [PubMed]

12. Ullmann, A.J.; Aguado, J.M.; Arikan-Akdagli, S.; Denning, D.W.; Groll, A.H.; Lagrou, K.; Lass-Flörl, C.; Lewis, R.E.; Munoz, P.; Verweij, P.E.; et al. Diagnosis and management of Aspergillus diseases: Executive summary of the 2017 ESCMID-ECMM-ERS guideline. *Clin. Microbiol. Infect.* **2018**, *24* (Suppl. 1), e1–e38. [CrossRef]

13. Patterson, T.F.; Thompson, G.R., 3rd; Denning, D.W.; Fishman, J.A.; Hadley, S.; Herbrecht, R.; Kontoyiannis, D.P.; Marr, K.A.; Morrison, V.A.; Nguyen, M.H.; Segal, B.H.; et al. Practice Guidelines for the Diagnosis and Management of Aspergillosis: 2016 Update by the Infectious Diseases Society of America. *Clin. Infect. Dis.* **2016**, *63*, e1–e60. [CrossRef] [PubMed]

14. Ascioglu, S.; Rex, J.H.; de Pauw, B.; Bennett, J.E.; Bille, J.; Crokaert, F.; Denning, D.W.; Donnelly, J.P.; Edwards, J.E.; Erjavec, Z.; et al. Defining opportunistic invasive fungal infections in immunocompromised patients with cancer and hematopoietic stem cell transplants: An international consensus. *Clin. Infect. Dis.* **2002**, *34*, 7–14. [CrossRef] [PubMed]

15. Stanzani, M.; Lewis, R.E.; Fiacchini, M.; Ricci, P.; Tumietto, F.; Viale, P.; Ambretti, S.; Baccarani, M.; Cavo, M.; Vianelli, N. A risk prediction score for invasive mold disease in patients with hematological malignancies. *PLoS ONE* **2013**, *8*, e75531. [CrossRef] [PubMed]

16. Caira, M.; Candoni, A.; Verga, L.; Busca, A.; Delia, M.; Nosari, A.; Caramatti, C.; Castagnola, C.; Cattaneo, C.; Fanci, R.; et al. Pre-chemotherapy risk factors for invasive fungal diseases: Prospective analysis of 1192 patients with newly diagnosed acute myeloid leukemia (SEIFEM 2010-a multicenter study). *Haematologica* **2015**, *100*, 284–292. [CrossRef] [PubMed]

17. Pagano, L.; Caira, M.; Candoni, A.; Offidani, M.; Fianchi, L.; Martino, B.; Pastore, D.; Picardi, M.; Bonini, A.; Chierichini, A.; et al. The epidemiology of fungal infections in patients with hematologic malignancies: The SEIFEM-2004 study. *Haematologica* **2006**, *91*, 1068–1075.

18. Pagano, L.; Girmenia, C.; Mele, L.; Ricci, P.; Tosti, M.E.; Nosari, A.; Buelli, M.; Picardi, M.; Allione, B.; Corvatta, L.; et al. Infections caused by filamentous fungi in patients with hematologic malignancies. A report of 391 cases by GIMEMA Infection Program. *Haematologica* **2001**, *86*, 862–870.

19. Lionakis, M.S.; Lewis, R.E.; Kontoyiannis, D.P. Breakthrough Invasive Mold Infections in the Hematology Patient: Current Concepts and Future Directions. *Clin. Infect. Dis.* **2018**, *67*, 1621–1630. [CrossRef]

20. Estey, E.H. Acute myeloid leukemia: 2019 update on risk-stratification and management. *Am. J. Hematol.* **2018**, *93*, 1267–1291. [CrossRef]

21. Falantes, J.F.; Calderón, C.; Márquez-Malaver, F.J.; Aguilar-Guisado, M.; Martín-Peña, A.; Martino, M.L.; Montero, I.; González, J.; Parody, R.; Pérez-Simón, J.A.; Espigado, I. Patterns of infection in patients with myelodysplastic syndromes and acute myeloid leukemia receiving azacitidine as salvage therapy. Implications for primary antifungal prophylaxis. *Clin. Lymphoma Myeloma Leuk.* **2014**, *14*, 80–86. [CrossRef] [PubMed]

22. Trubiano, J.A.; Dickinson, M.; Thursky, K.A.; Spelman, T.; Seymour, J.F.; Slavin, M.A.; Worth, L.J. Incidence, etiology and timing of infections following azacitidine therapy for myelodysplastic syndromes. *Leuk. Lymphoma* **2017**, *58*, 2379–2386. [CrossRef] [PubMed]

23. Pomares, H.; Arnan, M.; Sánchez-Ortega, I.; Sureda, A.; Duarte, R.F. Invasive fungal infections in AML/MDS patients treated with azacitidine: A risk worth considering antifungal prophylaxis? *Mycoses* **2016**, *59*, 516–519. [CrossRef] [PubMed]

24. Cornely, O.A.; Leguay, T.; Maertens, J.; Vehreschild, M.J.G.T.; Anagnostopoulos, A.; Castagnola, C.; Verga, L.; Rieger, C.; Kondakci, M.; Härter, G.; et al. Randomized comparison of liposomal amphotericin B versus placebo to prevent invasive mycoses in acute lymphoblastic leukaemia. *J. Antimicrob. Chemother.* **2017**, *72*, 2359–2367. [CrossRef] [PubMed]

25. Storring, J.M.; Minden, M.D.; Kao, S.; Gupta, V.; Schuh, A.C.; Schimmer, A.D.; Yee, K.W.L.; Kamel-Reid, S.; Chang, H.; Lipton, J.H.; et al. Treatment of adults with BCR-ABL negative acute lymphoblastic leukaemia with a modified paediatric regimen. *Br. J. Haematol.* **2009**, *146*, 76–85. [CrossRef] [PubMed]

26. Kadia, T.M.; Cortes, J.; Ravandi, F.; Jabbour, E.; Konopleva, M.; Benton, C.B.; Burger, J.; Sasaki, K.; Borthakur, G.; DiNardo, C.D.; et al. Cladribine and low-dose cytarabine alternating with decitabine as front-line therapy for elderly patients with acute myeloid leukaemia: A phase 2 single-arm trial. *Lancet Haematol* **2018**, *5*, e411–e421. [CrossRef]

27. Bassan, R.; Rossi, G.; Pogliani, E.M.; Di Bona, E.; Angelucci, E.; Cavattoni, I.; Lambertenghi-Deliliers, G.; Mannelli, F.; Levis, A.; Ciceri, F.; et al. Chemotherapy-phased imatinib pulses improve long-term outcome of adult patients with Philadelphia chromosome-positive acute lymphoblastic leukemia: Northern Italy Leukemia Group protocol 09/00. *J. Clin. Oncol.* **2010**, *28*, 3644–3652. [CrossRef]

28. Foà, R.; Vitale, A.; Vignetti, M.; Meloni, G.; Guarini, A.; De Propris, M.S.; Elia, L.; Paoloni, F.; Fazi, P.; Cimino, G.; et al. Dasatinib as first-line treatment for adult patients with Philadelphia chromosome-positive acute lymphoblastic leukemia. *Blood* **2011**, *118*, 6521–6528. [CrossRef]

29. Ottmann, O.G.; Druker, B.J.; Sawyers, C.L.; Goldman, J.M.; Reiffers, J.; Silver, R.T.; Tura, S.; Fischer, T.; Deininger, M.W.; Schiffer, C.A.; et al. A phase 2 study of imatinib in patients with relapsed or refractory Philadelphia chromosome–positive acute lymphoid leukemias. *Blood* **2002**, *100*, 1965–1971. [CrossRef]

30. Teng, J.C.; Slavin, M.A.; Teh, B.W.; Lingaratnam, S.M.; Ananda-Rajah, M.R.; Worth, L.J.; Seymour, J.F.; Thursky, K.A. Epidemiology of invasive fungal disease in lymphoproliferative disorders. *Haematologica* **2015**, *100*, e462–e466. [CrossRef]

31. Chamilos, G.; Lionakis, M.S.; Kontoyiannis, D.P. Call for Action: Invasive Fungal Infections Associated with Ibrutinib and Other Small Molecule Kinase Inhibitors Targeting Immune Signaling Pathways. *Clin. Infect. Dis.* **2018**, *66*, 140–148. [CrossRef] [PubMed]

32. Teh, B.W.; Tam, C.S.; Handunnetti, S.; Worth, L.J.; Slavin, M.A. Infections in patients with chronic lymphocytic leukaemia: Mitigating risk in the era of targeted therapies. *Blood Rev.* **2018**. [CrossRef] [PubMed]

33. Byrd, J.C.; Brown, J.R.; O'Brien, S.; Barrientos, J.C.; Kay, N.E.; Reddy, N.M.; Coutre, S.; Tam, C.S.; Mulligan, S.P.; Jaeger, U.; et al. Ibrutinib versus ofatumumab in previously treated chronic lymphoid leukemia. *N. Engl. J. Med.* **2014**, *371*, 213–223. [CrossRef] [PubMed]

34. Jabbour, E.; Kantarjian, H.M.; Saglio, G.; Steegmann, J.L.; Shah, N.P.; Boqué, C.; Chuah, C.; Pavlovsky, C.; Mayer, J.; Cortes, J.; et al. Early response with dasatinib or imatinib in chronic myeloid leukemia: 3-year follow-up from a randomized phase 3 trial (DASISION). *Blood* **2014**, *123*, 494–500. [CrossRef] [PubMed]

35. Polverelli, N.; Breccia, M.; Benevolo, G.; Martino, B.; Tieghi, A.; Latagliata, R.; Sabattini, E.; Riminucci, M.; Godio, L.; Catani, L.; et al. Risk factors for infections in myelofibrosis: Role of disease status and treatment. A multicenter study of 507 patients. *Am. J. Hematol.* **2017**, *92*, 37–41. [CrossRef]

36. Valdez, J.M.; Scheinberg, P.; Young, N.S.; Walsh, T.J. Infections in patients with aplastic anemia. *Semin. Hematol.* **2009**, *46*, 269–276. [CrossRef]

37. Valdez, J.M.; Scheinberg, P.; Nunez, O.; Wu, C.O.; Young, N.S.; Walsh, T.J. Decreased infection-related mortality and improved survival in severe aplastic anemia in the past two decades. *Clin. Infect. Dis.* **2011**, *52*, 726–735. [CrossRef]

38. Bodey, G.P. The changing face of febrile neutropenia-from monotherapy to moulds to mucositis. Fever and neutropenia: The early years. *J. Antimicrob. Chemother.* **2009**, *63* (Suppl. 1), i3–i13. [CrossRef]

39. Bodey, G.P. Fungal infections complicating acute leukemia. *J. Chronic Dis.* **1966**, *19*, 667–687. [CrossRef]

40. Gerson, S.L.; Talbot, G.H.; Hurwitz, S.; Strom, B.L.; Lusk, E.J.; Cassileth, P.A. Prolonged granulocytopenia: The major risk factor for invasive pulmonary aspergillosis in patients with acute leukemia. *Ann. Intern. Med.* **1984**, *100*, 345–351. [CrossRef]

41. De Pauw, B.; Walsh, T.J.; Donnelly, J.P.; Stevens, D.A.; Edwards, J.E.; Calandra, T.; Pappas, P.G.; Maertens, J.; Lortholary, O.; Kauffman, C.A.; et al. Revised definitions of invasive fungal disease from the European Organization for Research and Treatment of Cancer/Invasive Fungal Infections Cooperative Group and the National Institute of Allergy and Infectious Diseases Mycoses Study Group (EORTC/MSG) Consensus Group. *Clin. Infect. Dis.* **2008**, *46*, 1813–1821. [PubMed]

42. Mikulska, M.; Raiola, A.M.; Bruno, B.; Furfaro, E.; Van Lint, M.T.; Bregante, S.; Ibatici, A.; Del Bono, V.; Bacigalupo, A.; Viscoli, C. Risk factors for invasive aspergillosis and related mortality in recipients of

allogeneic SCT from alternative donors: An analysis of 306 patients. *Bone Marrow Transplant.* **2009**, *44*, 361–370. [CrossRef]

43. Garcia-Vidal, C.; Upton, A.; Kirby, K.A.; Marr, K.A. Epidemiology of invasive mold infections in allogeneic stem cell transplant recipients: Biological risk factors for infection according to time after transplantation. *Clin. Infect. Dis.* **2008**, *47*, 1041–1050. [CrossRef] [PubMed]

44. Lewis, R.E.; Georgiadou, S.P.; Sampsonas, F.; Chamilos, G.; Kontoyiannis, D.P. Risk factors for early mortality in haematological malignancy patients with pulmonary mucormycosis. *Mycoses* **2014**, *57*, 49–55. [CrossRef] [PubMed]

45. Graham, B.S.; Tucker, W.S., Jr. Opportunistic infections in endogenous Cushing's syndrome. *Ann. Intern. Med.* **1984**, *101*, 334–338. [CrossRef]

46. Lionakis, M.S.; Kontoyiannis, D.P. Glucocorticoids and invasive fungal infections. *Lancet* **2003**, *362*, 1828–1838. [CrossRef]

47. Taccone, F.S.; Van den Abeele, A.-M.; Bulpa, P.; Misset, B.; Meersseman, W.; Cardoso, T.; Paiva, J.-A.; Blasco-Navalpotro, M.; De Laere, E.; Dimopoulos, G.; et al. Epidemiology of invasive aspergillosis in critically ill patients: Clinical presentation, underlying conditions, and outcomes. *Crit. Care* **2015**, *19*, 7. [CrossRef]

48. Bassetti, M.; Garnacho-Montero, J.; Calandra, T.; Kullberg, B.; Dimopoulos, G.; Azoulay, E.; Chakrabarti, A.; Kett, D.; Leon, C.; Ostrosky-Zeichner, L.; et al. Intensive care medicine research agenda on invasive fungal infection in critically ill patients. *Intensive Care Med.* **2017**, 1–14. [CrossRef] [PubMed]

49. Marr, K.A.; Carter, R.A.; Boeckh, M.; Martin, P.; Corey, L. Invasive aspergillosis in allogeneic stem cell transplant recipients: Changes in epidemiology and risk factors. *Blood* **2002**, *100*, 4358–4366. [CrossRef]

50. Wald, A.; Leisenring, W.; van Burik, J.A.; Bowden, R.A. Epidemiology of *Aspergillus* infections in a large cohort of patients undergoing bone marrow transplantation. *J. Infect. Dis.* **1997**, *175*, 1459–1466. [CrossRef]

51. Ribaud, P.; Chastang, C.; Latgé, J.P.; Baffroy-Lafitte, L.; Parquet, N.; Devergie, A.; Espérou, H.; Sélimi, F.; Rocha, V.; Espérou, H.; et al. Survival and prognostic factors of invasive aspergillosis after allogeneic bone marrow transplantation. *Clin. Infect. Dis.* **1999**, *28*, 322–330. [CrossRef] [PubMed]

52. Hovi, L.; Saarinen-Pihkala, U.M.; Vettenranta, K.; Saxen, H. Invasive fungal infections in pediatric bone marrow transplant recipients: Single center experience of 10 years. *Bone Marrow Transplant.* **2000**, *26*, 999–1004. [CrossRef] [PubMed]

53. O'Donnell, M.R.; Schmidt, G.M.; Tegtmeier, B.R.; Faucett, C.; Fahey, J.L.; Ito, J.; Nademanee, A.; Niland, J.; Parker, P.; Smith, E.P. Prediction of systemic fungal infection in allogeneic marrow recipients: Impact of amphotericin prophylaxis in high-risk patients. *J. Clin. Oncol.* **1994**, *12*, 827–834. [CrossRef] [PubMed]

54. Girmenia, C.; Raiola, A.M.; Piciocchi, A.; Algarotti, A.; Stanzani, M.; Cudillo, L.; Pecoraro, C.; Guidi, S.; Iori, A.P.; Montante, B.; et al. Incidence and outcome of invasive fungal diseases after allogeneic stem cell transplantation: A prospective study of the Gruppo Italiano Trapianto Midollo Osseo (GITMO). *Biol. Blood Marrow Transplant.* **2014**, *20*, 872–880. [CrossRef] [PubMed]

55. Atalla, A.; Garnica, M.; Maiolino, A.; Nucci, M. Risk factors for invasive mold diseases in allogeneic hematopoietic cell transplant recipients. *Transpl. Infect. Dis.* **2015**, *17*, 7–13. [CrossRef] [PubMed]

56. Bittencourt, H.; Rocha, V.; Chevret, S.; Socié, G.; Espérou, H.; Devergie, A.; Dal Cortivo, L.; Marolleau, J.-P.; Garnier, F.; Ribaud, P.; Gluckman, E. Association of CD34 cell dose with hematopoietic recovery, infections, and other outcomes after HLA-identical sibling bone marrow transplantation. *Blood* **2002**, *99*, 2726–2733. [CrossRef] [PubMed]

57. Kuster, S.; Stampf, S.; Gerber, B.; Baettig, V.; Weisser, M.; Gerull, S.; Medinger, M.; Passweg, J.; Schanz, U.; Garzoni, C.; et al. Incidence and outcome of invasive fungal diseases after allogeneic hematopoietic stem cell transplantation: A Swiss transplant cohort study. *Transpl. Infect. Dis.* **2018**, e12981. [CrossRef] [PubMed]

58. Bochud, P.-Y.; Chien, J.W.; Marr, K.A.; Leisenring, W.M.; Upton, A.; Janer, M.; Rodrigues, S.D.; Li, S.; Hansen, J.A.; Zhao, L.P.; et al. Toll-like receptor 4 polymorphisms and aspergillosis in stem-cell transplantation. *N. Engl. J. Med.* **2008**, *359*, 1766–1777. [CrossRef] [PubMed]

59. Cunha, C.; Di Ianni, M.; Bozza, S.; Giovannini, G.; Zagarella, S.; Zelante, T.; D'Angelo, C.; Pierini, A.; Pitzurra, L.; Falzetti, F.; et al. Dectin-1 Y238X polymorphism associates with susceptibility to invasive aspergillosis in hematopoietic transplantation through impairment of both recipient- and donor-dependent mechanisms of antifungal immunity. *Blood* **2010**, *116*, 5394–5402. [CrossRef]

60. Cunha, C.; Aversa, F.; Lacerda, J.F.; Busca, A.; Kurzai, O.; Grube, M.; Löffler, J.; Maertens, J.A.; Bell, A.S.; Inforzato, A.; et al. Genetic PTX3 deficiency and aspergillosis in stem-cell transplantation. *N. Engl. J. Med.* **2014**, *370*, 421–432. [CrossRef]

61. Sucak, G.T.; Yegin, Z.A.; Ozkurt, Z.N.; Aki, S.Z.; Karakan, T.; Akyol, G. The role of liver biopsy in the workup of liver dysfunction late after SCT: Is the role of iron overload underestimated? *Bone Marrow Transplant.* **2008**, *42*, 461–467. [CrossRef] [PubMed]

62. Kontoyiannis, D.P.; Chamilos, G.; Lewis, R.E.; Giralt, S.; Cortes, J.; Raad, I.I.; Manning, J.T.; Han, X. Increased bone marrow iron stores is an independent risk factor for invasive aspergillosis in patients with high-risk hematologic malignancies and recipients of allogeneic hematopoietic stem cell transplantation. *Cancer* **2007**, *110*, 1303–1306. [CrossRef] [PubMed]

63. Stuehler, C.; Kuenzli, E.; Jaeger, V.K.; Baettig, V.; Ferracin, F.; Rajacic, Z.; Kaiser, D.; Bernardini, C.; Forrer, P.; Weisser, M.; et al. Immune reconstitution after allogeneic hematopoietic stem cell transplantation and association with occurrence and outcome of invasive aspergillosis. *J. Infect. Dis.* **2015**, *212*, 959–967. [CrossRef] [PubMed]

64. Kontoyiannis, D.P.; Marr, K.A.; Park, B.J. Prospective surveillance for invasive fungal infections in hematopoietic stem cell transplant recipients, 2001–2006: Overview of the Transplant-Associated Infection Surveillance Network (TRANSNET). *Clin. Infect. Dis.* **2010**, *50*, 1091–1100. [CrossRef] [PubMed]

65. Gil, L.; Kozlowska-Skrzypczak, M.; Mol, A.; Poplawski, D.; Styczynski, J.; Komarnicki, M. Increased risk for invasive aspergillosis in patients with lymphoproliferative diseases after autologous hematopoietic SCT. *Bone Marrow Transplant.* **2009**, *43*, 121–126. [CrossRef] [PubMed]

66. Teh, B.W.; Teng, J.C.; Urbancic, K.; Grigg, A.; Harrison, S.J.; Worth, L.J.; Slavin, M.A.; Thursky, K.A. Invasive fungal infections in patients with multiple myeloma: A multi-center study in the era of novel myeloma therapies. *Haematologica* **2015**, *100*, e28–e31. [CrossRef] [PubMed]

67. Peleg, A.Y.; Weerarathna, T.; McCarthy, J.S.; Davis, T.M.E. Common infections in diabetes: Pathogenesis, management and relationship to glycaemic control. *Diabetes. Metab. Res. Rev.* **2007**, *23*, 3–13. [CrossRef]

68. Ibrahim, A.S. Host cell invasion in mucormycosis: Role of iron. *Curr. Opin. Microbiol.* **2011**, *14*, 406–411. [CrossRef] [PubMed]

69. Artis, W.M.; Fountain, J.A.; Delcher, H.K.; Jones, H.E. A mechanism of susceptibility to mucormycosis in diabetic ketoacidosis: Transferrin and iron availability. *Diabetes* **1982**, *31*, 1109–1114. [CrossRef] [PubMed]

70. Corzo-León, D.E.; Satlin, M.J.; Soave, R.; Shore, T.B.; Schuetz, A.N.; Jacobs, S.E.; Walsh, T.J. Epidemiology and outcomes of invasive fungal infections in allogeneic haematopoietic stem cell transplant recipients in the era of antifungal prophylaxis: A single-centre study with focus on emerging pathogens. *Mycoses* **2015**, *58*, 325–336. [CrossRef] [PubMed]

71. Yong, M.K.; Slavin, M.A.; Kontoyiannis, D.P. Invasive fungal disease and cytomegalovirus infection: Is there an association? *Curr. Opin. Infect. Dis.* **2018**, *31*, 481. [CrossRef] [PubMed]

72. Martino, R.; Piñana, J.L.; Parody, R.; Valcarcel, D.; Sureda, A.; Brunet, S.; Briones, J.; Delgado, J.; Sánchez, F.; Rabella, N.; Sierra, J. Lower respiratory tract respiratory virus infections increase the risk of invasive aspergillosis after a reduced-intensity allogeneic hematopoietic SCT. *Bone Marrow Transplant.* **2009**, *44*, 749–756. [CrossRef] [PubMed]

73. Giménez, E.; Solano, C.; Nieto, J.; Remigia, M.J.; Clari, M.Á.; Costa, E.; Muñoz-Cobo, B.; Amat, P.; Bravo, D.; Benet, I.; Navarro, D. An investigation on the relationship between the occurrence of CMV DNAemia and the development of invasive aspergillosis in the allogeneic stem cell transplantation setting. *J. Med. Virol.* **2014**, *86*, 568–575. [CrossRef] [PubMed]

74. Yong, M.K.; Lewin, S.R.; Manuel, O. Immune Monitoring for CMV in Transplantation. *Curr. Infect. Dis. Rep.* **2018**, *20*, 4. [CrossRef] [PubMed]

75. Lamoth, F.; Calandra, T. Let's add invasive aspergillosis to the list of influenza complications. *Lancet Respir. Med.* **2018**, *6*, 733–735. [CrossRef]

76. Schauwvlieghe, A.F.A.D.; Rijnders, B.J.A.; Philips, N.; Verwijs, R.; Vanderbeke, L.; Van Tienen, C.; Lagrou, K.; Verweij, P.E.; Van de Veerdonk, F.L.; Gommers, D.; et al. Invasive aspergillosis in patients admitted to the intensive care unit with severe influenza: A retrospective cohort study. *Lancet Respir. Med.* **2018**, *6*, 782–792. [CrossRef]

77. Garnacho-Montero, J.; León-Moya, C.; Gutiérrez-Pizarraya, A.; Arenzana-Seisdedos, A.; Vidaur, L.; Guerrero, J.E.; Gordón, M.; Martín-Loeches, I.; Rodriguez, A. Clinical characteristics, evolution,

and treatment-related risk factors for mortality among immunosuppressed patients with influenza A (H1N1) virus admitted to the intensive care unit. *J. Crit. Care* **2018**, *48*, 172–177. [CrossRef]

78. Garcia-Vidal, C.; Barba, P.; Arnan, M.; Moreno, A.; Ruiz-Camps, I.; Gudiol, C.; Ayats, J.; Ortí, G.; Carratalà, J. Invasive aspergillosis complicating pandemic influenza A (H1N1) infection in severely immunocompromised patients. *Clin. Infect. Dis.* **2011**, *53*, e16–e19. [CrossRef]

79. Crum-Cianflone, N.F. Invasive Aspergillosis associated with severe influenza infections. *Open Forum Infect. Dis.* **2016**, *3*, ofw171. [CrossRef]

80. Guarner, J.; Falcón-Escobedo, R. Comparison of the pathology caused by H1N1, H5N1, and H3N2 influenza viruses. *Arch. Med. Res.* **2009**, *40*, 655–661. [CrossRef]

81. Panackal, A.A.; Li, H.; Kontoyiannis, D.P.; Mori, M.; Perego, C.A.; Boeckh, M.; Marr, K.A. Geoclimatic influences on invasive aspergillosis after hematopoietic stem cell transplantation. *Clin. Infect. Dis.* **2010**, *50*, 1588–1597. [CrossRef] [PubMed]

82. Tanaka, R.J.; Boon, N.J.; Vrcelj, K.; Nguyen, A.; Vinci, C.; Armstrong-James, D.; Bignell, E. In silico modeling of spore inhalation reveals fungal persistence following low dose exposure. *Sci. Rep.* **2015**, *5*, 13958. [CrossRef]

83. Kanamori, H.; Rutala, W.A.; Sickbert-Bennett, E.E.; Weber, D.J. Review of Fungal Outbreaks and Infection Prevention in Healthcare Settings during Construction and Renovation. *Clin. Infect. Dis.* **2015**, *61*, 433–444. [CrossRef]

84. Alberti, C.; Bouakline, A.; Ribaud, P.; Lacroix, C.; Rousselot, P.; Leblanc, T.; Derouin, F.; Aspergillus Study Group. Relationship between environmental fungal contamination and the incidence of invasive aspergillosis in haematology patients. *J. Hosp. Infect.* **2001**, *48*, 198–206. [CrossRef] [PubMed]

85. Cornet, M.; Levy, V.; Fleury, L.; Lortholary, J.; Barquins, S.; Coureul, M.H.; Deliere, E.; Zittoun, R.; Brücker, G.; Bouvet, A. Efficacy of prevention by high-efficiency particulate air filtration or laminar airflow against Aspergillus airborne contamination during hospital renovation. *Infect. Control Hosp. Epidemiol.* **1999**, *20*, 508–513. [CrossRef]

86. Portugal, R.D.; Garnica, M.; Nucci, M. Index to predict invasive mold infection in high-risk neutropenic patients based on the area over the neutrophil curve. *J. Clin. Oncol.* **2009**, *27*, 3849–3854. [CrossRef] [PubMed]

87. Garnica, M.; Sinhorelo, A.; Madeira, L.; Portugal, R.; Nucci, M. Diagnostic-driven antifungal therapy in neutropenic patients using the D-index and serial serum galactomannan testing. *Braz. J. Infect. Dis.* **2016**, *20*, 354–359. [CrossRef]

88. Takaoka, K.; Nannya, Y.; Shinohara, A.; Arai, S.; Nakamura, F.; Kurokawa, M. A novel scoring system to predict the incidence of invasive fungal disease in salvage chemotherapies for malignant lymphoma. *Ann. Hematol.* **2014**, *93*, 1637–1644. [CrossRef]

89. Sassi, C.; Stanzani, M.; Lewis, R.E. Computerized tomographic pulmonary angiography discriminates invasive mould disease of the lung from lymphoma. *Br. J. Haematol.* **2015**, *169*, 462. [CrossRef]

90. Montesinos, P.; Rodríguez-Veiga, R.; Boluda, B.; Martínez-Cuadrón, D.; Cano, I.; Lancharro, A.; Sanz, J.; Arilla, M.J.; López-Chuliá, F.; Navarro, I.; et al. Incidence and risk factors of post-engraftment invasive fungal disease in adult allogeneic hematopoietic stem cell transplant recipients receiving oral azoles prophylaxis. *Bone Marrow Transpl.* **2015**, *50*, 1465–1472. [CrossRef]

91. Hol, J.A.; Wolfs, T.F.W.; Bierings, M.B.; Lindemans, C.A.; Versluys, A.B.J.; Wildt de, A.; Gerhardt, C.E.; Boelens, J.J. Predictors of invasive fungal infection in pediatric allogeneic hematopoietic SCT recipients. *Bone Marrow Transpl.* **2014**, *49*, 95–101. [CrossRef] [PubMed]

92. Sorror, M.L.; Maris, M.B.; Storb, R.; Baron, F.; Sandmaier, B.M.; Maloney, D.G.; Storer, B. Hematopoietic cell transplantation (HCT)-specific comorbidity index: A new tool for risk assessment before allogeneic HCT. *Blood* **2005**, *106*, 2912–2919. [CrossRef] [PubMed]

93. Sorror, M.L.; Storb, R.F.; Sandmaier, B.M.; Maziarz, R.T.; Pulsipher, M.A.; Maris, M.B.; Bhatia, S.; Ostronoff, F.; Deeg, H.J.; Syrjala, K.L.; et al. Comorbidity-age index: A clinical measure of biologic age before allogeneic hematopoietic cell transplantation. *J. Clin. Oncol.* **2014**, *32*, 3249–3256. [CrossRef]

94. Busca, A.; Passera, R.; Maffini, E.; Festuccia, M.; Brunello, L.; Dellacasa, C.M.; Aydin, S.; Frairia, C.; Manetta, S.; Butera, S.; et al. Hematopoietic cell transplantation comorbidity index and risk of developing invasive fungal infections after allografting. *Bone Marrow Transpl.* **2018**, *53*, 1304–1310. [CrossRef] [PubMed]

95. Moons, K.G.M.; Altman, D.G.; Vergouwe, Y.; Royston, P. Prognosis and prognostic research: Application and impact of prognostic models in clinical practice. *BMJ* **2009**, *338*, 1487–1490. [CrossRef] [PubMed]

96. Duarte, R.F.; Sánchez-Ortega, I.; Cuesta, I.; Arnan, M.; Patiño, B.; Fernández de Sevilla, A.; Gudiol, C.; Ayats, J.; Cuenca-Estrella, M. Serum galactomannan-based early detection of invasive aspergillosis in hematology patients receiving effective anti-mold prophylaxis. *Clin. Infect. Dis.* **2014**, *59*, 1696–1702. [CrossRef]

97. Stanzani, M.; Sassi, C.; Lewis, R.E.; Tolomelli, G.; Bazzocchi, A.; Cavo, M.; Vianelli, N.; Battista, G. High resolution computed tomography angiography improves the radiographic diagnosis of invasive mold disease in patients with hematological malignancies. *Clin. Infect. Dis.* **2015**, *60*, 1603–1610. [CrossRef] [PubMed]

98. Moons, K.G.M.; Altman, D.G.; Reitsma, J.B.; Ioannidis, J.P.A.; Macaskill, P.; Steyerberg, E.W.; Vickers, A.J.; Ransohoff, D.F.; Collins, G.S. Transparent Reporting of a multivariable prediction model for Individual Prognosis or Diagnosis (TRIPOD): Explanation and elaboration. *Ann. Intern. Med.* **2015**, *162*, W1–W73. [CrossRef]

99. Altman, D.G.; Vergouwe, Y.; Royston, P.; Moons, K.G.M. Prognosis and prognostic research: Validating a prognostic model. *BMJ* **2009**, *338*, b605. [CrossRef]

100. White, P.L.; Parr, C.; Barnes, R.A. Predicting Invasive Aspergillosis in Hematology Patients by Combining Clinical and Genetic Risk Factors with Early Diagnostic Biomarkers. *J. Clin. Microbiol.* **2018**, *56*, e01122-17. [CrossRef]

Journal of
Fungi

MDPI

Review

Invasive Candidiasis in Infants and Children: Recent Advances in Epidemiology, Diagnosis, and Treatment

Thomas J. Walsh [1,*], Aspasia Katragkou [2], Tempe Chen [3], Christine M. Salvatore [4] and Emmanuel Roilides [5]

[1]   Departments of Medicine, Pediatrics, and Microbiology & immunology, Weill Cornell Medicine of Cornell University and New York Presbyterian Hospital, New York, NY 10065, USA

[2]   Department of Pediatrics, Division of Pediatric Infectious Diseases, Nationwide Children's Hospital and The Ohio State University School of Medicine, Columbus, OH 43205, USA; Aspasia.Katragkou@nationwidechildrens.org

[3]   Pediatric Infectious Diseases, Department of Pediatrics, Miller Children's Hospital and University of California Irvine, Long Beach, CA 90806, USA; TChen@memorialcare.org

[4]   Department of Pediatrics, Division of Pediatric Infectious Diseases, Weill Cornell Medicine of Cornell University and New York Presbyterian Hospital, New York, NY 10065, USA; chs2032@med.cornell.edu

[5]   Infectious Diseases Section, 3rd Department of Pediatrics, Faculty of Medicine, Aristotle University School of Health Sciences and Hippokration General Hospital, 54642 Thessaloniki, Greece; roilides@med.auth.gr

*   Correspondence: thw2003@med.cornell.edu; Fax: +1-212-746-8852

Received: 5 December 2018; Accepted: 17 January 2019; Published: 24 January 2019

Abstract: This paper reviews recent advances in three selected areas of pediatric invasive candidiasis: epidemiology, diagnosis, and treatment. Although the epidemiological trends of pediatric invasive candidiasis illustrate a declining incidence, this infection still carries a heavy burden of mortality and morbidity that warrants a high index of clinical suspicion, the need for rapid diagnostic systems, and the early initiation of antifungal therapy. The development of non-culture-based technologies, such as the T2Candida system and $(1{\rightarrow}3)$-β-D-glucan detection assay, offers the potential for early laboratory detection of candidemia and CNS candidiasis, respectively. Among the complications of disseminated candidiasis in infants and children, hematogenous disseminated *Candida* meningoencephalitis (HCME) is an important cause of neurological morbidity. Detection of $(1{\rightarrow}3)$-β-D-glucan in cerebrospinal fluid serves as an early diagnostic indicator and an important biomarker of therapeutic response. The recently reported pharmacokinetic data of liposomal amphotericin B in children demonstrate dose–exposure relationships similar to those in adults. The recently completed randomized clinical trial of micafungin versus deoxycholate amphotericin B in the treatment of neonatal candidemia provides further safety data for an echinocandin in this clinical setting.

Keywords: candidemia; *Candida* meningoencephalitis; $(1{\rightarrow}3)$-β-D-glucan; T2Candida; PCR; liposomal amphotericin B; micafungin; anidulafungin

---

1. Introduction

This paper reviews the recent advances in three selected areas of pediatric invasive candidiasis: epidemiology, diagnosis, and treatment, as presented in a lecture at the 20th Meeting of the International Immunocompromised Host Society. The paper reviews the nationwide secular trends of pediatric invasive candidiasis in the United States and Europe. Our review then further discusses new approaches to laboratory diagnosis and therapeutic monitoring while underscoring the continued need for bedside clinical evaluation. We then further review recent studies in pediatric antifungal pharmacology and therapeutics that provide new insights into safety, tolerability, pharmacokinetics, and efficacy for the management of invasive candidiasis.

2. Epidemiology

2.1. Secular Trends of Candidemia

Candidemia is the leading cause of invasive fungal infections in hospitalized children. Among the different populations of pediatric patients, the highest rates of candidemia have been recorded in neonates and infants <1 year of age [1–4]. However, candidemia in pediatric patients is associated with better therapeutic outcomes than in adults. For neonates and young infants, this improved outcome is associated with higher inpatient costs, in comparison with the costs associated with the treatment of adults. Additional comparative data pertaining to pediatric and adult secular trends are depicted at https://www.cdc.gov/fungal/diseases/candidiasis/invasive/statistics.html [5].

During the last decade, there has been a declining secular trend in the incidence of pediatric candidemia the United States and European Union [1–5]. The United States Centers for Diseases Control (CDC) initiated a population-based surveillance of four US metropolitan areas between 2009 and 2015 [5]. The overall incidence of candidemia in neonates decreased from 31.5 cases/100,000 births in 2009 to 10.7 and to 11.8 cases/100,000 births between 2012 and 2015, while the incidence in infants decreased from 52.1 cases/100,000 births in 2009 to 15.7 and to 17.5 between 2012 and 2015. The incidence of candidemia in non-infant children decreased similarly from 1.8 cases/100,000 births in 2009 to 0.8 cases/100,000 births in 2014.

Consistent with these data, there was a decline in the incidence of candidemia in patients who were <1 year in a population-based observational study conducted in Atlanta, Georgia, from approximately 60 per 100,000 person-years in 2008–2009, to less than 40 per 100,000 person-years in 2012–2013. Similarly, there was a decline of approximately 40 per 100,000 person-years in 2008–2009 to less than 20 per 100,000 person-years in 2012–2013. The secular trends in adults were relatively stable.

This decrease in the incidence of pediatric candidemia may be related to several factors regarding the care of central venous catheters [1,2]. These include hospital-wide implementation bundles, guiding insertion and the maintenance of central lines. These measures underscore the importance of using fully sterile barrier precautions, using chlorhexidine in the preparation of the skin during insertion of central lines, taking meticulous care of the catheter and its insertion site, and having daily discussions over the need for a central venous catheter.

2.2. Risk Factors

The risk factors for invasive candidiasis in neonates, particularly in prematurely born infants, warrant special consideration. In a study involving a prospective observational cohort of 1515 extremely low-birth-weight (ELBW) infants, which took place over three years at 19 centers of the US Eunice Kennedy Shriver National Institute of Child Health and Human Development (NICHD) Neonatal Research Network, Benjamin et al. quantified the risk factors predicting infection in high-risk premature infants [3]. Among the 1515 infants enrolled, 137 (9.0%) developed invasive candidiasis, documented by positive culture from one or more of the following sources: blood ($n = 96$); urine obtained by catheterization or suprapubic aspiration ($n = 52$); CSF ($n = 9$); other sterile body fluids ($n = 10$).

Among the different predictive models that have been developed for invasive candidiasis in neonates, a multivariable analysis of potentially modifiable risk factors associated with candidiasis identified the presence of an endotracheal tube, the presence of a central venous catheter, and a receipt of an intravenous lipid emulsion [3]. A second model predicted candidiasis at the time of blood cultures. Components of the history, physical exam, and initial laboratory evaluation that predicted candidiasis included vaginal delivery, week of gestational age, presence of *Candida*-like dermatitis observed during the physical exam, central venous catheter, lack of enteral feeding, hyperglycemia, number of days of antibiotic exposure in the week prior to culture, and thrombocytopenia [3]. The clinical prediction model had an area under the receiver operating characteristic curve of 0.79 and was superior to clinician judgment (0.70) in predicting neonatal invasive candidiasis.

In this groundbreaking study, invasive candidiasis was found to increase the risk of death in neonates; for example, 47 of 137 (34%) infants with candidiasis died, in comparison with 197 of 1378 (14%) patients without candidiasis ($p < 0.0001$) [3]. Mortality was the highest in the infants from whom *Candida* was isolated from multiple sources. For infant patients with positive urine and blood or positive urine and CSF, the rate of mortality was 16 of 28 (57%). Underscoring the significance of the recovery of *Candida* spp. from urine in neonates, mortality rate was similar in patients who had *Candida* spp. isolated only from blood and those with *Candida* isolated only from urine.

3. Diagnosis

3.1. Clinical Diagnosis

The bedside assessment of disseminated candidiasis begins with an understanding of the relative risks and a recognition of its clinical manifestations [6–10]. The clinical manifestations of invasive candidiasis include endophthalmitis (chorioretinal and vitreal lesions), hematogenous *Candida* meningoencephalitis (HCME) (seizures, intraventricular hemorrhages, developmental regression or delays, and CSF pleocytosis), endocarditis (murmurs, peripheral embolic manifestations, and congestive heart failure), hepatosplenic candidiasis (chronic disseminated candidiasis, persistent fever, left upper quadrant or right upper quadrant abdominal pain, and anorexia), acute disseminated candidiasis (multiple cutaneous lesions, diffuse myalgias, hypotension, and multiorgan failure), renal candidiasis (decreasing creatinine clearance, obstructive nephropathy, and renal bezoars), and osteoarticular infections (osteoarticular lesions that are unresponsive to empirical antibacterial therapy).

3.2. Laboratory Detection

Blood culture systems are relatively insensitive in the detection of deeply invasive candidiasis [11]. Non-culture-based methods, such as nucleic acid amplification systems, enzyme immunoassays for circulating mannans, and enzymatic systems for detection of $(1\rightarrow3)$-β-D-glucan, are emerging as important laboratory tools for the diagnosis of invasive candidiasis [12–16].

3.3. T2Candida for Detection of Candidemia

The T2Candida system was recently licensed by the US FDA and was designated as superior to conventional blood culture systems for the detection of candidemia. In detecting the five most commonly recovered medically important *Candida* spp. (*Candida albicans*, *Candida tropicalis*, *Candida parapsilosis*, *Candida glabrata*, and *Candida krusei*), the T2Candida system utilizes a T2 magnetic resonance technology coupled with pathogen-derived nucleic acid amplification to identify the five target pathogens within 2 to 5 h from the time of initiation of the assay. Studies in adults have demonstrated a more rapid time to detection, in comparison with that of conventional blood cultures [15,16].

Little is known about the diagnostic utility of the T2Candida system in pediatric patients. In a study conducted at Children's Hospital of Philadelphia, whole blood from 15 children with candidemia was collected immediately following a blood culture draw [14]. Given the need for conserving the blood volume in this pediatric study, the amount of blood required by the system was reduced by pipetting whole blood directly onto the T2Candida cartridge. The specimens were subsequently run on the T2Dx Instrument (T2 Biosystems). The T2Candida biosystem correctly identified 15 positive and nine negative results within 3 to 5 h. The authors concluded that the T2Candida system was able to efficiently diagnose candidemia in pediatric patients while using low-volume blood specimens.

3.4. CSF (1→3)-β-D-Glucan as a Biomarker for Detection and Therapeutic Monitoring of Candida Infections of the Central Nervous System

HCME in pediatric patients is a life-threatening infection that is fraught with the potential of serious neurologic morbidity if not recognized and treated early [6,17]. HCME is observed in neonates, as well as in children with B-cell acute lymphoblastic leukemia, acute myelogenous leukemia, and primary immunodeficiencies. Associated with seizures, intraventricular hemorrhage, cortical blindness, and neurocognitive impairment, as well as the loss of developmental milestones, the early diagnosis of HCME is difficult, and its recurrence following the completion of antifungal therapy is common.

Petraitiene et al. originally demonstrated that CSF (1→3)-β-D-glucan levels correlated with CNS tissue infection in experimental HCME [18]. The expression of CSF and plasma (1→3)-β-D-glucan in the non-neutropenic rabbit model of experimental HCME treated with micafungin and with amphotericin B was predictive of the clinical features of this infection. Consistent with clinical observations regarding the difficulty in establishing a microbiological diagnosis, despite a well-established infection throughout CNS tissues, only 8% of CSF cultures were positive in untreated control animals. By comparison, all 25 CSF samples from these animals were found to be positive for (1→3)-β-D-glucan (755 to 7,750 pg/mL) ($p < 0.001$).

Changes in CSF (1→3)-β-D-glucan levels were highly predictive of antifungal therapeutic response, while clearance of *C. albicans* from blood cultures was not predictive of the eradication of organisms from the CNS [17]. The levels of (1→3)-β-D-glucan in CSF significantly decreased in comparison to those in untreated control animals. The levels of CSF (1→3)-β-D-glucan correlated with therapeutic responses to micafungin in a dose-dependent pattern, with a residual fungal burden in the cerebral tissue ($r = 0.842$). Thus, CSF (1→3)-β-D-glucan levels were predictive biomarkers for the detection and the therapeutic monitoring of experimental HCME. Building upon these data, a clinical trial was designed in the attempt to improve the management of HCME in pediatric patients.

Salvatore, Chen, and colleagues measured (1→3)-β-D-glucan levels in serially collected samples of serum and CSF of pediatric patients (aged 0–18 years) with a diagnosis of probable or proven HCME and CNS aspergillosis [19]. Among the nine cases of fungal infections of the central nervous system, seven were caused by HCME. All patients at baseline had detectable (1→3)-β-D-glucan in their CSF. In the six patients who completed the therapy for HCME, the elevated CSF (1→3)-β-D-glucan levels decreased to <31 pg/mL. One patient, who was unable to complete the antifungal therapy, died as the result of an overwhelmingly disseminated candidiasis. Monitoring serial CSF (1→3)-β-D-glucan levels in HCME was critical in determining the length of therapy, which ranged from 3 to 6 months, on the basis of individualized assessments. Subsequent reports have confirmed the utility of measuring CSF (1→3)-β-D-glucan levels for initial diagnosis and therapeutic monitoring of HCME [20–22].

4. Treatment

4.1. Liposomal Amphotericin B in Immunocompromised Children

Liposomal Amphotericin B (L-AMB) is widely used in the treatment of invasive fungal infections in immunocompromised children; however, little is known about its safety and pharmacokinetics in this vulnerable patient population. Seibel and colleagues therefore conducted a study of the safety, tolerability, and pharmacokinetics of L-AMB in 40 immunocompromised children and adolescents in a sequential-dose-escalation, multidose clinical trial [23]. Ten to 13 patients between the ages of 1 and 17 years were enrolled into each of the four dosage cohorts: 2.5, 5.0, 7.5, or 10 mg/kg, to receive empirical antifungal therapy for persistent fever and neutropenia or for the treatment of documented invasive fungal infections.

Serum creatinine increased from a mean of 0.45 ± 0.04 mg/dL to 0.63 ± 0.06 mg/dL across all dosage groups ($p = 0.003$). There was a significant increase in serum creatinine in dosage cohorts of 5.0 and 10 mg/kg/day. A greater frequency of hypokalemia and vomiting was also observed in patients receiving 10 mg/kg. Among the 565 infusions, 63 (11%) infusion-related adverse effects occurred. Five patients experienced acute infusion-related reactions at both the 7.5 and the 10 mg/kg dosage levels.

L-AMB in this patient population exhibited nonlinear pharmacokinetics [24]. The area under the concentration–time curve from 0 to 24 h (AUC_{0-24}) on day 1 increased from 54.7 ± 32.9 to 430 ± 566 µg·h/mL in patients receiving 2.5 and 10.0 mg/kg/day. The pharmacokinetic data were best described by a 2-compartment model that incorporated weight and an exponential decay function describing volume of distribution. The population-based model also demonstrated a significant ($p = 0.004$) relationship between the mean AUC_{0-24} and the probability of nephrotoxicity, with an odds ratio of 2.37 (95% confidence interval, 1.84 to 3.22).

In summary, these data collectively support the use of a range of dosages comparable to those used in adult patients for the treatment of invasive fungal infections, with the understanding that azotemia may occur in direct relation to the AUC_{0-24}.

4.2. Micafungin in Neonates

Extensive preclinical studies in the treatment of experimental disseminated candidiasis [25–29] and clinical studies in pediatric patients [30–39] supported the investigation of micafungin in neonates in comparison with that of amphotericin B deoxycholate. Benjamin and colleagues compared the efficacy, safety and pharmacokinetics of intravenous micafungin at 10 mg/kg/d with intravenous amphotericin B deoxycholate at 1 mg/kg/d, in a phase 3, randomized, double-blind, multicenter, parallel-group, noninferiority trial, performed on infants >2–120 days of age with proven invasive candidiasis [40]. A total of 20 infants received micafungin, and 10 received amphotericin B deoxycholate. Although the study was terminated early because of low recruitment, fungal-free survival was observed in 12 out of the 20 [60%; 95% CI: 36–81%] infants treated with micafungin, versus 7 of the 10 (70%; 95% CI: 35–93%) infants treated with amphotericin B deoxycholate. The pharmacokinetic-model-derived mean area under the concentration-time curve (AUC) at steady state for micafungin was 399.3 ± 163.9 µg·h/mL, with an AUC pharmacodynamic target exposure of micafungin of 170 µg·h/mL. A population-based pharmacodynamic analysis supported a direct relationship between plasma exposure and the successful eradication of candidemia [41].

4.3. Anidulafungin in Pediatric Patients

Building upon earlier preclinical and clinical studies of anidulafungin [42,43], a recently published study reports on the safety and efficacy of anidulafungin in pediatric invasive candidiasis [44]. Anidulafungin was administered at a 3 mg/kg loading dose on day 1 and at a 1.5 mg/kg/d maintenance dose thereafter to patients between the ages of 2 and <18 years. Among 49 patients who received ≥1 dose of anidulafungin for a median 11 days (range of 1–35 days), all were reported to have a treatment-emergent adverse event (AE), such as, most commonly, diarrhea, vomiting, and fever. Treatment was discontinued due to AEs in four cases which were thought to be related to anidulafungin. Among the 48 patients with an isolate of a *Candida* spp., organisms were identified as *C. albicans* in 37.5%, *C. parapsilosis* in 25.0%, *C. tropicalis* in 14.6%, and *Candida lusitaniae* in 10.4%. One patient did not have an isolate of *Candida* sp. recovered. The global response success rate was 70.8%, while all-cause mortality was 8.2% at the end of intravenous therapy and 14.3% at 6-week follow-up. None of the deaths were considered to be treatment-related. The results of this study support the role of anidulafungin at the studied dosages for the treatment of pediatric invasive candidiasis.

5. Conclusions and Future Directions

During the past several years, important advances have been achieved in key areas of the epidemiology, laboratory diagnosis, and treatment of pediatric invasive candidiasis. There has been a clear downward trend in the incidence of pediatric invasive candidiasis. Nonetheless, pediatric invasive candidiasis remains an important cause of healthcare-associated sepsis and infectious morbidity. The bedside recognition of invasive candidiasis is challenging, and clinical manifestations

J. Fungi **2019**, *5*, 11

are usually non-specific. Advances in the laboratory diagnosis of candidemia and deeply invasive candidiasis are helping to improve the recognition of these serious infections. Among these advances is the T2Candida system, which has the ability to detect more cases of candidemia than conventional blood cultures within 3 to 5 h. Another important advance is the detection of CSF $(1\rightarrow3)$-β-D-glucan levels, which are highly sensitive in the diagnosis of HCME and can be serially monitored to guide the duration of and evaluate the response to the antifungal therapy.

Recently reported studies of L-AMB, micafungin, and anidulafungin in pediatric patients have been have been important advances in the management of invasive candidiasis in infants can children. A major body of preclinical and clinical studies has established the safety, pharmacokinetic, and efficacy profile of these potent antifungal agents in pediatric patients, including those patients with candidemia and other forms of deeply invasive candidiasis. As resistance to echinocandins and antifungal triazoles develops in *Candida* spp., new antifungal agents will be necessary to treat these emerging medically important pathogens. Among the new antifungal agents in development are the first-in-class molecules SCY-078, which is a triterpene inhibitor of $(1\rightarrow3)$-β-D-glucan synthase, and APX-001, which is an inhibitor of fungal glycosyl-phosphatidyl-inositol (GPI) biosynthesis [21]. As these agents are developed, well-defined pediatric studies will need to be designed and implemented for the management of pediatric invasive candidiasis and other mycoses.

Funding: Thomas J. Walsh was supported in the writing of this manuscript as a Scholar of Emerging Infectious Diseases of the Save Our Sick Kids Foundation.

Acknowledgments: This paper is based upon a presentation given at the 20th ICHS Symposium on Infections in the Immunocompromised Host, 17–19 June 2018, Athens, Greece.

Conflicts of Interest: Thomas J. Walsh has received research grants to his institution from Allergan, Amplyx, Astellas, Lediant, Medicines Company, Merck, Scynexis, and Tetraphase; he has served as consultant to Amplyx, Astellas, Allergan, ContraFect, Gilead, Lediant, Medicines Company, Merck, Methylgene, Pfizer, and Scynexis. Emmanuel Roilides has received research grants from Astellas, Gilead, and Pfizer Inc; he is a scientific advisor and member of speaker bureaus for Astellas, Gilead, Merck, and Pfizer. The other authors declare no conflicts of interest.

References

1. Mantadakis, E.; Pana, Z.D.; Zaoutis, T. Candidemia in children: Epidemiology, prevention and management. *Mycoses* **2018**, *61*, 614–622. [CrossRef] [PubMed]
2. Pana, Z.D.; Roilides, E.; Warris, A.; Groll, A.H.; Zaoutis, T. Epidemiology of Invasive Fungal Disease in Children. *J. Pediatr. Infect. Dis.* **2017**, *6*, S3–S11. [CrossRef] [PubMed]
3. Benjamin, D.K., Jr.; Stoll, B.J.; Gantz, M.G.; Walsh, M.C.; Sánchez, P.J.; Das, A.; Shankaran, S.; Higgins, R.D.; Auten, K.J.; Miller, N.A.; et al. Neonatal Candidiasis: Epidemiology, Risk Factors, and Clinical Judgment. *Pediatrics* **2010**, *126*, e865–e873. [CrossRef] [PubMed]
4. Aliaga, S.; Clark, R.H.; Clark, R.H.; Laughon, M.; Walsh, T.J.; Hope, W.; Benjamin, D.K.; Benjamin, D.K., Jr.; Smith, P.B. Decreasing incidence of candidiasis in infants in neonatal intensive care units. *Pediatrics* **2014**, *133*, 236–242. [CrossRef] [PubMed]
5. Invasive Candidiasis Statistics. Available online: https://www.cdc.gov/fungal/diseases/candidiasis/invasive/statistics.html (accessed on 1 December 2018).
6. Benjamin, D.K.; Poole, C.; Steinbach, W.J.; Rowen, J.L.; Walsh, T.J. Neonatal Candidemia and End-Organ Damage: A Critical Appraisal of the Literature Using Meta-analytic Techniques. *Pediatrics* **2003**, *112*, 634–640. [CrossRef] [PubMed]
7. Roilides, E. Invasive candidiasis in neonates and children. *Early Hum. Dev.* **2011**, *87*, S75–S76. [CrossRef] [PubMed]
8. Prasad, P.A.; Fisher, B.T.; Coffin, S.E.; Walsh, T.J.; McGowan, K.L.; Gross, R.; Zaoutis, T.E. Pediatric Risk Factors for Candidemia Secondary to Candida glabrata and Candida krusei Species. *J. Pediatr. Infect. Dis.* **2012**, *2*, 263–266. [CrossRef] [PubMed]
9. Fierro, J.L.; Prasad, P.; Fisher, B.; Gerber, J.; Coffin, S.E.; Walsh, T.J.; Zaoutis, T.E. Ocular manifestations of candidemia in a pediatric population. *Pediatr. Infect. Dis. J.* **2013**, *32*, 84–86. [CrossRef]

10. Gamaletsou, M.N.; Kontoyiannis, D.P.; Sipsas, N.V.; Moriyama, B.; Alexander, E.; Roilides, E.; Brause, B.; Walsh, T.J. Candida Osteomyelitis: Analysis of 207 Pediatric and Adult Cases (1970–2011). *Clin. Infect. Dis.* **2012**, *55*, 1338–1351. [CrossRef] [PubMed]

11. Berenguer, J.; Buck, M.; Witebsky, F.; Stock, F.; Pizzo, P.A.; Walsh, T.J. Lysis—Centrifugation blood cultures in the detection of tissue-proven invasive candidiasis disseminated versus single-organ infection. *Diagn. Microbiol. Infect. Dis.* **1993**, *17*, 103–109. [CrossRef]

12. Huppler, A.R.; Fisher, B.T.; Lehrnbecher, T.; Walsh, T.J.; Steinbach, W.J. Role of Molecular Biomarkers in the Diagnosis of Invasive Fungal Diseases in Children. *J. Pediatr. Infect. Dis.* **2017**, *6*, S32–S44. [CrossRef] [PubMed]

13. McCarthy, M.W.; Walsh, T.J. Molecular diagnosis of invasive mycoses of the central nervous system. *Expert Rev. Mol. Diagn.* **2016**, *17*, 129–139. [CrossRef] [PubMed]

14. Hamula, C.L.; Hughes, K.; Fisher, B.T.; Zaoutis, T.E.; Singh, I.R.; Velegraki, A. T2Candida Provides Rapid and Accurate Species Identification in Pediatric Cases of Candidemia. *Am. J. Clin. Pathol.* **2016**, *145*, 858–861. [CrossRef] [PubMed]

15. Mylonakis, E.; Clancy, C.J.; Ostrosky-Zeichner, L.; Garey, K.W.; Alangaden, G.J.; Vazquez, J.A.; Groeger, J.S.; Judson, M.A.; Vinagre, Y.-M.; Heard, S.O.; et al. T2 Magnetic Resonance Assay for the Rapid Diagnosis of Candidemia in Whole Blood: A Clinical Trial. *Clin. Infect. Dis.* **2015**, *60*, 892–899. [CrossRef]

16. Clancy, C.J.; Pappas, P.G.; Vazquez, J.; Judson, M.A.; Kontoyiannis, D.P.; Thompson, G.R., 3rd; Garey, K.W.; Reboli, A.; Greenberg, R.N.; Apewokin, S.; et al. Detecting Infections Rapidly and Easily for Candidemia Trial, Part 2 (DIRECT2): A Prospective, Multicenter Study of the T2Candida Panel. *Clin. Infect. Dis.* **2018**, *66*, 1678–1686. [CrossRef] [PubMed]

17. McCullers, J.A.; Vargas, S.L.; Flynn, P.M.; Razzouk, B.I.; Shenep, J.L. Candidal Meningitis in Children with Cancer. *Clin. Infect. Dis.* **2000**, *31*, 451–457. [CrossRef]

18. Petraitiene, R.; Petraitis, V.; Hope, W.W.; Mickiene, D.; Kelaher, A.M.; Murray, H.A.; Mya-San, C.; Hughes, J.E.; Cotton, M.P.; Bacher, J.; et al. Cerebrospinal fluid and plasma (1–>3)-beta-D-glucan as surrogate markers for detection and monitoring of therapeutic response in experimental hematogenous Candida meningoencephalitis. *Antimicrob. Agents Chemother.* **2008**, *52*, 4121–4129. [CrossRef]

19. Salvatore, C.M.; Chen, T.K.; Toussi, S.S.; DeLaMora, P.; Petraitiene, R.; Finkelman, M.A.; Walsh, T.J. (1→3)-β-D-glucan in Cerebrospinal Fluid as a Biomarker for Candida and Aspergillus Infections of the Central Nervous System in Pediatric Patients. *J. Pediatr. Infect. Dis. Soc.* **2016**, *5*, 277–286. [CrossRef]

20. Ceccarelli, G.; Ghezzi, M.C.; Raponi, G.; Brunetti, G.; Marsiglia, C.; Fallani, S.; Novelli, A.; Venditti, M. Voriconazole treatment of Candida tropicalis meningitis: Persistence of (1,3)-β-D-glucan in the cerebrospinal fluid is a marker of clinical and microbiological failure: A case report. *Medicine* **2016**, *95*, e4474. [CrossRef]

21. McCarthy, M.W.; Walsh, T.J. Drugs currently under investigation for the treatment of invasive candidiasis. *Expert Opin. Investig. Drugs* **2017**, *26*, 825–831. [CrossRef]

22. Farrugia, M.K.; Fogha, E.P.; Miah, A.R.; Yednock, J.; Palmer, H.C.; Guilfoose, J. Candida meningitis in an immunocompetent patient detected through (1→3)-beta-d-glucan. *Int. J. Infect. Dis.* **2016**, *51*, 25–26. [CrossRef] [PubMed]

23. Seibel, N.L.; Shad, A.T.; Bekersky, I.; Groll, A.H.; Gonzalez, C.; Wood, L.V.; Jarosinski, P.; Buell, D.; Hope, W.W.; Walsh, T.J.; et al. Safety, Tolerability, and Pharmacokinetics of Liposomal Amphotericin B in Immunocompromised Pediatric Patients. *Antimicrob. Agents Chemother.* **2016**, *61*, e01477-16. [CrossRef]

24. Lestner, J.M.; Groll, A.H.; Aljayyoussi, G.; Seibel, N.L.; Shad, A.; Gonzalez, C.; Wood, L.V.; Jarosinski, P.F.; Walsh, T.J.; Hope, W.W. Population Pharmacokinetics of Liposomal Amphotericin B in Immunocompromised Children. *Antimicrob. Agents Chemother.* **2016**, *60*, 7340–7346. [CrossRef] [PubMed]

25. Groll, A.H.; Mickiene, D.; Petraitis, V.; Petraitiene, R.; Ibrahim, K.H.; Piscitelli, S.C.; Bekersky, I.; Walsh, T.J. Compartmental Pharmacokinetics and Tissue Distribution of the Antifungal Echinocandin Lipopeptide Micafungin (FK463) in Rabbits. *Antimicrob. Agents Chemother.* **2001**, *45*, 3322–3327. [CrossRef]

26. Petraitis, V.; Petraitiene, R.; Groll, A.H.; Roussillon, K.; Hemmings, M.; Lyman, C.A.; Sein, T.; Bacher, J.; Bekersky, I.; Walsh, T.J.; et al. Comparative Antifungal Activities and Plasma Pharmacokinetics of Micafungin (FK463) against Disseminated Candidiasis and Invasive Pulmonary Aspergillosis in Persistently Neutropenic Rabbits. *Antimicrob. Agents Chemother.* **2002**, *46*, 1857–1869. [CrossRef] [PubMed]

27. Andes, D.R.; Diekema, D.; Pfaller, M.A.; Marchillo, K.; Bohrmueller, J. In Vivo Pharmacodynamic Target Investigation for Micafungin against Candida albicans and C. glabrata in a Neutropenic Murine Candidiasis Model. *Antimicrob. Agents Chemother.* **2008**, *52*, 3497–3503. [CrossRef] [PubMed]
28. Mickiene, D.; Petraitis, V.; Petraitiene, R.; Bacher, J.; Buell, D.; Heresi, G.; Kelaher, A.M.; Hughes, J.E.; Cotton, M.P.; Hope, W.W.; et al. The Pharmacokinetics and Pharmacodynamics of Micafungin in Experimental Hematogenous Candida Meningoencephalitis: Implications for Echinocandin Therapy in Neonates. *J. Infect. Dis.* **2008**, *197*, 163–171.
29. Simitsopoulou, M.; Chlichlia, K.; Kyrpitzi, D.; Walsh, T.J.; Roilides, E. Pharmacodynamic and Immunomodulatory Effects of Micafungin on Host Responses against Biofilms of Candida parapsilosis in Comparison to Those of Candida albicans. *Antimicrob. Agents Chemother.* **2018**, *62*. [CrossRef]
30. Seibel, N.L.; Schwartz, C.; Arrieta, A.; Flynn, P.; Shad, A.; Albano, E.; Keirns, J.; Lau, W.M.; Facklam, D.P.; Buell, D.N.; et al. Safety, Tolerability, and Pharmacokinetics of Micafungin (FK463) in Febrile Neutropenic Pediatric Patients. *Antimicrob. Agents Chemother.* **2005**, *49*, 3317–3324. [CrossRef]
31. Santos, R.P.; Sánchez, P.J.; Mejias, A.; Benjamin, D.K., Jr.; Walsh, T.J.; Patel, S.; Jafri, H.S. Successful medical treatment of cutaneous aspergillosis in a premature infant using liposomal amphotericin B, voriconazole and micafungin. *Pediatr. Infect. Dis. J.* **2007**, *26*, 364–366. [CrossRef]
32. Hope, W.; Seibel, N.L.; Schwartz, C.L.; Arrieta, A.; Flynn, P.; Shad, A.; Albano, E.; Keirns, J.J.; Buell, D.N.; Gumbo, T.; et al. Population Pharmacokinetics of Micafungin in Pediatric Patients and Implications for Antifungal Dosing. *Antimicrob. Agents Chemother.* **2007**, *51*, 3714–3719. [CrossRef] [PubMed]
33. Queiroz-Telles, F.; Berezin, E.; Leverger, G.; Freire, A.; van der Vyver, A.; Chotpitayasunondh, T.; Konja, J.; Diekmann-Berndt, H.; Koblinger, S.; Groll, A.H.; et al. Micafungin versus liposomal amphotericin B for pediatric patients with invasive candidiasis: Substudy of a randomized double-blind trial. *Pediatr. Infect. Dis. J.* **2008**, *27*, 820–826. [CrossRef] [PubMed]
34. Smith, P.B.; Walsh, T.J.; Hope, W.; Arrieta, A.; Takada, A.; Kovanda, L.L.; Kearns, G.L.; Kaufman, D.; Sawamoto, T.; Buell, D.N.; et al. Pharmacokinetics of an Elevated Dosage of Micafungin in Premature Neonates. *Pediatr. Infect. Dis. J.* **2009**, *28*, 412–415. [CrossRef] [PubMed]
35. Benjamin, D.K.; Arrieta, A.; Castro, L.; Sánchez, P.J.; Kaufman, D.; Arnold, L.J.; Kovanda, L.L.; Sawamoto, T.; Buell, D.N.; Hope, W.; et al. Safety and Pharmacokinetics of Repeat-Dose Micafungin in Young Infants. *Clin. Pharmacol. Ther.* **2009**, *87*, 93–99. [CrossRef] [PubMed]
36. Walsh, T.J.; Goutelle, S.; Jelliffe, R.W.; Golden, J.A.; Little, E.A.; DeVoe, C.; Mickiene, D.; Hayes, M.; Conte, J.E., Jr. Intrapulmonary Pharmacokinetics and Pharmacodynamics of Micafungin in Adult Lung Transplant Patients. *Antimicrob. Agents Chemother.* **2010**, *54*, 3451–3459. [CrossRef] [PubMed]
37. Hope, W.; Smith, P.B.; Arrieta, A.; Buell, D.N.; Roy, M.; Kaibara, A.; Walsh, T.J.; Cohen-Wolkowiez, M.; Benjamin, D.K., Jr. Population Pharmacokinetics of Micafungin in Neonates and Young Infants. *Antimicrob. Agents Chemother.* **2010**, *54*, 2633–2637. [CrossRef]
38. Arrieta, A.C.; Maddison, P.; Groll, A.H. Safety of Micafungin in Pediatric Clinical Trials. *Pediatr. Infect. Dis. J.* **2011**, *30*, e97–e102. [CrossRef]
39. Bochennek, K.; Balan, A.; Müller-Scholden, L.; Becker, M.; Farowski, F.; Müller, C.; Groll, A.H.; Lehrnbecher, T. Micafungin twice weekly as antifungal prophylaxis in paediatric patients at high risk for invasive fungal disease. *J. Antimicrob. Chemother.* **2015**, *70*, 1527–1530. [CrossRef]
40. Benjamin, D.K., Jr.; Kaufman, D.A.; Hope, W.W.; Smith, P.B.; Arrieta, A.; Manzoni, P.; Kovanda, L.L.; Lademacher, C.; Isaacson, B.; Jednachowski, D.; et al. A Phase 3 Study of Micafungin Versus Amphotericin B Deoxycholate in Infants with Invasive Candidiasis. *Pediatr. Infect. Dis. J.* **2018**, *37*, 992–998. [CrossRef]
41. Kovanda, L.L.; Walsh, T.J.; Benjamin, D.; Arrieta, A.; Kauffman, D.; Smith, P.B.; Manzoni, P.; Desai, A.V.; Kaibara, A.; Bonate, P.; et al. Exposure-response analysis of micafungin in neonatal candidiasis: Pooled analysis of two clinical trials. *Pediatr. Infect. Dis. J.* **2008**, *37*, 580–585. [CrossRef]
42. Groll, A.H.; Mickiene, D.; Petraitiene, R.; Petraitis, V.; Lyman, C.A.; Bacher, J.S.; Piscitelli, S.C.; Walsh, T.J. Pharmacokinetic and Pharmacodynamic Modeling of Anidulafungin (LY303366): Reappraisal of Its Efficacy in Neutropenic Animal Models of Opportunistic Mycoses Using Optimal Plasma Sampling. *Antimicrob. Agents Chemother.* **2001**, *45*, 2845–2855. [CrossRef] [PubMed]

43. Benjamin, D.K.; Driscoll, T.; Seibel, N.L.; González, C.E.; Roden, M.M.; Kilaru, R.; Clark, K.; Dowell, J.A.; Schranz, J.; Walsh, T.J.; et al. Safety and Pharmacokinetics of Intravenous Anidulafungin in Children with Neutropenia at High Risk for Invasive Fungal Infections. *Antimicrob. Agents Chemother.* **2006**, *50*, 632–638. [CrossRef] [PubMed]

44. Roilides, E.; Carlesse, F.; Leister-Tebbe, H.; Conte, U.; Yan, J.L.; Liu, P.; Tawadrous, M.; Aram, J.A.; Queiroz-Telles, F.; Anidulafungin A8851008 Pediatric Study Group. A Prospective, Open-label Study to Assess the Safety, Tolerability, and Efficacy of Anidulafungin in the Treatment of Invasive Candidiasis in Children 2 to <18 Years of Age. *Pediatr. Infect. Dis. J.* **2018**. [CrossRef]

Journal of
Fungi

MDPI

Review

Invasive Aspergillosis in Pediatric Leukemia Patients: Prevention and Treatment

Savvas Papachristou, Elias Iosifidis and Emmanuel Roilides *

Infectious Diseases Unit, 3rd Department of Pediatrics, Faculty of Medicine, Aristotle University School of Health Sciences, Konstantinoupoleos 49, 54642 Thessaloniki, Greece; savvas.gpap@gmail.com (S.P.); iosifidish@gmail.com (E.I.)
* Correspondence: roilides@med.auth.gr; Tel.: +30-3023-1089-2444

Received: 4 December 2018; Accepted: 5 February 2019; Published: 11 February 2019

Abstract: The purpose of this article is to review and update the strategies for prevention and treatment of invasive aspergillosis (IA) in pediatric patients with leukemia and in patients with hematopoietic stem cell transplantation. The major risk factors associated with IA will be described since their recognition constitutes the first step of prevention. The latter is further analyzed into chemoprophylaxis and non-pharmacologic approaches. Triazoles are the mainstay of anti-fungal prophylaxis while the other measures revolve around reducing exposure to mold spores. Three levels of treatment have been identified: (a) empiric, (b) pre-emptive, and (c) targeted treatment. Empiric is initiated in febrile neutropenic patients and uses mainly caspofungin and liposomal amphotericin B (LAMB). Pre-emptive is a diagnostic driven approach attempting to reduce unnecessary use of anti-fungals. Treatment targeted at proven or probable IA is age-dependent, with voriconazole and LAMB being the cornerstones in >2yrs and <2yrs age groups, respectively.

Keywords: *Aspergillus*; anti-fungal agents; hematological malignancies

1. Introduction

Aspergillosis can be present in an acute or chronic form [1]. Syndromes of clinical significance include invasive aspergillosis (IA), chronic and saprophytic aspergillosis, and allergic aspergillosis [2]. We focus on IA because it occurs in immuno-compromised hosts. It is associated with notable morbidity and mortality in pediatric patients suffering from immuno-compromising conditions [3–5], with one multi-center retrospective study recording at a 52.5% mortality rate [6]. IA in children has also been related to increased financial costs [7]. While immuno-compromised pediatric patients also display susceptibility to invasive fungal disease (IFD), like IA, differences from adults have been highlighted to pertain to several aspects of these infections [8–12]. These differences are summarized in Table 1.

During recent years, clinicians have been extrapolating evidence from adult studies of IA, due to the lack of respective pediatric data [6]. According to the 2017 European Society for Clinical Microbiology and Infectious Diseases (ESCMID), the European Confederation of Medical Mycology (ECMM) and the European Respiratory Society (ERS) Joint Clinical Guidelines, according to recent guidelines by the Fourth European Conference on Infections in Leukemia (ECIL-4), pediatric recommendations about intervention are based on efficacy data from phase 2 and 3 trials in adults, on pediatric pharmacokinetic (PK), dosing, safety, supportive efficacy data, and on regulatory approvals [8,13].

Table 1. Differences between pediatric and adult Invasive Aspergillosis.

Field of Difference
A) Comorbidities
-Biology
-Management
-Prognosis
B) High-risk populations
C) Epidemiology
D) Diagnostic techniques
-Performance
-Utility
E) Anti-fungal drugs
-Pharmacology
-Dosing scheme
F) Phase 3 clinical trials

In pediatric patients, risk factors for IA include primary immunodeficiencies and especially chronic granulomatous disease (CGD), secondary immunodeficiencies (associated with cancer chemotherapy and failure syndromes of the bone marrow), critical illness, chronic diseases of the airways, low birth-weight, and prematurity (the last two are related to neonatal patients) [3–5,14–16]. Additionally, immunosuppressive treatments—including corticosteroids in high doses and biologic agents interacting with immune pathways (like monoclonal antibodies targeting tumor necrosis factor alpha)—are also regarded as risk factors for IA [5,6,17]. Lastly, solid organ transplantation (SOT) is related to IA, which becomes more evident in the case of heart and/or lung recipients [1,18]. Nevertheless, the most significant risk factors are considered to be hematological malignancies and *hematopoietic stem cell transplantation* (HSCT) [1,6,19–21]. These two conditions constitute the two major risk factors that are commonly encountered in patients with IA [1,6,19–21]. Furthermore, the detection of IA in leukemia patients affects the decisions regarding the administration of chemotherapy [5,7]. More specifically, delayed delivery of chemotherapy decreases the risk for IA progression, on one hand, but, conversely, it renders the progression of the malignancy more likely [5]. This delicate balance makes it more urgent to address the management of this group of patients.

This article intends to review the current strategies for prevention and treatment of IA in pediatric leukemia patients. In the section of prevention, the following topics will be covered: (a) epidemiology and risk factors for IA in pediatric patients with leukemia, (b) anti-fungal prophylaxis, and (c) other preventive measures. Treatment will be subdivided into three main areas: (a) empiric treatment, (b) pre-emptive treatment, and (c) treatment for proven/probable IA. The latter will also include an analysis of the therapeutic approaches to invasive pulmonary aspergillosis (IPA) and the central nervous system (CNS) aspergillosis.

2. Prevention

2.1. Epidemiology and Risk Factors for Invasive Aspergillosis

The incidence of IA in pediatric patients with hematological malignancies has been estimated by several studies between 4.57% and 9.5% [7,20,22,23]. Identified routes of infection include the respiratory tract, the gastrointestinal tract, and the skin [24]. A retrospective multi-center study incorporating a diverse population [6] found lungs, skin, and paranasal sinuses as the most frequently affected foci of infection. Regarding microbiology, *Aspergillus fumigatus*, *Aspergillus flavus*, *Aspergillus terreus*, and *Aspergillus niger* were the predominant isolates (in order of frequency) in the previous study [6].

Recognizing pediatric patients with leukemia at risk for developing IA is the cornerstone of prevention. This will enable physicians to timely implement the appropriate strategies to reduce

J. Fungi **2019**, *5*, 14

modifiable risk factors and initiate anti-fungal prophylaxis in pediatric leukemia and HSCT patients at high risk for invasive *Aspergillus* spp. [8]. Risk factors for IA in the previously mentioned pediatric patients are summarized in Table 2.

Table 2. Risk factors for Invasive Aspergillosis in pediatric patients.

Leukemia Patients	HSCT Recipients
Severe and persistent neutropenia	Severe and persistent neutropenia
Corticosteroids in high-doses	Corticosteroids in high-doses
Mucosal damage	Mucosal damage
Increasing age	Increasing age
AML	Allogeneic transplant
ALL: relapse	GVHD
ALL: de novo	HLA discordance
ALL: high-risk	CMV coinfection
Refractoriness of acute leukemia	Respiratory virus coinfection
	Colonization by *Aspergillus* spp.
	T-cell depletion
	CD 34 selection
Ward-associated factors (local epidemiology, environmental conditions, contamination of hospital water supply systems, construction works)	Ward-associated factors (local epidemiology, environmental conditions, contamination of hospital water supply systems, construction works)

AML, acute myelogenous leukemia. ALL, acute lymphoblastic leukemia. HSCT, hematopoietic stem cell transplantation. GVHD, graft-versus-host disease. HLA, human leukocyte antigen. CMV, cytomegalovirus. References are provided in the text.

Generally, an IFD incidence >10% is considered high-risk [8]. Severe and persistent neutropenia, high-dose corticosteroid regimens, and damage to mucosal surfaces render these two groups of patients susceptible to IA [8,25,26]. A recent systematic review of publications since 1980, that addressed pediatric-specific factors for invasive fungal diseases (IFDs), indicated that increasing age is a risk factor in both groups [27]. In leukemia patients, the type of malignancy determines the risk, with acute myelogenous leukemia (AML) ranking first (3.7–28% risk), while relapse and de novo acute lymphoblastic leukemia (ALL) are associated with a 4–9% and a 0.6–2% risk for IA, respectively [1,20,21,28]. It should be noted, that according to other studies, the risk was nearly equal between AML and ALL patients [6], or even greater in ALL patients [7]. However, these observations could be attributed to the specific characteristics or limitations of the studies. Refractoriness among acute leukemia patients is also a significant risk factor for IA [2]. High-risk ALL is recognized as a risk factor, but the heterogeneity characterizing this group of patients was underlined by the International Pediatric Fever and Neutropenia Guideline Panel [27,29]. In HSCT recipients, an allogeneic transplant is associated with a greater risk for IA than an autologous one [2,30]. Specific risk factors in allogeneic HSCT include the development of graft-versus-host disease (GVHD), the extension of human leukocyte antigen (HLA) discordance, the presence of cytomegalovirus (CMV) or respiratory virus coinfection, and the colonization by *Aspergillus* spp. [1,28,31–33]. In addition, two strategies for reducing GVHD—T-cell depletion and CD34 selection—are also related to IA infection [2,32,34]. Despite the absence of a risk stratification model for IFDs in pediatrics, a differentiation between high-risk and low-risk patients has been attempted [27,29]. More specifically, AML, high-risk ALL, acute leukemia relapse, allogeneic HSCT, protracted granulocytopenia, and administration of corticosteroids in high doses are considered high-risk conditions [29]. All other conditions are low-risk [29]. Lastly, topics in the field of risk factors for further research include the role of lymphopenia in IFDs and IA and the development of a prediction model for IFDs [27].

Certain risk factors for IA in children with leukemia and HSCT are ward-associated. These include local epidemiology, environmental conditions, contamination of hospital water supply systems, and construction work [2,7,8,35–39].

2.2. Anti-Fungal Prophylaxis

Anti-fungal prophylaxis is divided into primary and secondary entities [8]. Primary is defined as the administration of antifungal agents to high-risk patients without infection, whereas secondary lacks a robust definition and occasionally coincides conceptually with treatment for proven/probable IFDs [8].

Initiation of primary prophylaxis for IA is justified due to the lack of efficacy of diagnostic tests and the dismal outcomes of this infection [5,13]. Anti-fungal agents used for primary prophylaxis include the triazoles itraconazole, voriconazole, and posaconazole, liposomal amphotericin B (LAMB) (in the systemic and aerosolized form) and the echinocandins micafungin and caspofungin. Fluconazole has no activity against molds and, thus, it is not used in prophylaxis against IA [8].

Itraconazole is active against both yeasts and molds and is administered at a per os (PO) dose of 2.5 mg/kg/12 h, provided that the patient's age is ≥2 years [8], while therapeutic drug monitoring (TDM) is necessary to achieve the dosing target of ≥0.5 mg/L [40]. Itraconazole has been studied in pediatric cancer and HSCT patients and is considered a reliable option [41,42] even though prospective studies of larger scale are required to reach further conclusions [43]. The use of this azole is restricted by adverse reactions, according to a meta-analysis [44]. It is not approved in EU for patients <18 years of age [8].

Voriconazole has been found to be superior to itraconazole, in terms of tolerability, for allogeneic HSCT in patients ≥12 years. However, the two agents were equally effective in preventing IFD [45]. According to the ECIL-4, the recommended voriconazole dose in pediatrics for ages 2–<12 years, or 12–14 years with a body weight <50 kg, is 9 mg/kg/12 h for the PO forms and 9 mg/kg/12 h the first day, which is followed by 8 mg/kg/12 h on subsequent days for the intravenous (IV) forms [8]. For ages 12–14 years with a body weight ≥50 kg, or ages ≥15 years, the recommended dose for the PO form is 200 mg/12 h and, for the IV form, 6 mg/kg/12 h the first day, which is followed by 4 mg/kg/12 h on subsequent days [8]. In pediatric patients, voriconazole exposure displays substantial variability [46] and, consequently, TDM is required to maintain the plasma concentration of 1–5 mg/L [8,47,48]. In the "Voriproph" study—the largest cohort study of voriconazole chemoprophylaxis in children—the use of this agent has been well tolerated [49]. An age of <2 years is a contraindication to the use of voriconazole [8].

Posaconazole is approved for use as a chemoprophylactic agent with both anti-mold and anti-yeast activity in pediatric patients with AML, GVHD after HSCT and HSCT with a long neutropenic period [10,50–52]. Posaconazole is appropriate only for children ≥13 years, based on scant PK data from two adult clinical trials, which also recruited a few patients older than 13, but younger than 18 years [1,53,54]. This anti-fungal agent is available in two PO formulations including an oral suspension and gastro-resistant tablets [55]. For tablets, the established dose is 300 mg/24 h, whereas, for suspension, it is 200 mg/8 h [50,55]. A recent, non-randomized, single-center study in pediatric patients with HSCT found that the tablets were more reliable than the suspension in terms of plasma trough levels [55]. TDM is, however, still required, when the oral suspension is used, to maintain plasma levels ≥0.5 mg/L [56]. Use of posaconazole in patients <13 years of age is contraindicated due to the lack of PK data, unstable plasma concentrations, lack of an IV preparation, and undependable PO absorption [50].

Liposomal amphotericin B (LAMB), in various dosing regimens, has been assessed in several pediatric studies with positive results regarding its safety, efficacy, and feasibility [57–59]. The IV form is reserved for patients intolerant or with contraindications to the use of triazoles [13]. The recommended dosing scheme is either 1 mg/kg/48 h, or 2.5 mg/kg two times per week [8]. The aerosolized form of LAMB is described in ECIL-4 guidelines as prophylaxis against pulmonary infections. However, the route is not approved and doses in patients younger than 18 years have not been established [8].

Micafungin has been compared to fluconazole in a phase III randomized, double-blind clinical trial including both adults and children with HSCT-associated neutropenia [60]. The results were in favor of micafungin in terms of efficacy as a prophylactic agent [60]. However, the scarcity of pediatric data, the lack of PO forms, and the high cost restrict the use of micafungin in anti-fungal prophylaxis [50]. According to ECIL-4 guidelines, the recommended dose is 1 mg/kg/24 h or 50 mg if the patient's weight ≥50 kg [8]. A dose of 2 mg/kg/24 h has been evaluated and found to be safe in children with allogeneic HSCT [61]. Micafungin is used in cases of intolerance or contra-indication to triazoles [13].

Caspofungin has also been evaluated in the setting of anti-fungal prophylaxis [50]. A randomized study including both adult and pediatric patients with AML and myelodysplastic syndrome (MDS) compared caspofungin to itraconazole and found similar efficiency and tolerability [62]. Studies including only pediatric patients are limited to two retrospective cohort studies [50,63,64]. The first compared caspofungin to LAMB in HSCT recipients and reported comparable efficiency [63]. The second study used micafungin as a comparator and recommended that caspofungin should not be preferred over micafungin [64]. Both studies recognized the need for the conduction of randomized clinical trials [63,64].

In summary, in high-risk, acute leukemia patients, the recommended agents for primary prophylaxis against IA include itraconazole, posaconazole, IV LAMB, aerosolized LAMB, micafungin, and voriconazole [8]. In allogeneic HSCT recipients, the options for IA prophylaxis are itraconazole or voriconazole, micafungin, LAMB, aerosolized LAMB, and posaconazole [8]. When GVHD develops, the recommended anti-fungals are posaconazole, voriconazole, itraconazole, IV LAMB, and micafungin [8]. All the above agents are ranked according to the strength of recommendation and quality of evidence and are summarized in Table 3. The grading system used is the one developed by the Infectious Diseases Society of America (IDSA) [8,65]. According to the previously mentioned system, each recommendation receives a letter (A, B, C, D, or E) reflecting its strength, followed by a Latin number (I, II or III) pertaining to the quality of evidence [65].

Table 3. Primary prophylaxis against Invasive Aspergillosis.

Drug	Route	Dosage	Indications for Use (Recommendation Ranking)	Refs
AMB formulations				
LAMB	IV	1 mg/kg/48 h, or 2.5 mg/kg two times per week	• High-risk acute leukemia patients (B-II) • Allogeneic HSCT recipients (C-III) • GVHD (no grading)	[8]
LAMB	Aerosolized	Not established	• High-risk acute leukemia patients (no grading) • Allogeneic HSCT recipients (no grading)	[8]
Azoles				
ITC	PO	2.5 mg/kg/24 h	• High-risk acute leukemia patients (B-I) • Allogeneic HSCT recipients (B-I) • GVHD (C-II)	[8]
VRC	PO	• 9 mg/kg/12 h (ages 2-<12, or 12–14 weighing <50 kg) • 200 mg/12 h (ages ≥15 years, or 12–14 weighing ≥50 kg) • 9 mg/kg/12 h the first day, followed by 8 mg/kg/12 h on next days (ages 2-<12, or 12–14 weighing <50 kg)	• High-risk acute leukemia patients (no grading) • Allogeneic HSCT recipients (B-I) • GVHD (B-I)	[8]
	IV	• 6 mg/kg/12 h the first day, followed by 4 mg/kg/12 h on next days (ages ≥15 years, or 12–14 weighing ≥50 kg)		

Table 3. *Cont.*

Drug	Route	Dosage	Indications for Use (Recommendation Ranking)	Refs
PSC	PO	• 200 mg/8 h (oral susp.) • 300 mg/24 h (tabl.)	• High-risk acute leukemia patients (B-I) • Allogeneic HSCT recipients (no grading) • GVHD (B-I)	[8]
		Echinocandins		
MFG	IV	• 1 mg/kg/24 h (max 50 mg if weight ≥50 kg) • 2 mg/kg/24 h	• High-risk acute leukemia patients (no grading) • Allogeneic HSCT recipients (C-I) • GVHD (no grading) • Allogeneic HSCT recipients	[8,61]

AMB, amphotericin B. LAMB, liposomal amphotericin B. IV, intravenous. HSCT, hematopoietic stem cell transplantation. GVHD, graft-versus-host disease. ITC, itraconazole. PO, per os. VRC, voriconazole. PSC, posaconazole. susp., suspension. tabl., tablets. MFG, micafungin.

Secondary chemoprophylaxis, as mentioned above, is a vague term, which cannot be easily differentiated from continued treatment against previous IA [8]. It should be administered for the entire period during which the risk factors for IA persist (such as granulocytopenia and immunosuppression) and it should be targeted against the previous isolates of *Aspergillus* spp. [8,13]. The "VOSIFI Study" has evaluated voriconazole in the setting of allogeneic HSCT in adult patients and has found that this agent is an efficient option when considering secondary prophylaxis [66]. The regimen consisting of LAMB followed by voriconazole has been evaluated in pediatric patients with acute leukemia and IPA [67]. Other options include itraconazole and caspofungin. However, supporting evidence is derived from adult studies [8].

2.3. Other Preventive Measures

Apart from the administration of drug prophylaxis, several infection control strategies can be applied to prevent IA among pediatric leukemia and HSCT patients. These strategies aim to decrease the exposure to sources of mold spores, which would, otherwise, increase the risk of IA [1].

The cornerstone is the construction of a "protective environment" for inpatients, which regulates room ventilation and involves a specific number of air exchanges/hour, the application of high-efficiency particulate air (HEPA) filters (with or without laminar airflow), appropriate room sealing, automatic-closing doors, pressure monitoring (to maintain a positive pressure differential between the ward and the outside), and directed airflow [2,68]. Furthermore, plants and flowers are not permitted in these rooms, but the installation of shower filters is recommended [13]. In hospitals with a limited number of "protective environments," strict criteria should be implemented regarding which patients will be accommodated in these wards [2,68]. Another option is the admission to a private ward with restricted connections [2].

In outpatients, the previously mentioned measures are not applicable and, thus, different recommendations have been developed for this patient group [69]. These include, among others, avoidance of construction areas, stagnant waters, and areas with increased moisture, avoidance of gardening and lawn mowing, appropriate checking of foods, and hand hygiene [69]. The efficacy against IA of surgical masks and N95 respirators has not been established [2].

In the case of construction or renovation works in the hospital, or in any adjacent sites, infection control strategies should be escalated and interdisciplinary committees should be established to ensure compliance with these strategies [68].

Environmental sampling for microbiological analysis is useful in cases of outbreaks even though its application as part of routine clinical practice has been questioned and there is a lack of data to support it [2,68]. Nevertheless, it is a useful tool to evaluate the function of the filters [13].

3. Treatment

3.1. Empiric Treatment

Empiric treatment should be started in individuals at high-risk for IA presenting with fever and neutropenia, which persist for a minimum of four days after the initiation of broad-spectrum anti-bacteria [1,8,29]. Another group for which empiric therapy is indicated includes neutropenic patients presenting with recurrent febrile episodes after defervescence following the administration of antibiotics [8]. This indication, however, has not been graded in the latest guidelines by ECIL-4. Moreover, ECIL-4 guidelines recommend initiation of empiric therapy in low-risk children with persistent fever accompanied by severe neutropenia and mucosal damage [8]. Nonetheless, according to the clinical practice guideline (CPG) developed by the International Pediatric Fever and Neutropenia Guideline Panel, empiric treatment should not be administered to low-risk pediatric patients with persistent fever and neutropenia [29]. This recommendation is supported by the results of a randomized, prospective study comparing either caspofungin or LAMB to no treatment in low-risk children [70]. In resource-poor settings with inadequate laboratory capabilities, empiric treatment has been associated with better results in individuals at high risk for IA [7].

The anti-fungal agents recommended for this approach are caspofungin and LAMB [8,29]. The first is administered at a loading dose of 70 mg/m^2, which is followed by 50 mg/m^2/24 h (maximum dose 70 mg/24 h) and the second at a dose of 1–3 mg/kg/24 h [8]. Use of these agents in pediatrics has been established by three randomized prospective trials [1]. The results of the first trial underlined the superiority of liposomal to deoxycholate amphotericin B (AMB) [1,71]. According to the second study, amphotericin B colloidal dispersion (ABCD) was superior to deoxycholate AMB in terms of adverse events, but similar in efficacy [1,72]. Lastly, the third study found that caspofungin performed similarly to LAMB in terms of efficiency and adverse effect rates [1,73]. Caspofungin at a dose of 50 mg/m^2/24 h in pediatric patients results in similar exposure levels with adult patients [74]. The use of voriconazole in empiric treatment has also been studied in a randomized, multi-center, open-label trial that compared this second-generation azole with LAMB [2,75]. The trial included both adult and pediatric patients and the results were indicative of a comparable response rate between voriconazole and LAMB groups in high-risk patients [2,75]. In another study, though, oral voriconazole was not preferred over deoxycholate AMB in patients presenting with gastrointestinal symptoms or in those receiving vincristine [7]. Identification of new anti-fungals to be used in empiric treatment remains a "research gap" [29].

Lastly, the duration of empiric anti-fungal schemes is a field that requires further investigation [29]. The 2017 ESCMID-ECMM-ERS Joint Clinical Guidelines recommend that the administration of caspofungin or LAMB should be carried on until defervescence and recovery of the neutrophil count [13,71,73,76].

3.2. Pre-Emptive Treatment

Empiric treatment has the disadvantage of exposing most patients to unnecessary use of anti-fungal drugs, due to the low specificity of fever as a symptom of IA [2]. Furthermore, the low sensitivity of fever as a criterion in the diagnosis of IA might delay the initiation of treatment. On the other hand, establishing a definite diagnosis in these patients is challenging and the risk of a dismal outcome is high [5,77]. Thus, the need for the adoption of a pre-emptive approach is highlighted [5,77].

Pre-emptive treatment is an alternative to empiric treatment that uses clinical and noninvasive, non-culture diagnostic methods to further assess the risk of IFDs and IA in febrile patients with neutropenia or in asymptomatic patients [2,8]. The diagnostic methods utilized in this approach include imaging techniques—mainly computed tomography—and microbiological markers, such as galactomannan (GM) antigen, (1,3)-β-D glucan (BDG), and *Aspergillus* polymerase chain reaction (PCR) [2,8].

Data regarding the application of pre-emptive therapy and the use of the associated biomarkers in children are limited [2,8]. Furthermore, these biomarkers do not perform optimally and consistently in pediatrics [78]. More specifically, the GM antigen displays variable sensitivity and specificity, low positive prognostic value (PPV), and a high false-positive rate [78,79]. The high negative prognostic value (NPV) of this biomarker applies only for *Aspergillus* spp. and not for other pathogens [78,79]. The sensitivity and specificity of BDG ranged from 50% to 82% and from 46% to 82%, respectively [78]. According to the International Pediatric Fever and Neutropenia Guideline Panel, its use is not recommended in the context of empiric anti-fungal treatment [29]. Lastly, the lack of standardization and the false-positive rates limit the use of PCR as well [78]. According to ECIL-4 guidelines, however, there is consensus in the favor of the use of the pre-emptive approach in pediatric patients, which has not received any grading [8]. A recent randomized, multi-center study compared pre-emptive with empiric treatment in pediatric cancer patients with fever and neutropenia with the exception of HSCT recipients [80]. The two methods were found to be comparable in terms of efficiency [80].

3.3. Targeted Treatment for Proven/Probable Invasive Aspergillosis

The European Organization for Research and Treatment of Cancer/Invasive Fungal Infections Cooperative Group (EORTC) and the National Institute of Allergy and Infectious Diseases Mycoses Study Group (MSG) provided the definitions for proven and probable IA. Proven IA requires evidence of *Aspergillus* to be identified in tissue by microscopic examination or culture, whereas probable entails a combination of patient risk factors, clinical manifestations, and mycological criteria [77]. Nevertheless, establishing a diagnosis is a challenging and time-consuming task, which may postpone treatment initiation and, thus, the above definitions should be used only for studies and not for routine clinical practice [13,77]. In the current review, the definitions by EORTC/MSG are used.

Targeted treatment for IA includes anti-fungal medications and adjunctive measures [2]. Similarities between pediatric and adult patients do exist, but there are several critical differences, such as the dosing scheme [2]. Targeted anti-fungal drugs (summarized in Table 4) can be further divided into the first and second-line, with the latter being reserved for unresponsive patients or for cases of intolerance to the adverse events [1,8]. Strength of recommendation and quality of evidence are in line with the IDSA grading system [8,65].

Table 4. Anti-fungal drugs for proven/probable Invasive Aspergillosis.

Drug	Route	Dosage	Indications for Use (Recommendation Ranking)	Refs
AMB formulations				
LAMB	IV	• 3 mg/kg/24 h	• First-line treatment (B-I) (especially for ages <2 years) • Second-line treatment (B-I) (for cases with VRC intolerance, or for settings with azole resistance)	[1,8,13]
		• ≥5 mg/kg/24 h	• Second-line treatment for CNS Aspergillosis	[5,81]
ABLC	IV	5 mg/kg/24 h	• First-line treatment (B-II) • Second-line treatment (B-II)	[8]
Azoles				
ITC	PO	2.5 mg/kg/12 h (for ages ≥2 years)	Second-line treatment (no grading) (not approved for ages <18 years)	[8]

Table 4. *Cont.*

Drug	Route	Dosage	Indications for Use (Recommendation Ranking)	Refs
VRC	PO IV	• 9 mg/kg/12 h (ages 2-<12, or 12–14 weighing <50 kg) • 200 mg/12 h (ages ≥15 years, or 12–14 weighing ≥50 kg) • 9 mg/kg/12 h the first day, followed by 8 mg/kg/12 h on next days (ages 2-<12, or 12–14 weighing <50 kg) • 6 mg/kg/12 h the first day, followed by 4 mg/kg/12 h on next days (ages ≥15 years, or 12–14 weighing ≥50 kg)	• First-line treatment (A-I) (not approved for ages <2 years) • Second-line treatment (A-I)	[8]
PSC	PO	800 mg/24 h divided in 2–4 doses (oral susp.)	Second-line treatment (no grading) (not approved for ages <18 years by the EU, approved by the FDA for ages ≥13 years)	[2,8]
ISA	IV PO	Not established	• Evaluation of PK for ages 1–18 years (ClinicalTrials.gov NCT03241550) • Evaluation of PK for ages 6–18 years (ClinicalTrials.gov NCT03241550)	[50]
Echinocandins				
CAS	IV	70 mg/m² loading dose the first day, followed by 50 mg/m²/24 h the next days (maximum 70 mg)	Second-line treatment (A-II)	[8]
MFG	IV	2–4 mg/24 h (100–200 mg/24 h if patient's weight ≥50 kg)	Second-line treatment (no grading) (non-approved indication by the EU, approved by the FDA for ages ≥4 months)	[1,2,8]
AFG	IV	3 mg/kg loading dose, followed by 1.5 mg/kg/24 h	Not approved by the FDA for ages <18 years	[2,50,82]

AMB, amphotericin B. LAMB, liposomal amphotericin B. IV, intravenous. VRC, voriconazole. CNS, central nervous system. ABLC, amphotericin B lipid complex. ITC, itraconazole. PO, per os. PSC, posaconazole. susp., suspension. EU, European Union. FDA, Food and Drug Administration. ISA, isavuconazole. PK, pharmacokinetics. CAS, caspofungin. MFG, micafungin. AFG, anidulafungin.

3.3.1. First-Line Anti-Fungal Drugs

First-line agents are voriconazole, LAMB, and amphotericin B lipid complex (ABLC) [1,8].

Adequate data and experience have rendered voriconazole the cornerstone of IA treatment in children of all ages, apart from neonates and children <2 years [2,13]. A randomized clinical trial underlying its superiority against deoxycholate AMB in patients ≥12 years of age has played a significant role in establishing the use of this azole [83–85]. The PK of voriconazole is linear in pediatric patients, in contrast to the nonlinear pattern observed in adults and further population analyses of this parameter have facilitated the development of the dosing scheme [2,86,87]. For ages 2–<12 years, or 12–14 years with a body weight <50 kg, the dose is 9 mg/kg/12 h for the PO forms and 9 mg/kg/12 h the first day, followed by 8 mg/kg/12 h on subsequent days for the intravenous (IV) forms [8]. For ages 12–14 years with a body weight ≥50 kg, or ages ≥15 years, it is 200 mg/12 h for the PO form and for the IV form 6 mg/kg/12 h the first day, followed by 4 mg/kg/12 h on subsequent days [8]. According to a meta-analysis that highlighted the value of voriconazole TDM, therapeutic plasma levels increased the probability of a successful result, whereas greater concentrations were likely to cause toxicity [85,88]. TDM is, therefore, recommended, with an optimal range of 1–5 mg/L [8,47,48].

Liposomal amphotericin B (LAMB) is recommended at a dose of 3 mg/kg/24 h, IV [8]. The latter dose has been compared to a regimen of 10 mg/kg/24 h in the "AmBiLoad trial." However, the higher dose has not been associated with greater efficacy [89]. Moreover, in the previous randomized trial, no direct comparison to voriconazole has been attempted [13,89]. LAMB is indicated predominantly for children <2 years and neonates, for whom voriconazole has not been approved, and for settings with increased prevalence of azole-resistance [1,13]. When compared to deoxycholate AMB, LAMB is less nephrotoxic and has fewer infusion-related toxic reactions, but its use is limited due to its high cost [90–92].

Lastly, ABLC is another option for the first-line targeted treatment of IA, which is administered at 5 mg/kg/24 h in one IV dose [8]. However, recommendations for the use of ABLC stem from experience in naïve patients who received the agent in terms of second-line therapy [8].

3.3.2. Second-Line Anti-Fungal Drugs

Second-line agents include caspofungin, micafungin, itraconazole, and posaconazole [8]. The anti-fungal drugs described in the section of first-line treatment may also be used [8]. Voriconazole is reserved for patients naive to this agent and LAMB is used in cases of unresponsiveness or intolerance to the former, as well as in patients naive to AMB [1,8].

Caspofungin is the most preferred echinocandin for pediatric IFDs, based on results from the Antibiotic Resistance and Prescribing in European Children ("ARPEC") study [50,93]. The recommended IV dose is calculated, according to the body surface area (BSA), and is defined as a 70 mg/m^2 loading dose during the first day, which is followed by 50 mg/m^2/24 h on subsequent days (maximum 70 mg) [2,8,74]. Caspofungin is approved for pediatric use both in the United States of America (USA), by the Food and Drug Administration (FDA), and in Europe [2,8]. Two prospective studies verified the efficacy and safety of caspofungin [50]. The first evaluated the drug in the setting of salvage therapy for IA with encouraging results that showed 45% of patients demonstrated complete or partial clinical response [94]. The second study showed that the use of caspofungin in children from 6 months to 17 years of age was efficient against IA in a consistent manner with adult studies [95]. Lastly, caspofungin has been confirmed to be a feasible alternative choice for the treatment of children with IFDs, based on results from a systematic review and meta-analysis [96]. The authors, though, acknowledge the need for further research on this topic [96].

Micafungin—for therapeutic purposes—is administered at a dose of 2 to 4 mg/kg/24 h, IV, which reached a maximum of 100 to 200 mg/24 h if the patient weights ≥50 kg [1,8]. ECIL-4 stated that its use as a second-line agent for IA is a non-approved indication, due to a lack of robust evidence [8]. The FDA has approved the use of micafungin in pediatric patients ≥4 months of age [2]. Dose adjustments need to be considered in children ≤8 years, due to the fact that micafungin clearance increases with decreasing age, which necessitates higher doses in young age groups [2,97]. When compared to triazoles in a meta-analysis evaluating the treatment of IFDs in hematologic patients with neutropenia, micafungin has been associated with higher efficacy, fewer severe adverse events, but displayed similar all-cause mortality [50,98].

Itraconazole, despite being the first in its class exhibiting anti-Aspergillus activity, has fallen into disuse, due to several limitations including unpredictable bioavailability, interactions with chemotherapeutic drugs such as cyclophosphamide, and interactions with drugs that cause QTc prolongation [2,82,85,99]. Using this azole as second-line therapy is not approved in individuals <18 years of age [8] and it is reserved for less critical cases of IA [2]. When used in patients ≥2 years old, the dose of the PO form of itraconazole is 2.5 mg/kg/12 h with subsequent TDM, to aim for plasma concentrations ≥0.5mg/L [1,8,40].

Posaconazole exists in three different formulations: oral suspension, gastro-resistant tablet, and IV solution [85]. It is not approved in the European Union (EU) for children <18 years of age [8], but has received approval by the FDA for patients ≥13 years for both PO formulations and for patients ≥18 years for the IV form [2]. Based on limited PK data in pediatric patients ≥13 years of age, the dose of oral suspension of posaconazole is 800 mg/24 h divided in 2 to 4 doses [1,8]. TDM is required to maintain plasma concentrations ≥0.7–1.5 mg/L [8,56,100]. No need for TDM exists for the PO gastro-resistant tablets [50]. Posaconazole may be used for salvage therapy in cases of refractoriness or intolerance to previous agents among pediatric patients aged 13 years or older [8,84,85,100].

It should be noted that treatment for proven/probable IA in pediatrics is age-dependent. More specifically, the options for children ≥2 years of age are voriconazole, LAMB, ABLC, ABCD, caspofungin, itraconazole, and posaconazole (for ages ≥13 years), whereas, for children <2 years old, LAMB, ABLC, and caspofungin may be used [5,13]. For neonates, the only option is LAMB [13].

3.3.3. Novel Anti-Fungal Drugs

A brief review has to be made regarding the role that the newer agents of echinocandins and triazoles play in treating IA. These drugs are anidulafungin and isavuconazole, respectively.

Anidulafungin has not received FDA approval yet for use in patients aged <18 years [2,50]. The safety and PK parameters of anidulafungin in neutropenic pediatric patients at risk for IA have been assessed in a multi-center, dose-escalation study [101]. Anidulafungin displayed good tolerability and the regimen of 3 mg/kg loading dose, which was followed by 1.5 mg/kg/24 h that resulted in concentration profiles consistent with the adult dose of 100 mg/24 hours. This is the preferred one for IA [101]. The same regimen has been recently evaluated in a single-center study from Argentina, which underlined the safety and efficiency of this agent [102]. Lastly, the results of an open-label, non-comparative, pediatric study for the use of anidulafungin in invasive candidiasis (IC) have also been in favor of the safety of this drug in children, when administered in the previously mentioned doses [50,103].

Isavuconazole is a novel triazole with an extended spectrum of activity [104]. Its use in adult patients has been established by the "SECURE" trial—a phase 3, multi-center, randomized, double-blind, trial—which highlighted the safety and the non-inferiority of isavuconazole compared to voriconazole for the treatment of IA [85,105]. In pediatric patients, the PK and safety of both oral and IV forms of isavuconazole are currently being evaluated by a phase 1, open-label, multi-center study against no comparators (ClinicalTrials.gov NCT03241550) [50]. Additionally, limited experience has also been reported regarding the successful use of isavuconazole in pediatric hematology-oncology patients with invasive mucormycosis [50,106].

3.3.4. Combination Therapies and Duration of Treatment

The role of combination therapies in pediatric cancer and HSCT patients with IA has not been completely elucidated and requires further assessment [5,85]. However, use of a combination of anti-fungals in such patients has been reported by two multi-center cohort studies [85] including one retrospective [6] and one prospective [107]. Moreover, ECIL-4 provides a recommendation for the use of such therapy—both in the setting of first and second-line treatment for IA—but this is based on poor evidence [8]. Safety and efficacy data have arisen from a study evaluating combination therapy with caspofungin in 40 hematologic pediatric patients with IA [8,108]. In both ECIL-4 and IDSA 2016 Guidelines, the recommended combination is that of a polyene or triazole with an echinocandin [2,8].

There is no consensus regarding the optimal duration for which patients with proven/probable IA need to be treated and, thus, this parameter should be individualized [5,8]. Duration of treatment has been documented to range from 3 to more than 50 weeks [13,83,89,105,109]. Treatment may be discontinued after clinical improvement, microbiological response, and recovery from GVHD [5,8,13].

3.3.5. Breakthrough Infection and *Aspergillus* Resistance

Treatment of breakthrough infection occurring in patients who have received anti-mold prophylaxis includes salvage therapy, changing the antifungal drug class, awareness regarding local epidemiology patterns, and verification of serum triazole levels [2]. Salvage therapy is also used for the refractory or progressive aspergillosis and involves a switch in the class of antifungals, the addition of a second agent, the correction of underlying immunosuppression, and surgery [2,5].

Another essential topic that needs to be addressed is *Aspergillus* resistance. The previously mentioned anti-fungal agents are not active against all *Aspergillus* spp. [13]. Some species may display intrinsic resistance to azoles and polyenes [13,110], while others have acquired resistance to azoles [13,111]. For instance, *Aspergillus calidoustus* and *Aspergillus terreus* exhibit intrinsic resistance to azoles and AMB, respectively [13,110], whereas *Aspergillus fumigatus* may develop acquired resistance to azoles [13,112]. However, anti-fungal susceptibility testing is not recommended for the routine management of the initial infection [2]. More specifically, in patients naïve to azoles or in regions with no documented resistance, anti-fungal susceptibility testing should not be performed [13]. However, such testing is indicated in cases that do not respond to initial treatment, or if there is clinical suspicion of azole resistance [13]. In patients with infections due to *Aspergillus fumigatus* with documented azole-resistance, a group of experts recommended a switch to LAMB or a combination of voriconazole with an echinocandin [112,113]. In cases of environmental resistance to azoles, the latter regimen switch is recommended only if the respective rates are >10% [13]. Azole resistance of *Aspergillus fumigatus* has also been reported in pediatric patients and should be taken into consideration in cases that are unresponsive to azole treatment [114].

3.3.6. Adjunctive Measures

Apart from anti-fungal drugs, treatment for proven/probable IA also includes several adjunctive measures. Colony-stimulating factors (CSFs) may be administered either as prophylaxis, to reduce the duration of granulocytopenia, or as therapy for patients with IA and neutropenia [2]. Granulocyte transfusions may be an option in cases of severe and prolonged neutropenia [8]. An updated Cochrane review, which evaluates the efficacy and safety of this method in the setting of prophylaxis, concluded that the risk of fungemia decreased (low-grade evidence) due to the transfusions, but data was inadequate to support any differences in infection-associated mortality or adverse events [115]. Recombinant interferon gamma (IFN-γ) may be used in cases of severe or refractory IA [2]. Lastly, adoptive transfer of pathogen-specific T lymphocytes, which have derived from donors, is under study for use in IA [85,116,117].

Surgical treatment is reserved for cases of accessible localized lesions [2]. Indications include sinus aspergillosis, localized cutaneous aspergillosis, CNS aspergillosis, pulmonary disease that is localized or adjacent to great vessels/pericardium, or has invaded the pleural space and chest wall, or has caused uncontrolled bleeding [2]. Surgery is contraindicated for unstable patients, or for patients with disseminated disease [6]. In each case, decisions should be individualized [2].

3.3.7. Management of Selected Localized Infections

The following part of this review will be devoted to the management of IPA and CNS aspergillosis in pediatric patients with leukemia and HSCT. These clinical manifestations represent the most common site of infection [6] and one of the most serious complications [118], respectively.

Invasive pulmonary aspergillosis, when suspected, should prompt early initiation of treatment, due to the fact that this practice restricts the development of the disease and due to the unreliability of diagnostic testing [2,89,119]. When the diagnosis is established, the patient should be further evaluated for other foci of IA, such as the CNS [13]. Optimal duration of treatment has not been identified, but a minimum of six to 12 weeks should be applied [2]. Lastly, surgical intervention is reserved for cases with lesions adjacent to great vessels, or vital organs, lesions associated with unmanageable hemoptysis, or lesion causing bone erosions [2].

Aspergillosis of the CNS is most frequently associated with hematologic cancer since the underlying condition in patients >1 year of age [120]. Infection of the CNS has been observed in 6% of children with leukemia with IA [8,121]. The advent of AMB lipid formulations and anti-mold azoles and the progress achieved in early diagnosis have reduced mortality rates from 82.8% to 39.5%, before and after 1990, respectively [121]. Treatment principles of CNS aspergillosis are early diagnosis, initiation of appropriate anti-mold drugs, evaluation of the indications for surgery and decreasing immunosuppression [2,122]. The cornerstone of pharmacotherapy in children has been AMB, with the deoxycholate form having been tolerated in ages <3 months [15,123] and the lipid formulations of AMB having been reserved for older children [15]. Currently, voriconazole is the established first-line treatment of CNS aspergillosis [5,8,81,83,121,124] and the next choice is LAMB at high doses (≥5mg/kg/24 h) [5,125]. TDM for voriconazole is necessary and younger children may require higher doses to reach therapeutic levels [121,126]. However, there is limited evidence regarding the levels of voriconazole in the CSF [121]. Data is also scarce about the use of adjunctive immunotherapy (CSFs, cytokines) in pediatric patients with CNS aspergillosis [120]. Surgical intervention may be indicated in patients with localized lesions [120]. Lastly, the use of corticosteroids and the intrathecal administration of anti-fungals are not recommended [2,127].

4. Conclusions

It has become evident that IA is a major issue in immuno-compromised pediatric patients, especially in those with leukemia and in HSCT recipients. Management of this infection consists of two main components, which includes prevention and treatment. The role of primary anti-fungal prophylaxis is highlighted particularly due to the insufficiency of diagnostic tests. Several agents have been evaluated in this setting including triazoles, polyenes, and echinocandins. Empiric and pre-emptive treatments are two approaches that can be initiated before establishing a definitive diagnosis. The mainstay of targeted treatment is voriconazole for children older than two years of age and LAMB in the younger age group. Further research is required in the field of pediatric IA management in order to reach the evidence quantity and quality of the respective field in adults.

Funding: This research received no external funding.

Conflicts of Interest: Elias Roilides has received research grants from Astellas, Gilead, and Pfizer Inc. and is a scientific advisor and member of speaker bureaux for Astellas, Gilead, Merck, and Pfizer Inc. The other authors declare no conflict of interest.

References

1. Frange, P.; Bougnoux, M.-E.; Lanternier, F.; Neven, B.; Moshous, D.; Angebault, C.; Lortholary, O.; Blanche, S. An update on pediatric invasive aspergillosis. *Med. Mal. Infect.* **2015**, *45*, 189–198. [CrossRef] [PubMed]
2. Patterson, T.F.; Thompson, G.R., 3rd; Denning, D.W.; Fishman, J.A.; Hadley, S.; Herbrecht, R.; Kontoyiannis, D.P.; Marr, K.A.; Morrison, V.A.; Nguyen, M.H.; et al. Practice Guidelines for the Diagnosis and Management of Aspergillosis: 2016 Update by the Infectious Diseases Society of America. *Clin. Infect. Dis.* **2016**, *63*, e1–e60. [CrossRef] [PubMed]
3. Steinbach, W.J. Pediatric aspergillosis: Disease and treatment differences in children. *Pediatr. Infect. Dis. J.* **2005**, *24*, 358–364. [CrossRef] [PubMed]
4. Walsh, T.J.; Gonzalez, C.; Lyman, C.A.; Chanock, S.J.; Pizzo, P.A. Invasive fungal infections in children: Recent advances in diagnosis and treatment. *Adv. Pediatr. Infect. Dis.* **1996**, *11*, 187–290. [PubMed]
5. Tragiannidis, A.; Roilides, E.; Walsh, T.J.; Groll, A.H. Invasive aspergillosis in children with acquired immunodeficiencies. *Clin. Infect. Dis.* **2012**, *54*, 258–267. [CrossRef] [PubMed]
6. Burgos, A.; Zaoutis, T.E.; Dvorak, C.C.; Hoffman, J.A.; Knapp, K.M.; Nania, J.J.; Prasad, P.; Steinbach, W.J. Pediatric invasive aspergillosis: A multicenter retrospective analysis of 139 contemporary cases. *Pediatrics* **2008**, *121*, e1286-94. [CrossRef]
7. Jain, S.; Kapoor, G. Invasive aspergillosis in children with acute leukemia at a resource-limited oncology center. *J. Pediatr. Hematol. Oncol.* **2015**, *37*, e1–e5. [CrossRef]
8. Groll, A.H.; Castagnola, E.; Cesaro, S.; Dalle, J.-H.; Engelhard, D.; Hope, W.; Roilides, E.; Styczynski, J.; Warris, A.; Lehrnbecher, T. Fourth European Conference on Infections in Leukaemia (ECIL-4): Guidelines for diagnosis, prevention, and treatment of invasive fungal diseases in paediatric patients with cancer or allogeneic haemopoietic stem-cell transplantation. *Lancet. Oncol.* **2014**, *15*, e327-40. [CrossRef]

9. Sung, L.; Phillips, R.; Lehrnbecher, T. Time for paediatric febrile neutropenia guidelines—children are not little adults. *Eur. J. Cancer* **2011**, *47*, 811–813. [CrossRef]

10. Groll, A.H.; Tragiannidis, A. Update on antifungal agents for paediatric patients. *Clin. Microbiol. Infect.* **2010**, *16*, 1343–1353. [CrossRef]

11. Dornbusch, H.J.; Manzoni, P.; Roilides, E.; Walsh, T.J.; Groll, A.H. Invasive fungal infections in children. *Pediatr. Infect. Dis. J.* **2009**, *28*, 734–737. [CrossRef] [PubMed]

12. Lestner, J.M.; Smith, P.B.; Cohen-Wolkowiez, M.; Benjamin, D.K.J.; Hope, W.W. Antifungal agents and therapy for infants and children with invasive fungal infections: A pharmacological perspective. *Br. J. Clin. Pharmacol.* **2013**, *75*, 1381–1395. [CrossRef] [PubMed]

13. Ullmann, A.J.; Aguado, J.M.; Arikan-Akdagli, S.; Denning, D.W.; Groll, A.H.; Lagrou, K.; Lass-Florl, C.; Lewis, R.E.; Munoz, P.; Verweij, P.E.; et al. Diagnosis and management of Aspergillus diseases: Executive summary of the 2017 ESCMID-ECMM-ERS guideline. *Clin. Microbiol. Infect.* **2018**, *24*, e1–e38. [CrossRef] [PubMed]

14. Antachopoulos, C.; Walsh, T.J.; Roilides, E. Fungal infections in primary immunodeficiencies. *Eur. J. Pediatr.* **2007**, *166*, 1099–1117. [CrossRef] [PubMed]

15. Groll, A.H.; Jaeger, G.; Allendorf, A.; Herrmann, G.; Schloesser, R.; von Loewenich, V. Invasive pulmonary aspergillosis in a critically ill neonate: Case report and review of invasive aspergillosis during the first 3 months of life. *Clin. Infect. Dis.* **1998**, *27*, 437–452. [CrossRef] [PubMed]

16. Groll, A.H.; Shah, P.M.; Mentzel, C.; Schneider, M.; Just-Nuebling, G.; Huebner, K. Trends in the postmortem epidemiology of invasive fungal infections at a university hospital. *J. Infect.* **1996**, *33*, 23–32. [CrossRef]

17. Nedel, W.L.; Kontoyiannis, D.P.; Pasqualotto, A.C. Aspergillosis in patients treated with monoclonal antibodies. *Rev. Iberoam. Micol.* **2009**, *26*, 175–183. [CrossRef] [PubMed]

18. Pappas, P.G.; Alexander, B.D.; Andes, D.R.; Hadley, S.; Kauffman, C.A.; Freifeld, A.; Anaissie, E.J.; Brumble, L.M.; Herwaldt, L.; Ito, J.; et al. Invasive fungal infections among organ transplant recipients: Results of the Transplant-Associated Infection Surveillance Network (TRANSNET). *Clin. Infect. Dis.* **2010**, *50*, 1101–1111. [CrossRef] [PubMed]

19. Abbasi, S.; Shenep, J.L.; Hughes, W.T.; Flynn, P.M. Aspergillosis in children with cancer: A 34-year experience. *Clin. Infect. Dis.* **1999**, *29*, 1210–1219. [CrossRef] [PubMed]

20. Groll, A.H.; Kurz, M.; Schneider, W.; Witt, V.; Schmidt, H.; Schneider, M.; Schwabe, D. Five-year-survey of invasive aspergillosis in a paediatric cancer centre. Epidemiology, management and long-term survival. *Mycoses* **1999**, *42*, 431–442. [CrossRef] [PubMed]

21. Zaoutis, T.E.; Heydon, K.; Chu, J.H.; Walsh, T.J.; Steinbach, W.J. Epidemiology, outcomes, and costs of invasive aspergillosis in immunocompromised children in the United States, 2000. *Pediatrics* **2006**, *117*, e711-6. [CrossRef] [PubMed]

22. Rubio, P.M.; Sevilla, J.; Gonzalez-Vicent, M.; Lassaletta, A.; Cuenca-Estrella, M.; Diaz, M.A.; Riesco, S.; Madero, L. Increasing incidence of invasive aspergillosis in pediatric hematology oncology patients over the last decade: A retrospective single centre study. *J. Pediatr. Hematol. Oncol.* **2009**, *31*, 642–646. [CrossRef] [PubMed]

23. Kaya, Z.; Gursel, T.; Kocak, U.; Aral, Y.Z.; Kalkanci, A.; Albayrak, M. Invasive fungal infections in pediatric leukemia patients receiving fluconazole prophylaxis. *Pediatr. Blood Cancer* **2009**, *52*, 470–475. [CrossRef] [PubMed]

24. Thomas, K.E.; Owens, C.M.; Veys, P.A.; Novelli, V.; Costoli, V. The radiological spectrum of invasive aspergillosis in children: A 10-year review. *Pediatr. Radiol.* **2003**, *33*, 453–460. [CrossRef] [PubMed]

25. Dvorak, C.C.; Fisher, B.T.; Sung, L.; Steinbach, W.J.; Nieder, M.; Alexander, S.; Zaoutis, T.E. Antifungal prophylaxis in pediatric hematology/oncology: New choices & new data. *Pediatr. Blood Cancer* **2012**, *59*, 21–26. [CrossRef] [PubMed]

26. Tragiannidis, A.; Dokos, C.; Lehrnbecher, T.; Groll, A.H. Antifungal chemoprophylaxis in children and adolescents with haematological malignancies and following allogeneic haematopoietic stem cell transplantation: Review of the literature and options for clinical practice. *Drugs* **2012**, *72*, 685–704. [CrossRef] [PubMed]

27. Fisher, B.T.; Robinson, P.D.; Lehrnbecher, T.; Steinbach, W.J.; Zaoutis, T.E.; Phillips, B.; Sung, L. Risk Factors for Invasive Fungal Disease in Pediatric Cancer and Hematopoietic Stem Cell Transplantation: A Systematic Review. *J. Pediatric Infect. Dis. Soc.* **2018**, *7*, 191–198. [CrossRef]

28. Crassard, N.; Hadden, H.; Pondarre, C.; Hadden, R.; Galambrun, C.; Piens, M.A.; Pracros, J.P.; Souillet, G.; Basset, T.; Berthier, J.C.; et al. Invasive aspergillosis and allogeneic hematopoietic stem cell transplantation in children: A 15-year experience. *Transpl. Infect. Dis.* **2008**, *10*, 177–183. [CrossRef]

29. Lehrnbecher, T.; Robinson, P.; Fisher, B.; Alexander, S.; Ammann, R.A.; Beauchemin, M.; Carlesse, F.; Groll, A.H.; Haeusler, G.M.; Santolaya, M.; et al. Guideline for the Management of Fever and Neutropenia in Children With Cancer and Hematopoietic Stem-Cell Transplantation Recipients: 2017 Update. *J. Clin. Oncol.* **2017**, *35*, 2082–2094. [CrossRef]

30. Kontoyiannis, D.P.; Marr, K.A.; Park, B.J.; Alexander, B.D.; Anaissie, E.J.; Walsh, T.J.; Ito, J.; Andes, D.R.; Baddley, J.W.; Brown, J.M.; et al. Prospective surveillance for invasive fungal infections in hematopoietic stem cell transplant recipients, 2001-2006: Overview of the Transplant-Associated Infection Surveillance Network (TRANSNET) Database. *Clin. Infect. Dis.* **2010**, *50*, 1091–1100. [CrossRef]

31. Lass-Florl, C. The changing face of epidemiology of invasive fungal disease in Europe. *Mycoses* **2009**, *52*, 197–205. [CrossRef] [PubMed]

32. Marr, K.A.; Carter, R.A.; Boeckh, M.; Martin, P.; Corey, L. Invasive aspergillosis in allogeneic stem cell transplant recipients: Changes in epidemiology and risk factors. *Blood* **2002**, *100*, 4358–4366. [CrossRef] [PubMed]

33. Segal, B.H. Aspergillosis. *N. Engl. J. Med.* **2009**, *360*, 1870–1884. [CrossRef] [PubMed]

34. van Burik, J.-A.H.; Carter, S.L.; Freifeld, A.G.; High, K.P.; Godder, K.T.; Papanicolaou, G.A.; Mendizabal, A.M.; Wagner, J.E.; Yanovich, S.; Kernan, N.A. Higher risk of cytomegalovirus and aspergillus infections in recipients of T cell-depleted unrelated bone marrow: Analysis of infectious complications in patients treated with T cell depletion versus immunosuppressive therapy to prevent graft-versus-host. *Biol. Blood Marrow Transplant.* **2007**, *13*, 1487–1498. [CrossRef] [PubMed]

35. Hope, W.W.; Castagnola, E.; Groll, A.H.; Roilides, E.; Akova, M.; Arendrup, M.C.; Arikan-Akdagli, S.; Bassetti, M.; Bille, J.; Cornely, O.A.; et al. ESCMID* guideline for the diagnosis and management of Candida diseases 2012: Prevention and management of invasive infections in neonates and children caused by Candida spp. *Clin. Microbiol. Infect.* **2012**, *18*, 38–52. [CrossRef] [PubMed]

36. Panackal, A.A.; Li, H.; Kontoyiannis, D.P.; Mori, M.; Perego, C.A.; Boeckh, M.; Marr, K.A. Geoclimatic influences on invasive aspergillosis after hematopoietic stem cell transplantation. *Clin. Infect. Dis. Off. Publ. Infect. Dis. Soc. Am.* **2010**, *50*, 1588–1597. [CrossRef] [PubMed]

37. Anaissie, E.J.; Stratton, S.L.; Dignani, M.C.; Lee, C.-K.; Mahfouz, T.H.; Rex, J.H.; Summerbell, R.C.; Walsh, T.J. Cleaning patient shower facilities: A novel approach to reducing patient exposure to aerosolized Aspergillus species and other opportunistic molds. *Clin. Infect. Dis.* **2002**, *35*, E86–E88. [CrossRef]

38. Anaissie, E.J.; Stratton, S.L.; Dignani, M.C.; Summerbell, R.C.; Rex, J.H.; Monson, T.P.; Spencer, T.; Kasai, M.; Francesconi, A.; Walsh, T.J. Pathogenic Aspergillus species recovered from a hospital water system: A 3-year prospective study. *Clin. Infect. Dis.* **2002**, *34*, 780–789. [CrossRef]

39. Warris, A.; Klaassen, C.H.W.; Meis, J.F.G.M.; De Ruiter, M.T.; De Valk, H.A.; Abrahamsen, T.G.; Gaustad, P.; Verweij, P.E. Molecular epidemiology of *Aspergillus fumigatus* isolates recovered from water, air, and patients shows two clusters of genetically distinct strains. *J. Clin. Microbiol.* **2003**, *41*, 4101–4106. [CrossRef]

40. Glasmacher, A.; Hahn, C.; Molitor, E.; Marklein, G.; Sauerbruch, T.; Schmidt-Wolf, I.G. Itraconazole trough concentrations in antifungal prophylaxis with six different dosing regimens using hydroxypropyl-beta-cyclodextrin oral solution or coated-pellet capsules. *Mycoses* **1999**, *42*, 591–600. [CrossRef]

41. Simon, A.; Besuden, M.; Vezmar, S.; Hasan, C.; Lampe, D.; Kreutzberg, S.; Glasmacher, A.; Bode, U.; Fleischhack, G. Itraconazole prophylaxis in pediatric cancer patients receiving conventional chemotherapy or autologous stem cell transplants. *Support. Care cancer Off. J. Multinatl. Assoc. Support. Care Cancer* **2007**, *15*, 213–220. [CrossRef] [PubMed]

42. Doring, M.; Blume, O.; Haufe, S.; Hartmann, U.; Kimmig, A.; Schwarze, C.-P.; Lang, P.; Handgretinger, R.; Muller, I. Comparison of itraconazole, voriconazole, and posaconazole as oral antifungal prophylaxis in pediatric patients following allogeneic hematopoietic stem cell transplantation. *Eur. J. Clin. Microbiol. Infect. Dis.* **2014**, *33*, 629–638. [CrossRef] [PubMed]

43. Grigull, L.; Kuehlke, O.; Beilken, A.; Sander, A.; Linderkamp, C.; Schmid, H.; Seidemann, K.; Sykora, K.W.; Schuster, F.R.; Welte, K. Intravenous and oral sequential itraconazole antifungal prophylaxis in paediatric stem cell transplantation recipients: A pilot study for evaluation of safety and efficacy. *Pediatr. Transplant.* **2007**, *11*, 261–266. [CrossRef] [PubMed]

44. Vardakas, K.Z.; Michalopoulos, A.; Falagas, M.E. Fluconazole versus itraconazole for antifungal prophylaxis in neutropenic patients with haematological malignancies: A meta-analysis of randomised-controlled trials. *Br. J. Haematol.* **2005**, *131*, 22–28. [CrossRef] [PubMed]

45. Marks, D.I.; Pagliuca, A.; Kibbler, C.C.; Glasmacher, A.; Heussel, C.-P.; Kantecki, M.; Miller, P.J.S.; Ribaud, P.; Schlamm, H.T.; Solano, C.; et al. Voriconazole versus itraconazole for antifungal prophylaxis following allogeneic haematopoietic stem-cell transplantation. *Br. J. Haematol.* **2011**, *155*, 318–327. [CrossRef] [PubMed]

46. Gastine, S.; Lehrnbecher, T.; Muller, C.; Farowski, F.; Bader, P.; Ullmann-Moskovits, J.; Cornely, O.A.; Groll, A.H.; Hempel, G. Pharmacokinetic Modeling of Voriconazole To Develop an Alternative Dosing Regimen in Children. *Antimicrob. Agents Chemother.* **2018**, *62*. [CrossRef] [PubMed]

47. Park, W.B.; Kim, N.H.; Kim, K.H.; Lee, S.H.; Nam, W.-S.; Yoon, S.H.; Song, K.H.; Choe, P.G.; Kim, N.J.; Jang, I.J.; et al. The effect of therapeutic drug monitoring on safety and efficacy of voriconazole in invasive fungal infections: A randomized controlled trial. *Clin. Infect. Dis.* **2012**, *55*, 1080–1087. [CrossRef]

48. Troke, P.F.; Hockey, H.P.; Hope, W.W. Observational study of the clinical efficacy of voriconazole and its relationship to plasma concentrations in patients. *Antimicrob. Agents Chemother.* **2011**, *55*, 4782–4788. [CrossRef]

49. Pana, Z.D.; Kourti, M.; Vikelouda, K.; Vlahou, A.; Katzilakis, N.; Papageorgiou, M.; Doganis, D.; Petrikkos, L.; Paisiou, A.; Koliouskas, D.; et al. Voriconazole Antifungal Prophylaxis in Children With Malignancies: A Nationwide Study. *J. Pediatr. Hematol. Oncol.* **2018**, *40*, 22–26. [CrossRef]

50. Iosifidis, E.; Papachristou, S.; Roilides, E. Advances in the Treatment of Mycoses in Pediatric Patients. *J. Fungi (Basel, Switzerland)* **2018**, *4*. [CrossRef]

51. Cecinati, V.; Guastadisegni, C.; Russo, F.G.; Brescia, L.P. Antifungal therapy in children: An update. *Eur. J. Pediatr.* **2013**, *172*, 437–446. [CrossRef] [PubMed]

52. Cesaro, S.; Milano, G.M.; Aversa, F. Retrospective survey on the off-label use of posaconazole in pediatric hematology patients. *Eur. J. Clin. Microbiol. Infect. Dis.* **2011**, *30*, 595–596. [CrossRef] [PubMed]

53. Cornely, O.A.; Maertens, J.; Winston, D.J.; Perfect, J.; Ullmann, A.J.; Walsh, T.J.; Helfgott, D.; Holowiecki, J.; Stockelberg, D.; Goh, Y.T.; et al. Posaconazole vs. fluconazole or itraconazole prophylaxis in patients with neutropenia. *N. Engl. J. Med.* **2007**, *356*, 348–359. [CrossRef] [PubMed]

54. Ullmann, A.J.; Lipton, J.H.; Vesole, D.H.; Chandrasekar, P.; Langston, A.; Tarantolo, S.R.; Greinix, H.; Morais de Azevedo, W.; Reddy, V.; Boparai, N.; et al. Posaconazole or fluconazole for prophylaxis in severe graft-versus-host disease. *N. Engl. J. Med.* **2007**, *356*, 335–347. [CrossRef] [PubMed]

55. Doring, M.; Cabanillas Stanchi, K.M.; Queudeville, M.; Feucht, J.; Blaeschke, F.; Schlegel, P.; Feuchtinger, T.; Lang, P.; Muller, I.; Handgretinger, R.; et al. Efficacy, safety and feasibility of antifungal prophylaxis with posaconazole tablet in paediatric patients after haematopoietic stem cell transplantation. *J. Cancer Res. Clin. Oncol.* **2017**, *143*, 1281–1292. [CrossRef]

56. Jang, S.H.; Colangelo, P.M.; Gobburu, J.V.S. Exposure-response of posaconazole used for prophylaxis against invasive fungal infections: Evaluating the need to adjust doses based on drug concentrations in plasma. *Clin. Pharmacol. Ther.* **2010**, *88*, 115–119. [CrossRef]

57. Bochennek, K.; Tramsen, L.; Schedler, N.; Becker, M.; Klingebiel, T.; Groll, A.H.; Lehrnbecher, T. Liposomal amphotericin B twice weekly as antifungal prophylaxis in paediatric haematological malignancy patients. *Clin. Microbiol. Infect.* **2011**, *17*, 1868–1874. [CrossRef]

58. Mehta, P.; Vinks, A.; Filipovich, A.; Vaughn, G.; Fearing, D.; Sper, C.; Davies, S. High-dose weekly AmBisome antifungal prophylaxis in pediatric patients undergoing hematopoietic stem cell transplantation: A pharmacokinetic study. *Biol. Blood Marrow Transplant.* **2006**, *12*, 235–240. [CrossRef]

59. Kolve, H.; Ahlke, E.; Fegeler, W.; Ritter, J.; Jurgens, H.; Groll, A.H. Safety, tolerance and outcome of treatment with liposomal amphotericin B in paediatric patients with cancer or undergoing haematopoietic stem cell transplantation. *J. Antimicrob. Chemother.* **2009**, *64*, 383–387. [CrossRef]

60. van Burik, J.-A.H.; Ratanatharathorn, V.; Stepan, D.E.; Miller, C.B.; Lipton, J.H.; Vesole, D.H.; Bunin, N.; Wall, D.A.; Hiemenz, J.W.; Satoi, Y.; et al. Micafungin versus fluconazole for prophylaxis against invasive fungal infections during neutropenia in patients undergoing hematopoietic stem cell transplantation. *Clin. Infect. Dis.* **2004**, *39*, 1407–1416. [CrossRef]

61. Yoshikawa, K.; Nakazawa, Y.; Katsuyama, Y.; Hirabayashi, K.; Saito, S.; Shigemura, T.; Tanaka, M.; Yanagisawa, R.; Sakashita, K.; Koike, K. Safety, tolerability, and feasibility of antifungal prophylaxis with micafungin at 2 mg/kg daily in pediatric patients undergoing allogeneic hematopoietic stem cell transplantation. *Infection* **2014**, *42*, 639–647. [CrossRef] [PubMed]

62. Mattiuzzi, G.N.; Alvarado, G.; Giles, F.J.; Ostrosky-Zeichner, L.; Cortes, J.; O'brien, S.; Verstovsek, S.; Faderl, S.; Zhou, X.; Raad, I.I.; et al. Open-label, randomized comparison of itraconazole versus caspofungin for prophylaxis in patients with hematologic malignancies. *Antimicrob. Agents Chemother.* **2006**, *50*, 143–147. [CrossRef] [PubMed]

63. Doring, M.; Hartmann, U.; Erbacher, A.; Lang, P.; Handgretinger, R.; Muller, I. Caspofungin as antifungal prophylaxis in pediatric patients undergoing allogeneic hematopoietic stem cell transplantation: A retrospective analysis. *BMC Infect. Dis.* **2012**, *12*, 151. [CrossRef] [PubMed]

64. Maximova, N.; Schillani, G.; Simeone, R.; Maestro, A.; Zanon, D. Comparison of Efficacy and Safety of Caspofungin Versus Micafungin in Pediatric Allogeneic Stem Cell Transplant Recipients: A Retrospective Analysis. *Adv. Ther.* **2017**, *34*, 1184–1199. [CrossRef] [PubMed]

65. Kish, M.A. Guide to development of practice guidelines. *Clin. Infect. Dis.* **2001**, *32*, 851–854. [CrossRef]

66. Cordonnier, C.; Rovira, M.; Maertens, J.; Olavarria, E.; Faucher, C.; Bilger, K.; Pigneux, A.; Cornely, O.A.; Ullmann, A.J.; Bofarull, R.M.; et al. Voriconazole for secondary prophylaxis of invasive fungal infections in allogeneic stem cell transplant recipients: Results of the VOSIFI study. *Haematologica* **2010**, *95*, 1762–1768. [CrossRef]

67. Allinson, K.; Kolve, H.; Gumbinger, H.G.; Vormoor, H.J.; Ehlert, K.; Groll, A.H. Secondary antifungal prophylaxis in paediatric allogeneic haematopoietic stem cell recipients. *J. Antimicrob. Chemother.* **2008**, *61*, 734–742. [CrossRef]

68. Tomblyn, M.; Chiller, T.; Einsele, H.; Gress, R.; Sepkowitz, K.; Storek, J.; Wingard, J.R.; Young, J.-A.H.; Boeckh, M.J. Guidelines for preventing infectious complications among hematopoietic cell transplantation recipients: A global perspective. *Biol. Blood Marrow Transplant.* **2009**, *15*, 1143–1238. [CrossRef]

69. Partridge-Hinckley, K.; Liddell, G.M.; Almyroudis, N.G.; Segal, B.H. Infection control measures to prevent invasive mould diseases in hematopoietic stem cell transplant recipients. *Mycopathologia* **2009**, *168*, 329–337. [CrossRef]

70. Caselli, D.; Cesaro, S.; Ziino, O.; Ragusa, P.; Pontillo, A.; Pegoraro, A.; Santoro, N.; Zanazzo, G.; Poggi, V.; Giacchino, M.; et al. A prospective, randomized study of empirical antifungal therapy for the treatment of chemotherapy-induced febrile neutropenia in children. *Br. J. Haematol.* **2012**, *158*, 249–255. [CrossRef]

71. Prentice, H.G.; Hann, I.M.; Herbrecht, R.; Aoun, M.; Kvaloy, S.; Catovsky, D.; Pinkerton, C.R.; Schey, S.A.; Jacobs, F.; Oakhill, A.; et al. A randomized comparison of liposomal versus conventional amphotericin B for the treatment of pyrexia of unknown origin in neutropenic patients. *Br. J. Haematol.* **1997**, *98*, 711–718. [CrossRef] [PubMed]

72. Sandler, E.S.; Mustafa, M.M.; Tkaczewski, I.; Graham, M.L.; Morrison, V.A.; Green, M.; Trigg, M.; Abboud, M.; Aquino, V.M.; Gurwith, M.; et al. Use of amphotericin B colloidal dispersion in children. *J. Pediatr. Hematol. Oncol.* **2000**, *22*, 242–246. [CrossRef] [PubMed]

73. Maertens, J.A.; Madero, L.; Reilly, A.F.; Lehrnbecher, T.; Groll, A.H.; Jafri, H.S.; Green, M.; Nania, J.J.; Bourque, M.R.; Wise, B.A.; et al. A randomized, double-blind, multicenter study of caspofungin versus liposomal amphotericin B for empiric antifungal therapy in pediatric patients with persistent fever and neutropenia. *Pediatr. Infect. Dis. J.* **2010**, *29*, 415–420. [CrossRef] [PubMed]

74. Walsh, T.J.; Adamson, P.C.; Seibel, N.L.; Flynn, P.M.; Neely, M.N.; Schwartz, C.; Shad, A.; Kaplan, S.L.; Roden, M.M.; Stone, J.A.; et al. Pharmacokinetics, safety, and tolerability of caspofungin in children and adolescents. *Antimicrob. Agents Chemother.* **2005**, *49*, 4536–4545. [CrossRef] [PubMed]

75. Walsh, T.J.; Pappas, P.; Winston, D.J.; Lazarus, H.M.; Petersen, F.; Raffalli, J.; Yanovich, S.; Stiff, P.; Greenberg, R.; Donowitz, G.; et al. Voriconazole compared with liposomal amphotericin B for empirical antifungal therapy in patients with neutropenia and persistent fever. *N. Engl. J. Med.* **2002**, *346*, 225–234. [CrossRef] [PubMed]

76. Caselli, D.; Paolicchi, O. Empiric antibiotic therapy in a child with cancer and suspected septicemia. *Pediatr. Rep.* **2012**, *4*, e2. [CrossRef]

77. De Pauw, B.; Walsh, T.J.; Donnelly, J.P.; Stevens, D.A.; Edwards, J.E.; Calandra, T.; Pappas, P.G.; Maertens, J.; Lortholary, O.; Kauffman, C.A.; et al. Revised definitions of invasive fungal disease from the European Organization for Research and Treatment of Cancer/Invasive Fungal Infections Cooperative Group and the National Institute of Allergy and Infectious Diseases Mycoses Study Group (EORTC/MSG) C. *Clin. Infect. Dis.* **2008**, *46*, 1813–1821. [CrossRef]

78. Huppler, A.R.; Fisher, B.T.; Lehrnbecher, T.; Walsh, T.J.; Steinbach, W.J. Role of Molecular Biomarkers in the Diagnosis of Invasive Fungal Diseases in Children. *J. Pediatric Infect. Dis. Soc.* **2017**, *6*, S32–S44. [CrossRef]

79. Lehrnbecher, T.; Robinson, P.D.; Fisher, B.T.; Castagnola, E.; Groll, A.H.; Steinbach, W.J.; Zaoutis, T.E.; Negeri, Z.F.; Beyene, J.; Phillips, B.; et al. Galactomannan, beta-D-Glucan, and Polymerase Chain Reaction-Based Assays for the Diagnosis of Invasive Fungal Disease in Pediatric Cancer and Hematopoietic Stem Cell Transplantation: A Systematic Review and Meta-Analysis. *Clin. Infect. Dis.* **2016**, *63*, 1340–1348. [CrossRef]

80. Santolaya, M.E.; Alvarez, A.M.; Acuna, M.; Aviles, C.L.; Salgado, C.; Tordecilla, J.; Varas, M.; Venegas, M.; Villarroel, M.; Zubieta, M.; et al. Efficacy of pre-emptive versus empirical antifungal therapy in children with cancer and high-risk febrile neutropenia: A randomized clinical trial. *J. Antimicrob. Chemother.* **2018**, *73*, 2860–2866. [CrossRef]

81. Felton, T.; Troke, P.F.; Hope, W.W. Tissue penetration of antifungal agents. *Clin. Microbiol. Rev.* **2014**, *27*, 68–88. [CrossRef] [PubMed]

82. Marr, K.A.; Leisenring, W.; Crippa, F.; Slattery, J.T.; Corey, L.; Boeckh, M.; McDonald, G.B. Cyclophosphamide metabolism is affected by azole antifungals. *Blood* **2004**, *103*, 1557–1559. [CrossRef] [PubMed]

83. Herbrecht, R.; Denning, D.W.; Patterson, T.F.; Bennett, J.E.; Greene, R.E.; Oestmann, J.-W.; Kern, W.V.; Marr, K.A.; Ribaud, P.; Lortholary, O.; et al. Voriconazole versus amphotericin B for primary therapy of invasive aspergillosis. *N. Engl. J. Med.* **2002**, *347*, 408–415. [CrossRef] [PubMed]

84. Walsh, T.J.; Anaissie, E.J.; Denning, D.W.; Herbrecht, R.; Kontoyiannis, D.P.; Marr, K.A.; Morrison, V.A.; Segal, B.H.; Steinbach, W.J.; Stevens, D.A.; et al. Treatment of aspergillosis: Clinical practice guidelines of the Infectious Diseases Society of America. *Clin. Infect. Dis.* **2008**, *46*, 327–360. [CrossRef] [PubMed]

85. Wattier, R.L.; Ramirez-Avila, L. Pediatric Invasive Aspergillosis. *J. fungi (Basel, Switzerland)* **2016**, *2*. [CrossRef] [PubMed]

86. Walsh, T.J.; Karlsson, M.O.; Driscoll, T.; Arguedas, A.G.; Adamson, P.; Saez-Llorens, X.; Vora, A.J.; Arrieta, A.C.; Blumer, J.; Lutsar, I.; et al. Pharmacokinetics and safety of intravenous voriconazole in children after single- or multiple-dose administration. *Antimicrob. Agents Chemother.* **2004**, *48*, 2166–2172. [CrossRef] [PubMed]

87. Friberg, L.E.; Ravva, P.; Karlsson, M.O.; Liu, P. Integrated population pharmacokinetic analysis of voriconazole in children, adolescents, and adults. *Antimicrob. Agents Chemother.* **2012**, *56*, 3032–3042. [CrossRef]

88. Luong, M.-L.; Al-Dabbagh, M.; Groll, A.H.; Racil, Z.; Nannya, Y.; Mitsani, D.; Husain, S. Utility of voriconazole therapeutic drug monitoring: A meta-analysis. *J. Antimicrob. Chemother.* **2016**, *71*, 1786–1799. [CrossRef]

89. Cornely, O.A.; Maertens, J.; Bresnik, M.; Ebrahimi, R.; Ullmann, A.J.; Bouza, E.; Heussel, C.P.; Lortholary, O.; Rieger, C.; Boehme, A.; et al. Liposomal amphotericin B as initial therapy for invasive mold infection: A randomized trial comparing a high-loading dose regimen with standard dosing (AmBiLoad trial). *Clin. Infect. Dis.* **2007**, *44*, 1289–1297. [CrossRef]

90. Girois, S.B.; Chapuis, F.; Decullier, E.; Revol, B.G.P. Adverse effects of antifungal therapies in invasive fungal infections: Review and meta-analysis. *Eur. J. Clin. Microbiol. Infect. Dis.* **2006**, *25*, 138–149. [CrossRef]

91. Pappas, P.G.; Rex, J.H.; Sobel, J.D.; Filler, S.G.; Dismukes, W.E.; Walsh, T.J.; Edwards, J.E. Guidelines for treatment of candidiasis. *Clin. Infect. Dis.* **2004**, *38*, 161–189. [CrossRef] [PubMed]

92. Filioti, I.; Iosifidis, E.; Roilides, E. Therapeutic strategies for invasive fungal infections in neonatal and pediatric patients. *Expert Opin. Pharmacother.* **2008**, *9*, 3179–3196. [CrossRef] [PubMed]

93. Lestner, J.M.; Versporten, A.; Doerholt, K.; Warris, A.; Roilides, E.; Sharland, M.; Bielicki, J.; Goossens, H. Systemic antifungal prescribing in neonates and children: Outcomes from the Antibiotic Resistance and Prescribing in European Children (ARPEC) Study. *Antimicrob. Agents Chemother.* **2015**, *59*, 782–789. [CrossRef] [PubMed]

94. Maertens, J.; Raad, I.; Petrikkos, G.; Boogaerts, M.; Selleslag, D.; Petersen, F.B.; Sable, C.A.; Kartsonis, N.A.; Ngai, A.; Taylor, A.; et al. Efficacy and safety of caspofungin for treatment of invasive aspergillosis in patients refractory to or intolerant of conventional antifungal therapy. *Clin. Infect. Dis.* **2004**, *39*, 1563–1571. [CrossRef] [PubMed]

95. Zaoutis, T.E.; Jafri, H.S.; Huang, L.-M.; Locatelli, F.; Barzilai, A.; Ebell, W.; Steinbach, W.J.; Bradley, J.; Lieberman, J.M.; Hsiao, C.-C.; et al. A prospective, multicenter study of caspofungin for the treatment of documented Candida or Aspergillus infections in pediatric patients. *Pediatrics* **2009**, *123*, 877–884. [CrossRef] [PubMed]

96. Rosanova, M.T.; Bes, D.; Serrano Aguilar, P.; Cuellar Pompa, L.; Sberna, N.; Lede, R. Efficacy and safety of caspofungin in children: Systematic review and meta-analysis. *Arch. Argent. Pediatr.* **2016**, *114*, 305–312. [CrossRef] [PubMed]

97. Seibel, N.L.; Schwartz, C.; Arrieta, A.; Flynn, P.; Shad, A.; Albano, E.; Keirns, J.; Lau, W.M.; Facklam, D.P.; Buell, D.N.; et al. Safety, tolerability, and pharmacokinetics of Micafungin (FK463) in febrile neutropenic pediatric patients. *Antimicrob. Agents Chemother.* **2005**, *49*, 3317–3324. [CrossRef]

98. Lee, C.-H.; Lin, J.-C.; Ho, C.-L.; Sun, M.; Yen, W.-T.; Lin, C. Efficacy and safety of micafungin versus extensive azoles in the prevention and treatment of invasive fungal infections for neutropenia patients with hematological malignancies: A meta-analysis of randomized controlled trials. *PLoS ONE* **2017**, *12*, e0180050. [CrossRef]

99. de Repentigny, L.; Ratelle, J.; Leclerc, J.M.; Cornu, G.; Sokal, E.M.; Jacqmin, P.; De Beule, K. Repeated-dose pharmacokinetics of an oral solution of itraconazole in infants and children. *Antimicrob. Agents Chemother.* **1998**, *42*, 404–408. [PubMed]

100. Walsh, T.J.; Raad, I.; Patterson, T.F.; Chandrasekar, P.; Donowitz, G.R.; Graybill, R.; Greene, R.E.; Hachem, R.; Hadley, S.; Herbrecht, R.; et al. Treatment of invasive aspergillosis with posaconazole in patients who are refractory to or intolerant of conventional therapy: An externally controlled trial. *Clin. Infect. Dis.* **2007**, *44*, 2–12. [CrossRef] [PubMed]

101. Benjamin, D.K.J.; Driscoll, T.; Seibel, N.L.; Gonzalez, C.E.; Roden, M.M.; Kilaru, R.; Clark, K.; Dowell, J.A.; Schranz, J.; Walsh, T.J. Safety and pharmacokinetics of intravenous anidulafungin in children with neutropenia at high risk for invasive fungal infections. *Antimicrob. Agents Chemother.* **2006**, *50*, 632–638. [CrossRef] [PubMed]

102. Rosanova, M.T.; Sarkis, C.; Escarra, F.; Epelbaum, C.; Sberna, N.; Carnovale, S.; Figueroa, C.; Bologna, R.; Lede, R. Anidulafungin in children: Experience in a tertiary care children's hospital in Argentina. *Arch. Argent. Pediatr.* **2017**, *115*, 374–376. [CrossRef] [PubMed]

103. Roilides, E.; Carlesse, F.; Leister-Tebbe, H.; Conte, U.; Yan, J.L.; Liu, P.; Tawadrous, M.; Aram, J.A.; Queiroz-Telles, F. A Prospective, Open-label Study to Assess the Safety, Tolerability, and Efficacy of Anidulafungin in the Treatment of Invasive Candidiasis in Children 2 to <18 Years of Age. *Pediatr. Infect. Dis. J.* **2018**. [CrossRef]

104. Miceli, M.H.; Kauffman, C.A. Isavuconazole: A New Broad-Spectrum Triazole Antifungal Agent. *Clin. Infect. Dis.* **2015**, *61*, 1558–1565. [CrossRef] [PubMed]

105. Maertens, J.A.; Raad, I.I.; Marr, K.A.; Patterson, T.F.; Kontoyiannis, D.P.; Cornely, O.A.; Bow, E.J.; Rahav, G.; Neofytos, D.; Aoun, M.; et al. Isavuconazole versus voriconazole for primary treatment of invasive mould disease caused by Aspergillus and other filamentous fungi (SECURE): A phase 3, randomised-controlled, non-inferiority trial. *Lancet (London, England)* **2016**, *387*, 760–769. [CrossRef]

106. Barg, A.A.; Malkiel, S.; Bartuv, M.; Greenberg, G.; Toren, A.; Keller, N. Successful treatment of invasive mucormycosis with isavuconazole in pediatric patients. *Pediatr. Blood Cancer* **2018**, *65*, e27281. [CrossRef] [PubMed]

107. Wattier, R.L.; Dvorak, C.C.; Hoffman, J.A.; Brozovich, A.A.; Bin-Hussain, I.; Groll, A.H.; Castagnola, E.; Knapp, K.M.; Zaoutis, T.E.; Gustafsson, B.; et al. A Prospective, International Cohort Study of Invasive Mold Infections in Children. *J. Pediatric Infect. Dis. Soc.* **2015**, *4*, 313–322. [CrossRef] [PubMed]

108. Cesaro, S.; Giacchino, M.; Locatelli, F.; Spiller, M.; Buldini, B.; Castellini, C.; Caselli, D.; Giraldi, E.; Tucci, F.; Tridello, G.; et al. Safety and efficacy of a caspofungin-based combination therapy for treatment of proven or probable aspergillosis in pediatric hematological patients. *BMC Infect. Dis.* **2007**, *7*, 28. [CrossRef] [PubMed]

109. Marr, K.A.; Schlamm, H.T.; Herbrecht, R.; Rottinghaus, S.T.; Bow, E.J.; Cornely, O.A.; Heinz, W.J.; Jagannatha, S.; Koh, L.P.; Kontoyiannis, D.P.; et al. Combination antifungal therapy for invasive aspergillosis: A randomized trial. *Ann. Intern. Med.* **2015**, *162*, 81–89. [CrossRef] [PubMed]

110. Van Der Linden, J.W.M.; Warris, A.; Verweij, P.E. Aspergillus species intrinsically resistant to antifungal agents. *Med. Mycol.* **2011**, *49*, S82–S89. [CrossRef] [PubMed]

111. Anderson, J.B. Evolution of antifungal-drug resistance: Mechanisms and pathogen fitness. *Nat. Rev. Microbiol.* **2005**, *3*, 547–556. [CrossRef] [PubMed]

112. Verweij, P.E.; Chowdhary, A.; Melchers, W.J.G.; Meis, J.F. Azole Resistance in *Aspergillus fumigatus*: Can We Retain the Clinical Use of Mold-Active Antifungal Azoles? *Clin. Infect. Dis.* **2016**, *62*, 362–368. [CrossRef] [PubMed]

113. Verweij, P.E.; Ananda-Rajah, M.; Andes, D.; Arendrup, M.C.; Bruggemann, R.J.; Chowdhary, A.; Cornely, O.A.; Denning, D.W.; Groll, A.H.; Izumikawa, K.; et al. International expert opinion on the management of infection caused by azole-resistant *Aspergillus fumigatus*. *Drug Resist. Updat.* **2015**, *21–22*, 30–40. [CrossRef] [PubMed]

114. Thors, V.S.; Bierings, M.B.; Melchers, W.J.G.; Verweij, P.E.; Wolfs, T.F.W. Pulmonary aspergillosis caused by a pan-azole-resistant *Aspergillus fumigatus* in a 10-year-old boy. *Pediatr. Infect. Dis. J.* **2011**, *30*, 268–270. [CrossRef] [PubMed]

115. Estcourt, L.J.; Stanworth, S.; Doree, C.; Blanco, P.; Hopewell, S.; Trivella, M.; Massey, E. Granulocyte transfusions for preventing infections in people with neutropenia or neutrophil dysfunction. *Cochrane database Syst. Rev.* **2015**, CD005341. [CrossRef] [PubMed]

116. Perruccio, K.; Tosti, A.; Burchielli, E.; Topini, F.; Ruggeri, L.; Carotti, A.; Capanni, M.; Urbani, E.; Mancusi, A.; Aversa, F.; et al. Transferring functional immune responses to pathogens after haploidentical hematopoietic transplantation. *Blood* **2005**, *106*, 4397–4406. [CrossRef]

117. Papadopoulou, A.; Kaloyannidis, P.; Yannaki, E.; Cruz, C.R. Adoptive transfer of Aspergillus-specific T cells as a novel anti-fungal therapy for hematopoietic stem cell transplant recipients: Progress and challenges. *Crit. Rev. Oncol. Hematol.* **2016**, *98*, 62–72. [CrossRef]

118. Lin, S.J.; Schranz, J.; Teutsch, S.M. Aspergillosis case-fatality rate: Systematic review of the literature. *Clin. Infect. Dis.* **2001**, *32*, 358–366. [CrossRef]

119. Greene, R.E.; Schlamm, H.T.; Oestmann, J.-W.; Stark, P.; Durand, C.; Lortholary, O.; Wingard, J.R.; Herbrecht, R.; Ribaud, P.; Patterson, T.F.; et al. Imaging findings in acute invasive pulmonary aspergillosis: Clinical significance of the halo sign. *Clin. Infect. Dis.* **2007**, *44*, 373–379. [CrossRef]

120. Dotis, J.; Iosifidis, E.; Roilides, E. Central nervous system aspergillosis in children: A systematic review of reported cases. *Int. J. Infect. Dis.* **2007**, *11*, 381–393. [CrossRef]

121. Palmisani, E.; Barco, S.; Cangemi, G.; Moroni, C.; Dufour, C.; Castagnola, E. Need of voriconazole high dosages, with documented cerebrospinal fluid penetration, for treatment of cerebral aspergillosis in a 6-month-old leukaemic girl. *J. Chemother.* **2017**, *29*, 42–44. [CrossRef] [PubMed]

122. McCarthy, M.; Rosengart, A.; Schuetz, A.N.; Kontoyiannis, D.P.; Walsh, T.J. Mold infections of the central nervous system. *N. Engl. J. Med.* **2014**, *371*, 150–160. [CrossRef] [PubMed]

123. Starke, J.R.; Mason, E.O.J.; Kramer, W.G.; Kaplan, S.L. Pharmacokinetics of amphotericin B in infants and children. *J. Infect. Dis.* **1987**, *155*, 766–774. [CrossRef] [PubMed]

124. Schwartz, S.; Ruhnke, M.; Ribaud, P.; Corey, L.; Driscoll, T.; Cornely, O.A.; Schuler, U.; Lutsar, I.; Troke, P.; Thiel, E. Improved outcome in central nervous system aspergillosis, using voriconazole treatment. *Blood* **2005**, *106*, 2641–2645. [CrossRef] [PubMed]

125. Groll, A.H.; Giri, N.; Petraitis, V.; Petraitiene, R.; Candelario, M.; Bacher, J.S.; Piscitelli, S.C.; Walsh, T.J. Comparative efficacy and distribution of lipid formulations of amphotericin B in experimental *Candida albicans* infection of the central nervous system. *J. Infect. Dis.* **2000**, *182*, 274–282. [CrossRef] [PubMed]

126. Bartelink, I.H.; Wolfs, T.; Jonker, M.; de Waal, M.; Egberts, T.C.G.; Ververs, T.T.; Boelens, J.J.; Bierings, M. Highly variable plasma concentrations of voriconazole in pediatric hematopoietic stem cell transplantation patients. *Antimicrob. Agents Chemother.* **2013**, *57*, 235–240. [CrossRef] [PubMed]
127. Stevens, D.A.; Shatsky, S.A. Intrathecal amphotericin in the management of coccidioidal meningitis. *Semin. Respir. Infect.* **2001**, *16*, 263–269. [CrossRef]

Review

Immunomodulation as a Therapy for *Aspergillus* Infection: Current Status and Future Perspectives

Chris D. Lauruschkat, Hermann Einsele and Juergen Loeffler *

Department of Internal Medicine II, University Hospital Wuerzburg, WÜ4i, Building C11, 97080 Wuerzburg, Germany; Lauruschka_c@ukw.de (C.D.L.); Einsele_h@ukw.de (H.E.)
* Correspondence: Loeffler_j@ukw.de; Tel.: +49-931-2013-6412; Fax: +49-931-2013-6409

Received: 15 November 2018; Accepted: 10 December 2018; Published: 14 December 2018

Abstract: Invasive aspergillosis (IA) is the most serious life-threatening infectious complication of intensive remission induction chemotherapy and allogeneic stem cell transplantation in patients with a variety of hematological malignancies. *Aspergillus fumigatus* is the most commonly isolated species from cases of IA. Despite the various improvements that have been made with preventative strategies and the development of antifungal drugs, there is an urgent need for new therapeutic approaches that focus on strategies to boost the host's immune response, since immunological recovery is recognized as being the major determinant of the outcome of IA. Here, we aim to summarize current knowledge about a broad variety of immunotherapeutic approaches against IA, including therapies based on the transfer of distinct immune cell populations, and the administration of cytokines and antibodies.

Keywords: Immunotherapy; invasive aspergillosis; *Aspergillus fumigatus*; fungal infections; innate immunity; adaptive immunity; cell therapy; cytokine therapy

1. Introduction

Within the last decade, the filamentous fungus *Aspergillus fumigatus* (*A. fumigatus*) has underlined its role as one of the most clinically relevant fungal pathogens. Conidia of this saprobic fungus can be isolated ubiquitously. Because of this high abundance, hundreds of spores of *Aspergillus* are inhaled daily by each individual [1]. Mucociliary clearance and phagocytic cells in the lung prevent disease in immunocompetent individuals. This includes alveolar macrophages, the major resident cells in the lung alveoli, which most efficiently engulf conidia in the lung [2].

The most severe disease caused by *Aspergillus* is invasive aspergillosis (IA). Major risk factors include immunosuppression, neutropenia, lymphopenia, and depletion of T cells [3]. Thus, IA occurs almost exclusively in immunocompromised patients. The incidence in allogeneic stem cell transplantation (allo-SCT) patients ranges from 4%–10%. Although many *Aspergilli* cause IA, *A. fumigatus* is responsible for more than 90% of all systemic *Aspergillus* infections [4].

Innate immunity is of major importance for the defense against *A. fumigatus*. In contrast to most bacterial pathogens, *A. fumigatus* undergoes major morphological changes during the early phase of infection. In the alveoli, inert spores swell, germinate, and grow into lung tissue, becoming subsequently angioinvasive and lastly undergoing hematogenous dissemination [5]. Cells of the innate immunity recognize the different fungal morphologies by distinct pattern recognition receptors (e.g., TLR2, TLR4, and dectin-1), which induce cell-specific and general defense mechanisms [6]. Upon stimulation with *A. fumigatus* in vitro, polymorphonuclear neutrophils (PMNs) release reactive oxygen intermediates and form neutrophil extracellular traps. Dendritic cells (DCs) release inflammatory cytokines (e.g., TNF-α and IL-1) and chemokines (e.g., IL-8 and CXCL10), which attract and activate other innate immune cell populations and build a bridge between the innate and the adaptive immunity by processing fungal antigens and presenting them to T cells. Natural killer (NK) cells degranulate and secrete cytotoxic proteins (perforin, granzymes) in response to *A. fumigatus*, causing fungal damage, and produce Th1 cytokines and chemokines, again attracting and activating other innate immune cell populations [7–9].

Unfortunately, there is still a lack of reliable diagnostic and therapeutic tools, resulting in high mortality rates of up to 90%, depending on the patient cohort and the localization of the fungus [10]. Therapy of *Aspergillus* infection remains limited to only a handful of antifungal agents. Voriconazole is the drug of choice for primary therapy of IA, with isavuconazole and the liposomal formulation of amphotericin B serving as alternatives. Echinocandins (e.g., anidulafungin) and other mold-active azoles (e.g., itraconazole and posaconazole) remain for salvage therapy [11]. Recently, triazole-resistant *A. fumigatus* strains have increasingly been isolated from patients. These strains emerge most likely due to the extensive use of azole fungicidals in agriculture and massively hinder antifungal treatment [12].

The following pages describe how different options of immune modulation have or will become alternatives to treat *Aspergillus* infection. In the case of immunocompromised patients, this usually involves therapeutic enhancement of immunity, including cell therapy approaches such as the transfusion of cells of the innate (granulocytes, dendritic cells, natural killer cells) and adaptive immune systems (T cells) as well as the administration of different cytokines, chemokines, and antibodies (Figure 1).

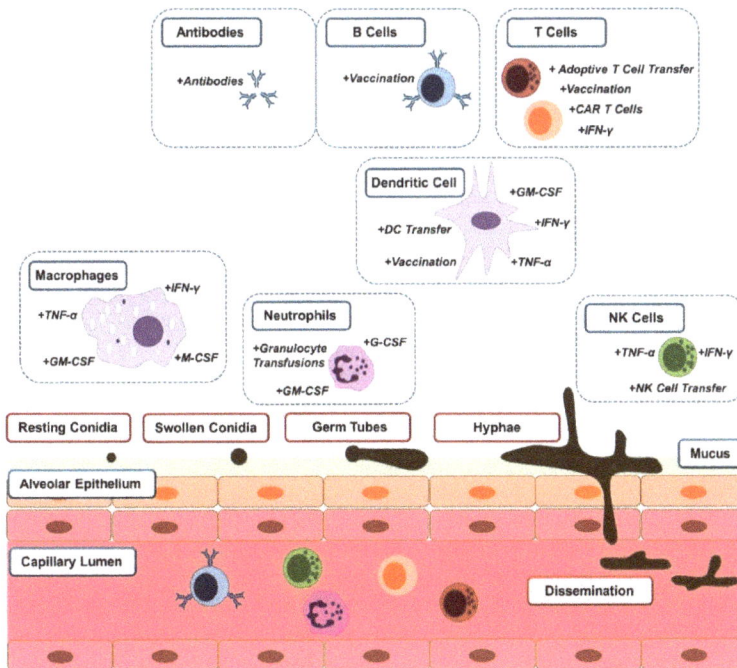

Figure 1. Cells of the innate and adaptive immune systems interact with different morphotypes of *Aspergillus fumigatus*. Macrophages clear resting and swollen conidia. Neutrophils attack all morphotypes of the fungus, while natural killer (NK) cells react to germ tubes and hyphae. Dendritic cells bridge the innate immune system to the adaptive immune system, which orchestrates fungal clearance of all morphotypes. Immunotherapeutic treatment options supporting these cells of the innate and adaptive immune systems are indicated next to the cell type they affect. Abbreviations: dendritic cell (DC); chimeric antigen receptor (CAR); granulocyte-macrophage colony stimulating factor (GM-CSF); granulocyte CSF (G-CSF); macrophage CSF (M-CSF); interferon (IFN) γ; tumor necrosis factor (TNF) α.

2. Cell Therapy

2.1. Granulocyte Transfusion

PMNs are able to engulf fungus, release antimicrobial peptides, and form extracellular traps [13,14]. After allo-SCT, especially during neutropenia, the ability of the immune system to effectively clear fungus is severely limited. Numerous studies have evaluated the transfusion of high numbers of neutrophils to patients during neutropenia in the past decades. Allogeneic granulocyte transfusions (GTs) dramatically increase neutrophil counts, which is speculated to reverse the increased susceptibility to infections in allo-SCT patients [15,16]. GTs have shown low toxicity in allo-SCT patients and are considered to be safe [16,17].

In a phase I/II clinical trial, allogeneic neutrophil transfusion in combination with dexamethasone and granulocyte colony stimulating factors (G-CSF) increased neutrophil counts and response; however, none of the five patients suffering from aspergillosis survived [16]. Mousset et al. reported that in hematological patients, who received either prophylactic or interventional GTs, *Aspergillus* infection could be controlled in 17 out of 22 cases. This result, however, was limited by the inclusion of possible IA cases into the study population and the trial's nonrandomized nature [18]. In contrast, in a randomized phase III clinical trial in which GTs were given to neutropenic patients, no difference in 100-day survival of fungal infections was found. The authors of the study did not discriminate between different fungal infections; nonetheless, 49 of the 55 cases were *Aspergillus* infections [17]. In a randomized multicenter controlled study, 58 neutropenic subjects were treated with GTs plus G-CSF and dexamethasone in addition to standard microbial treatment. This arm of the trial was compared to 56 neutropenic patients on standard microbial treatment alone. No difference in infections between the groups was found. Both the control and treatment groups included only three proven aspergillosis cases [19], which made it difficult to draw conclusions for *Aspergillus* infections. Moreover, to improve the limited life span of transfused granulocytes, granulocyte progenitors for transfusion could be used. Bitmansour et al. have shown protection of neutropenic mice from *A. fumigatus* infection by granulocyte/monocyte progenitors [20,21].

In summary, although GTs have a lot of potential and new trials should be performed to further clarify the effect of GTs, no recommendation of treating allo-SCT patients with GTs is currently given [22]. The limited success of GT transfusions up to date might be a result of the transfusion of too low granulocyte numbers in some patients [19] and points to the necessity of overcoming multifactorial dysfunctions of the immune system after allo-SCT in order to prevent and clear IA.

2.2. Dendritic Cells (DCs)

DCs connect innate and adaptive immunity. They recognize fungus by pattern recognition receptors and process fungal antigens. After activation, they secrete cytokines and chemokines and migrate to the lymph nodes. Here they present these antigens to specific T cells, which in turn are activated and primed. Ex vivo DCs stimulated with *Aspergillus* antigens induce protective immune responses to the fungus after transfusion to the patient due to activation of *Aspergillus*-specific T cells and secretion of cytokines and chemokines, which support the clearance of the fungus by both the innate and adaptive immune systems [23–25].

DC stimulation with unmethylated CpG oligodeoxynucleotides as an adjuvant in combination with one of the major *A. fumigatus* allergens, Asp f 16, induced a protective Th1 response in a hematopoietic stem cell transplantation (HSCT) mouse model of IA [26]. The same protective Th1 response was found when DCs were stimulated by *A. fumigatus* conidia and transfected with IL-12 in a similar murine model [27]. In addition, DCs that had been transduced with IL-12 and stimulated by *A. fumigatus* were administered to neutropenic mice in a model for IA. The treatment led to less mortality and decreased fungal burden due to a strong Th1 response [28]. Asp f 16-stimulated DCs were more effective in generating a cytotoxic T lymphocyte response against *Aspergillus* when antigen presentation of DCs was succeeded by a second antigen presentation using Epstein–Barr virus-transformed B lymphoblastoid cell lines. This method was described as more effective in generating Asp f 16-specific cytotoxic T lymphocytes (CTLs), and therefore would require less initial blood volume of the patient compared to DC stimulation alone [29].

In conclusion, an ex vivo stimulation of DCs and subsequent administration to the patient is cost inefficient, difficult to scale, and labor intensive. It shows, however, the therapeutic potential of fungal vaccination [30].

2.3. Natural Killer Cell Therapy

NK cells participate in the control of numerous pathogens, including viruses and fungi [31]. They have been shown to interact with *A. fumigatus*, *Cryptococcus neoformans*, *Candida albicans*, and Mucorales [32].

NK cells directly interact with *A. fumigatus* through the neural cell adhesion molecule (NCAM-1, CD56), and this interaction leads to the secretion of CC chemokine ligands CCL3, 4, and 5 [33]. After contact with *A. fumigatus*, NK cells become activated and release soluble factors such as perforin and granzyme, which mediate antifungal activity [34].

Higher reactive oxygen species (ROS) production and NK cell counts were associated with better control of IA in allo-SCT patients [35]. Referring to this study, Fernández-Ruiz et al. investigated NK cell counts of solid organ transplant recipients and correlated them to fungal infections. During the median follow-up period of 504.5 days, 10 out of 396 patients suffered from invasive fungal infection (IFI), and 4/10 IFI cases were classified as IA. Higher NK cell counts one month after transplantation decreased the incidence of fungal infections. In vivo, NK cells were the most significant contributor to IFN-γ secretion during the early stages of *Aspergillus* infection in the lungs of neutropenic mice. NK cell depletion resulted in higher mortality. In turn, fungal clearance was increased by transferring activated NK cells of wild-type mice to IFN-γ-deficient or wild-type neutropenic mice. The transfer of NK cells of IFN-γ-deficient mice into the same murine models, however, showed no effect [36]. Moreover, *Aspergillus niger* growth was partly inhibited due to increased NK cell activity in a murine model [37].

These results suggest that allogeneic NK cell transfer might be beneficial for the prevention of IA. Allogeneic NK cell transfer and transfer of the cell line NK92 after irradiation, which is already FDA-approved for clinical testing in certain types of cancer, were used in clinical studies and have a good safety profile in patients [38–42]. Nonetheless, they have to show their efficacy against IA in future studies.

2.4. Adoptive T Cell Transfer

The protective effect of *Aspergillus*-specific CD4$^+$ cells of the Th1 lineage has been shown throughout the literature [43]. After allo-SCT, the adaptive immune system reconstitutes much slower than the innate immune system. Only a few *Aspergillus*-specific T cells can be measured 9–12 months after allo-SCT [44]. Therefore, an artificial increase of these specific T cells might help to clear *Aspergillus* in immunocompromised patients. For adoptive T cell transfer, T cells are isolated from the patient and stimulated with defined antigens. Consequently, T cell populations that are specific for the antigens are activated and proliferate. In turn, high numbers of these specific T cells are injected into the patient, where they recognize their target and aid the immune system in its elimination [45]. While the benefits of adoptive T cell transfer were illustrated in viral infections after transplantation, the development of similar techniques for the transfer of fungus-specific T cells lags behind [45].

One major obstacle to successful specific T cell transfer for *Aspergillus* in allo-SCT patients is the generation of an adequate number of *Aspergillus*-specific T cells with sufficient purity, using Good Manufacturing Practice (GMP) guidelines. Because of the fast progression of IA, the enrichment process needs to be as fast as possible. Many groups have worked toward complying with these requirements [46–49]. Bacher et al. reported a GMP-compliant protocol, in which they were able to enrich *Aspergillus*-specific T cells 200-fold in the T cell population. After isolation, they depleted cytotoxic and regulatory T cells, stimulated the remaining T cells with a GMP-grade *A. fumigatus* lysate, and isolated *A. fumigatus*-specific T cells with the help of the T cell activation marker CD137. This protocol is being used in an ongoing clinical trial (EudraCT Nr.2013-002914-11) [50]. However, another group has demonstrated that immunosuppressants such as cyclosporine A, methylprednisolone, as well as mycophenolic acid, lowered the number and activation of *Aspergillus*-specific protective Th1 cells. These immunosuppressants are frequently used after allo-SCT, complicating the application of adoptive T cell transfer in allo-SCT patients [51].

Up to this point, to our knowledge, there is only one clinical trial testing the efficacy and safety of adoptive T cell transfer in invasive fungal diseases. In this study, 10 patients with IA were treated by adoptive T cell transfer, while 13 IA patients in the control group did not receive a cell transfusion. The IA clearance rate was 90% in the treatment group compared to 53% in the control group. Infused cells did not cause graft versus host disease (GvHD) and showed a high IFN-γ to IL-10 ratio, indicating Th1 priming in the first three weeks after infusion. In contrast, patients in the control group had only a few naturally occurring *Aspergillus*-specific T cells 9–12 months after transplantation, exhibiting a nonprotective Th2 profile. In addition, patients receiving adoptive T cell transfer showed significantly decreased galactomannan antigenemia in comparison to the levels in the control group [44]. In order to shorten the time in between diagnosis and the transfusion of the cell product, off-the-shelf T cells specific for certain viruses were developed. They were found to be safe and only rarely caused mild GvHD (grade 1) [52]. Production of off-the-shelf T cells for transfusing IA patients is desirable.

Current findings have indicated that not only CD4$^+$ T cell responses are important against *Aspergillus*, but cytotoxic CD8$^+$ T cells might play an important role as well [53,54]. CTLs stimulated with Asp f 16 were able to induce increased Th1 responses. The transfer of Asp f 16-specific CTLs resulted in higher survival in a murine model of IA [55]. In consequence, adoptive T cell transfer for IA might not only be limited to CD4$^+$ cells.

Adoptive T cell transfer is a promising tool for the fight against IA. GMP-compliant protocols for the production of sufficient numbers of *Aspergillus*-specific T cells are available, and off-the-shelf cell products against the infection might be developed soon. The results of the first clinical trial have been very promising, and there should be strong incentives for additional clinical trials.

2.5. Chimeric Antigen Receptors (CARs)

Chimeric antigen receptor (CAR) T cells are one of the most promising immunotherapeutic tools available and have shown their efficacy in primary clinical trials of B cell malignancies [56]. The FDA has already approved CAR T cell therapies for the treatment of acute lymphoblastic leukemia (ALL) and B cell lymphoma in appropriate patient groups [57]. CARs are artificially designed receptors that are introduced into T cells. MHC unrestricted antigen recognition and the capability to recognize glycoproteins and lipids are only two of the many advantages of this approach [58,59]. CARs consist of two extracellular, one transmembrane, and one intracellular element. The task of the extracellular targeting element is to recognize the target. The single chain variable fragment (scFv) region of a monoclonal antibody targeting the desired antigen is usually used, but other targeting elements such as extracellular parts of naturally occurring receptors are also tested. The targeting element is linked to a spacer or linker, which gives the targeting element the flexibility to bind to the intended target [60]. The most common transmembrane domain used in CARs is the transmembrane spanning region of CD28 [61]. The intracellular domain assures signal delivery, resulting in the activation of the cell. In first-generation CARs, CD3-ζ was used for signal transduction. In second-generation CARs, a costimulatory domain, most commonly CD28 and 4-1BB, was added to CD3-ζ, resulting in better persistence of CAR T cells [62] (Figure 2).

The success of CAR T cells in B cell malignancies led to the attempt to use CARs for infections like aspergillosis. The group of Cooper et al. swapped the CD19 targeting element of a second-generation CAR, which is currently being evaluated in a clinical trial, with the extracellular part of Dectin-1 [63,64]. Dectin-1 is a naturally occurring receptor of the innate immune system that is not expressed on T cells. Its ligand ß-glucan is a polysaccharide found on the surface of many fungi, including *Aspergillus*. In addition to the extracellular part of Dectin-1, the Dectin-1 CAR (D-CAR) consisted of an IgG4 spacer, a CD28 transmembrane domain, and an intracellular domain of CD28 and CD3-ζ. The authors demonstrated that the D-CAR was activated by ß-glucan and inhibited the growth of *A. fumigatus*. In addition, IFN-γ concentration increased after stimulation, and the perforin/granzyme pathway was likely activated [63]. Interestingly, steroid treatment did not inhibit the antifungal activity of the D-CAR.

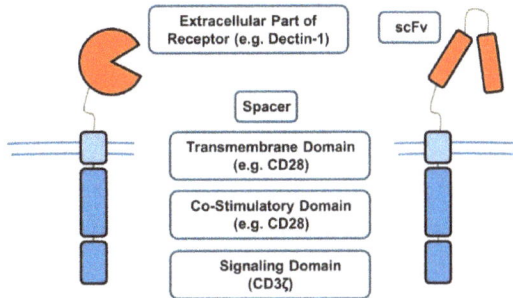

Figure 2. Second-generation chimeric antigen receptors (CARs) include a spacer, transmembrane domain, costimulatory domain, and signaling domain, as well as a targeting element. The two depicted CARs differ in their targeting element, which is usually the single chain variable fragment (scFv) of an antibody, but can also consist of the extracellular part of a receptor.

In light of this report, CAR T cells might not only be helpful for the treatment of B cell malignancies, but also for *Aspergillus* infections. However, there is only one report about the efficacy of CAR T cells against *Aspergillus* available, and more data needs to be generated. Additional CAR constructs containing alternative costimulatory domains and new targeting elements could be more efficient and should be evaluated. Moreover, the autologous generation of sufficient numbers of CAR T cells takes from just over one up to several weeks, which might lose critical time in an acute infection like IA [60].

3. Cytokine Therapy

One approach to fight aspergillosis after allo-SCT is to strengthen the immune system by administering cytokines. The most discussed cytokines are available as recombinant forms approved by the FDA, resulting in a more efficient evaluation on new patient cohorts. Cytokine therapies aimed at innate or both innate and adaptive immune systems have been assessed in several studies.

3.1. Colony Stimulating Factors

CSF treatment is aimed at increasing the capacity of the innate immune system to clear *Aspergillus*. This might be achieved by a faster reconstitution of innate immune cells mitigating risk factors such as neutropenia as well as increasing the activity of these cells against the fungus.

3.1.1. G-CSF

G-CSF increases neutrophil proliferation as well as maturation and is FDA-approved [65]. After stem cell transplantation, it is frequently administered during febrile episodes of neutropenia. Even though G-CSF does not decrease mortality caused by infections, it reduces time of neutropenia and febrile neutropenia-related hospitalization periods [66]. In an *A. fumigatus* mouse model, the addition of G-CSF to the antifungal caspofungin or caspofungin combined with amphotericin B-intralipid, resulted in higher survival rates of up to 78.9%, decreased fungal burden in organs, and reduced serum galactomannan [67]. It also increased neutrophil counts and led to a four-fold higher killing of *A. fumigatus* conidia by PMNs compared to untreated controls [68].

Research shows that G-CSF shortens neutropenia in patients, but more studies investigating the effects of G-CSF on the prevalence and outcomes of *Aspergillus* infections have to be conducted [69,70].

3.1.2. M-CSF

In contrast to G-CSF and granulocyte-macrophage CSF (GM-CSF), macrophage CSF (M-CSF) is not FDA-approved. Its main function is the stimulation of macrophage growth [71]. In a clinical phase I/II trial, the regular antifungal therapy of 46 bone marrow transplant patients was supplemented by recombinant human M-CSF. While the clinical outcome of patients infected with various *Candida* species improved compared to historical controls, no positive effect on patients suffering from aspergillosis was observed [72]. Treating transplanted mice with M-CSF before *A. fumigatus* challenge not only reduced fungal organ burden, but also increased survival rates from 10% in saline-treated animals to 60% [73]. Prophylactic administration of M-CSF to neutropenic rabbits in a model of pulmonary aspergillosis lowered pulmonary injury and increased survival, most likely due to increased macrophage numbers and phagocytosis activity [74].

Only a few experiments have been performed with M-CSF. M-CSF treatment has shown some promise in animal models for aspergillosis and should be evaluated further for the treatment of patients after allo-SCT.

3.1.3. GM-CSF

Like G-CSF, GM-CSF is FDA-approved, but has a broader effect on immune cells. It plays a role in the differentiation of dendritic cells, as well as macrophages, and stimulates the proliferation and activation of many cell types, including neutrophils, macrophages, eosinophils, and dendritic cells [75,76]. Therefore, GM-CSF treatment increases numbers of tissue macrophages, circulating monocytes, neutrophils, and platelets, as well as eosinophils [77,78].

After allo-SCT, GM-CSF administration is considered safe [79]. In a prospective multicenter randomized phase IV clinical trial, 206 allogeneic stem cell patients were prophylactically administered with either G-CSF or GM-CSF alone, or a combination of both. Although GM-CSF and GM-CSF + G-CSF decreased combined 600-day IFI-related mortality and yeast incidence, no benefit for IA incidence was found [78].

GM-CSF might partly mitigate the effect of certain immunosuppressive drugs but inhibit the ability of the immune system to clear *Aspergillus*. Brummer et al. illustrated that GM-CSF prevents the immunosuppressive effects of dexamethasone on murine bronchoalveolar macrophages, leading to increased killing of *A. fumigatus* conidia [80]. Supportively, GM-CSF exposure lowered the fungal burden in the lung among cyclophosphamide immunosuppressed mice in a model of pulmonary aspergillosis [81]. Macrophage suppression by the corticosteroid cortisone acetate was also inhibited by GM-CSF. This effect lasted for more than a week after treatment in a murine model. In addition, GM-CSF has been shown to counteract corticosteroid-induced downregulation of pro-inflammatory cytokines such as TNF-α, which are crucial to early defense mechanisms of the innate immune system against *Aspergillus* conidia [82].

Even though GM-CSF shortens the time of neutropenia, which is the major risk factor for IA, no benefit concerning incidence or course of IA in larger patient cohorts has been found to date. Again, this demonstrates that addressing one of the many dysfunctions of the immune system after allo-SCT might not be sufficient to prevent or clear IA. However, GM-CSF potential to partly reverse undesired immunosuppressive effects after allo-SCT, which hamper infection control, might be an additional advantage of GM-CSF administration.

3.2. IFN-γ

A strong Th1 response is essential to clear *Aspergillus* [83–85]. In order to increase the Th1 response of patients, FDA-approved forms of IFN-γ might be administered. In vivo, IFN-γ is secreted by T and NK cells. It has the capacity to induce protective responses of the innate and adaptive immune systems against *Aspergillus* [43]. Numerous clinical studies have investigated the benefit of supplementing antifungal therapy with IFN-γ.

Case reports describing the positive effect of adjunctive IFN-γ administration on aspergillosis have been published [86–89]. In a randomized prospective placebo-controlled double-blinded clinical study of 128 patients undergoing chronic granulomatous disease, decreased frequencies of infections were observed after frequent IFN-γ administration compared to controls. However, only one patient in the IFN-γ-treated group and four patients in the placebo group suffered from aspergillosis, which highly limited the predictive value of the study. The study was also limited by the short follow-up period of only 10 months [90]. A case series of IFIs on renal transplant patients included three patients suffering from disseminated IA. All three cases were cured after six weeks of combined amphotericin B and IFN-γ treatment [91]. In a prospective case series of eight patients, including three aspergillosis cases, Delsing et al. showed increased ability of peripheral blood mononuclear cells (PBMCs) to produce pro-inflammatory cytokines IL-1β and TNF-α, Th17-stimulating cytokines IL-17 and IL-22, and heightened HLA-DR expression after combined IFN-γ and antifungal treatment, all of which play an important role in protecting the host from IA. While lymphocyte and monocyte numbers were increased, granulocyte numbers were slightly decreased [92]. In order to reverse the drop in granulocyte numbers, the addition of IFN-γ with a granulocyte count-increasing cytokine such as GM-CSF might be beneficial. Combination therapy of IFN-γ and GM-CSF, supporting antifungal treatment in two HIV-negative and one HIV-positive patient suffering from progressive pulmonary aspergillosis, showed promising results. Peripheral leukocyte numbers increased and Th1 response was strengthened. This resulted in an improved control of the fungal infection [93]. In vitro, pre-incubation of human PMNs with IFN-γ and GM-CSF led to enhanced *Aspergillus flavus* hyphal damage and increased release of oxygen radicals by PMNs. This effect disappeared when pre-incubating PMNs with only either one of the cytokines [94]. Another report demonstrated that IFN-γ treatment of PMNs and PBMCs resulted in increased hyphal damage of *A. fumigatus* [95].

Even though adjunctive IFN-γ treatment has many potential advantages and is well tolerated in allo-SCT patients [96], the evidence supporting the use in patients is still weak [22]. More trials need to be conducted.

3.3. TNF-α

TNF-α is one of the most important cytokines in the defense against *Aspergillus* [97]. Comparable to IFN-γ, addition of TNF-α stimulates PMNs, which in turn increase oxygen radical release and cause enhanced hyphal damage against *A. fumigatus* in vitro. Although intracellular killing of *A. fumigatus* conidia by alveolar macrophages was not increased, phagocytosis was enhanced [98]. Administration of TNF-α to immunosuppressed mice in a model for pulmonary aspergillosis increased survival [99]. A time-dependent increase in TNF-α levels of the lung was correlated with higher migration of PMNs to the lung and increased survival of neutropenic and non-neutropenic mice after challenge with *A. fumigatus* conidia. In turn, blocking of TNF-α resulted in higher mortality and fungal lung burden in neutropenic mice. Prophylactic treatment of neutropenic mice with TNF-α increased their survival [100]. However, the major limitation of using TNF-α in the treatment of aspergillosis is its serious toxicity after systemic administration, including hepatotoxicity, nephrotoxicity, and neurotoxicity [101].

4. Other Immunotherapeutic Approaches

4.1. Vaccination

Successful vaccination elicits an adaptive immune response to a pathogen, leading to the generation of memory cells, which are able to fight subsequent infections with the same pathogen much more efficiently. There are different forms of vaccination available. First of all, inactivated whole-cell vaccines can be used; however, they have known limitations. They are complex and therefore difficult to standardize, and usually only elicit weak immune responses [102]. Live vaccines are more immunogenic, but are considered unsafe in immunocompromised patients, as they can potentially cause disease [103]. Subunit vaccines, which consist of purified elements, in combination with an adjuvant, could be the best method available in order to vaccinate immunocompromised patients, as they are easy to standardize and also considered safe in this cohort [53].

Subcutaneous vaccination using a hyphal sonicate protected immunocompromised mice in a model of IA [104]. The vaccination of mice with heat-killed *S. cerevisiae* before *A. fumigatus* challenge increased survival and decreased fungal organ burden. The major limitation of this study was the usage of immunocompetent mice without any immunosuppression. The usefulness of this approach in an immunocompromised setting cannot be predicted [105]. In another study, mice were vaccinated by intranasal inhalation of either filtrates of viable *A. fumigatus*, viable *A. fumigatus*, or heat-inactivated *A. fumigatus*. Thereafter, mice were immunosuppressed and challenged with the fungus. The filtrate and the live fungus vaccination were able to prolong survival and induced a protective Th1 response. In contrast, no prolonged protection was found in mice vaccinated with heat-inactivated fungus, which provoked a Th2 response [106]. Furthermore, vaccination with recombinant *Aspergillus* antigens Asp f 3, Asp f 9, Asp f 16 (all major allergens), Gel1 (a protein associated with cell wall morphogenesis), and Pep1 (an extracellular endopeptidase) resulted in protective effects in murine models of aspergillosis [107,108]. These antigens could potentially be used as the basis of subunit vaccines. The same is true for mannans, which can be found in the cell wall of *Aspergillus*. Liu et al. vaccinated immunocompetent mice with mannans derived from *C. albicans*, leading to increased survival rates after challenge with *A. fumigatus* conidia. Mortality was further decreased by the addition of bovine serum albumin (BSA) to the mannans [109].

Even though vaccine development in immunocompromised patients is difficult because of their weakened immune system, advances have been made to improve vaccination strategies in this patient cohort. However, T and B cell counts, as well as functionality, have to be at least partially restored in order to elicit a protective response. After allo-SCT, the reconstitution of the adaptive immune system takes several months [110]. Therefore, *Aspergillus* vaccination might not be effective in *Aspergillus* infections early after allo-SCT. It is still difficult and costly to develop vaccines against fungal pathogens, and in contrast to other microbes, no fungal vaccine has been licensed yet. Vaccines for other fungi, such as NDV-3A for *C. albicans*, which was used in a recent promising clinical trial, might lead to increased interest in the development of vaccines for *Aspergillus* [102,111].

4.2. Antibodies

Two decades ago, the humoral response was thought to play little to no role in the defense against fungi. More recent findings, however, show that this dogma needs to be revised. Humoral responses, in fact, are important for the host defense against fungal infection, including *Aspergillus* [112]. Patients with Good syndrome, a disease characterized by hypogammaglobulinemia, show increased incidences of fungal infections, including aspergillosis [113]. Anti-*Aspergillus* antibodies bind to swollen conidia and germ tubes, activating the classical pathway of the complement system. Complement activation leads to the killing of *Aspergillus* by neutrophils [114].

The efficacy of different kinds of antibodies against *Aspergillus* has been evaluated. One approach is to target polysaccharides found on the cell wall of fungi. The monoclonal antibody (mAb) 2G8 targets the cell wall polysaccharide laminarin, which consists of ß-glucan. It has demonstrated antifungal effects, including activity against *A. fumigatus* [115]. A different method is the usage of anti-idiotypic mAbs to yeast killer toxin, found in *Pichia anomala* and *Williopsis mrakii*, which displays antimicrobial effects. Administration of these mAbs to immunocompromised mice infected with *A. fumigatus* decreased fungal growth and increased survival [116]. Radioimmunotherapy has the potential to be another antibody-based immunotherapeutic strategy against *Aspergillus*. In this approach, an antibody directed against the fungus is tagged with radionuclides in order to deliver a lethal dose of radiation to the fungus [117].

The administration of antibodies might strengthen the ability of allo-SCT patients to prevent or clear IA in absence of a fully functional adaptive immune system. In contrast to adoptive T cell or CAR T cell transfer, in which autologous T cells have to be generated for each single patient over the period of weeks, one type of antibody would be instantly available for all affected patients, which might be a major advantage in acute infections. The research on the humoral response against *Aspergillus* and the implementation of immunotherapeutic strategies based on these findings is still in its early stages. However, the initial results generated are promising, and more data should be collected.

5. Summary and Outlook

The treatment of IA patients with standard antifungal drugs faces numerous challenges. No new class of antifungal drugs has been invented for over a decade, the number of fungal isolates resistant to azoles has been steadily increasing, and the side effects of conventional antifungal drugs are still considered to be severe. Immunotherapeutic approaches hold promise for improving antifungal therapy in order to decrease high mortality rates. In general, immunotherapy protocols treating *Aspergillus* infections are still exploratory, cost-intensive, might be accompanied by severe side effects, and involve complex as well as time-intensive genetic and cellular manipulations before use. Different immunotherapeutic strategies have been investigated for their efficacy, safety, and their potential to overcome these challenges. The most promising candidates should be evaluated in well-designed clinical trials. Up to this point, the low prevalence of IA has first resulted in the clinical evaluation of these exploratory methods in small patient cohorts with low statistical power; and second, the analysis of a treatment's efficacy has often been assessed in combined IFIs. Various IFIs are known to differ in their pathology and susceptibility to certain treatments. Thus, multicenter clinical trials for IA should be performed.

There are promising weapons against *Aspergillus* on the horizon. In the future, this fight might involve the use of new NK CAR technology, a tool that can be used as an allogeneic "off-the-shelf" product [118]. It might also include checkpoint inhibitors. These molecules disable inhibiting receptors on immune cells and therefore increase their activity. They have demonstrated their efficacy in cancer research and might attenuate the clinical progression of IA [119]. Another promising approach is the usage of neutrophil-dendritic cell hybrids (PMN-DCs), which are cells with the microbicide function of PMNs and the capacity of DCs to stimulate adaptive immunity [120]. In addition, as drug development is expensive, and only a few drug candidates reach market maturity, the repurposing of approved drugs for potential use in IA might be worthwhile. Drugs such as auranofin and ebselen have shown activity against *Aspergillus* in vitro. Both drugs block the thioredoxin reductase pathway, which is essential for cells to manipulate disulfide bonds. This pathway is different in humans compared to bacteria, as well as fungi, and therefore might be a suitable target [121,122]. Moreover, combining novel immunotherapeutic approaches with antifungals might yield positive synergistic effects. For example, echinocandins are drugs that uncover immunologically active epitopes in the fungal cell wall. Many immunotherapeutic strategies such as D-CARs or ß-glucan-specific antibodies target these epitopes, which might result in more efficient fungal clearance [119].

In conclusion, the fight against IA still relies heavily on conventional antifungal drugs. Immunotherapy has made a lot of progress in the last decade and might be used as an adjuvant therapy or even on its own in the future. In order to bring these new treatment strategies to the bedside, well-designed multicenter clinical trials are of the upmost importance.

Author Contributions: C.D.L., H.E., and J.L. wrote the manuscript. C.D.L designed the figures.

Funding: This work was supported by the Bavarian Ministry of Economics, Media, Energy and Technology (grant number BayBIO-1606-003, "T-cell based diagnostic monitoring of invasive aspergillosis in haematological patients") and the "Deutsche Forschungsgemeinschaft" (Collaborative Research Center / Transregio 124 "Pathogenic fungi and their human host: Networks of interaction – FungiNet", [project A2 to HE and JL].

Acknowledgments: We thank Lukas Page for proofreading the manuscript.

Conflicts of Interest: The authors declare no conflicts of interest.

References

1. Taccone, F.S.; Van den Abeele, A.M.; Bulpa, P.; Misset, B.; Meersseman, W.; Cardoso, T.; Paiva, J.A.; Blasco-Navalpotro, M.; De Laere, E.; Dimopoulos, G.; et al. Epidemiology of invasive aspergillosis in critically ill patients: Clinical presentation, underlying conditions, and outcomes. *Crit. Care* **2015**, *19*, 7. [CrossRef] [PubMed]

2. Filler, S.G.; Sheppard, D.C. Fungal invasion of normally non-phagocytic host cells. *PLoS Pathog.* **2006**, *2*, e129. [CrossRef] [PubMed]

3. White, P.L.; Posso, R.B.; Barnes, R.A. Analytical and Clinical Evaluation of the PathoNostics AsperGenius Assay for Detection of Invasive Aspergillosis and Resistance to Azole Antifungal Drugs Directly from Plasma Samples. *J. Clin. Microbiol.* **2017**, *55*, 2356–2366. [CrossRef] [PubMed]

4. Hohl, T.M. Immune responses to invasive aspergillosis: New understanding and therapeutic opportunities. *Curr. Opin. Infect. Dis.* **2017**, *30*, 364–371. [CrossRef] [PubMed]

5. Aimanianda, V.; Bayry, J.; Bozza, S.; Kniemeyer, O.; Perruccio, K.; Elluru, S.R.; Clavaud, C.; Paris, S.; Brakhage, A.A.; Kaveri, S.V.; et al. Surface hydrophobin prevents immune recognition of airborne fungal spores. *Nature* **2009**, *460*, 1117–1121. [CrossRef] [PubMed]

6. Bidula, S.; Schelenz, S. A Sweet Response to a Sour Situation: The Role of Soluble Pattern Recognition Receptors in the Innate Immune Response to Invasive Aspergillus fumigatus Infections. *PLoS Pathog.* **2016**, *12*, e1005637. [CrossRef]

7. Garth, J.M.; Reeder, K.M.; Godwin, M.S.; Mackel, J.J.; Dunaway, C.W.; Blackburn, J.P.; Steele, C. IL-33 Signaling Regulates Innate IL-17A and IL-22 Production via Suppression of Prostaglandin E_2 during Lung Fungal Infection. *J. Immunol.* **2017**. [CrossRef]

8. Espinosa, V.; Rivera, A. First line of defense: Innate cell-mediated control of pulmonary aspergillosis. *Front. Microbiol.* **2016**, *7*, 272. [CrossRef]

9. Khanna, N.; Stuehler, C.; Lünemann, A.; Wójtowicz, A.; Bochud, P.Y.; Leibundgut-Landmann, S. Host response to fungal infections—How immunology and host genetics could help to identify and treat patients at risk. *Swiss Med. Wkly.* **2016**, *146*. [CrossRef]

10. Heinz, W.J.; Vehreschild, J.J.; Buchheidt, D. Diagnostic workout to assess early response indicators in invasive pulmonary aspergillosis in adult patients with hematologic malignancies. *Mycoses* **2018**. [CrossRef]

11. Leroux, S.; Ullmann, A.J. Management and diagnostic guidelines for fungal diseases in infectious diseases and clinical microbiology: Critical appraisal. *Clin. Microbiol. Infect. Off. Publ. Eur. Soc. Clin. Microbiol. Infect. Dis.* **2013**, *19*, 1115–1121. [CrossRef]

12. Chowdhary, A.; Sharma, C.; Meis, J.F. Azole-Resistant Aspergillosis: Epidemiology, Molecular Mechanisms, and Treatment. *J. Infect. Dis.* **2017**, *216*, S436–S444. [CrossRef] [PubMed]

13. Hickey, M.J.; Kubes, P. Intravascular immunity: The host-pathogen encounter in blood vessels. *Nat. Rev. Immunol.* **2009**, *9*, 364–375. [CrossRef]

14. Shoham, S.; Levitz, S.M. The immune response to fungal infections. *Br. J. Haematol.* **2005**, *129*, 569–582. [CrossRef] [PubMed]

15. Hubel, K.; Carter, R.A.; Liles, W.C.; Dale, D.C.; Price, T.H.; Bowden, R.A.; Rowley, S.D.; Chauncey, T.R.; Bensinger, W.I.; Boeckh, M. Granulocyte transfusion therapy for infections in candidates and recipients of HPC transplantation: A comparative analysis of feasibility and outcome for community donors versus related donors. *Transfusion* **2002**, *42*, 1414–1421. [CrossRef] [PubMed]

16. Price, T.H.; Bowden, R.A.; Boeckh, M.; Bux, J.; Nelson, K.; Liles, W.C.; Dale, D.C. Phase I/II trial of neutrophil transfusions from donors stimulated with G-CSF and dexamethasone for treatment of patients with infections in hematopoietic stem cell transplantation. *Blood* **2000**, *95*, 3302–3309. [PubMed]

17. Seidel, M.G.; Peters, C.; Wacker, A.; Northoff, H.; Moog, R.; Boehme, A.; Silling, G.; Grimminger, W.; Einsele, H. Randomized phase III study of granulocyte transfusions in neutropenic patients. *Bone Marrow Transplant.* **2008**, *42*, 679–684. [CrossRef]
18. Mousset, S.; Hermann, S.; Klein, S.A.; Bialleck, H.; Duchscherer, M.; Bomke, B.; Wassmann, B.; Bohme, A.; Hoelzer, D.; Martin, H. Prophylactic and interventional granulocyte transfusions in patients with haematological malignancies and life-threatening infections during neutropenia. *Ann. Hematol.* **2005**, *84*, 734–741. [CrossRef]
19. Price, T.H.; Boeckh, M.; Harrison, R.W.; McCullough, J.; Ness, P.M.; Strauss, R.G.; Nichols, W.G.; Hamza, T.H.; Cushing, M.M.; King, K.E.; et al. Efficacy of transfusion with granulocytes from G-CSF/dexamethasone-treated donors in neutropenic patients with infection. *Blood* **2015**, *126*, 2153–2161. [CrossRef]
20. BitMansour, A.; Burns, S.M.; Traver, D.; Akashi, K.; Contag, C.H.; Weissman, I.L.; Brown, J.M. Myeloid progenitors protect against invasive aspergillosis and Pseudomonas aeruginosa infection following hematopoietic stem cell transplantation. *Blood* **2002**, *100*, 4660–4667. [CrossRef]
21. BitMansour, A.; Cao, T.M.; Chao, S.; Shashidhar, S.; Brown, J.M. Single infusion of myeloid progenitors reduces death from Aspergillus fumigatus following chemotherapy-induced neutropenia. *Blood* **2005**, *105*, 3535–3537. [CrossRef] [PubMed]
22. Patterson, T.F.; Thompson, G.R., 3rd; Denning, D.W.; Fishman, J.A.; Hadley, S.; Herbrecht, R.; Kontoyiannis, D.P.; Marr, K.A.; Morrison, V.A.; Nguyen, M.H.; et al. Practice Guidelines for the Diagnosis and Management of Aspergillosis: 2016 Update by the Infectious Diseases Society of America. *Clin. Infect. Dis.* **2016**, *63*, e1–e60. [CrossRef] [PubMed]
23. Ramirez-Ortiz, Z.G.; Means, T.K. The role of dendritic cells in the innate recognition of pathogenic fungi (A. fumigatus, C. neoformans and C. albicans). *Virulence* **2012**, *3*, 635–646. [CrossRef] [PubMed]
24. Thakur, R.; Anand, R.; Tiwari, S.; Singh, A.P.; Tiwary, B.N.; Shankar, J. Cytokines induce effector T-helper cells during invasive aspergillosis; what we have learned about T-helper cells? *Front. Microbiol.* **2015**, *6*, 429. [CrossRef] [PubMed]
25. Zelante, T.; Wong, A.Y.; Ping, T.J.; Chen, J.; Sumatoh, H.R.; Vigano, E.; Hong Bing, Y.; Lee, B.; Zolezzi, F.; Fric, J.; et al. CD103(+) Dendritic Cells Control Th17 Cell Function in the Lung. *Cell Rep.* **2015**, *12*, 1789–1801. [CrossRef] [PubMed]
26. Bozza, S.; Gaziano, R.; Lipford, G.B.; Montagnoli, C.; Bacci, A.; Di Francesco, P.; Kurup, V.P.; Wagner, H.; Romani, L. Vaccination of mice against invasive aspergillosis with recombinant Aspergillus proteins and CpG oligodeoxynucleotides as adjuvants. *Microbes Infect.* **2002**, *4*, 1281–1290. [CrossRef]
27. Bozza, S.; Perruccio, K.; Montagnoli, C.; Gaziano, R.; Bellocchio, S.; Burchielli, E.; Nkwanyuo, G.; Pitzurra, L.; Velardi, A.; Romani, L. A dendritic cell vaccine against invasive aspergillosis in allogeneic hematopoietic transplantation. *Blood* **2003**, *102*, 3807–3814. [CrossRef]
28. Shao, H.J.; Chen, L.; Su, Y.B. DNA fragment encoding human IL-1beta 163-171 peptide enhances the immune responses elicited in mice by DNA vaccine against foot-and-mouth disease. *Vet. Res. Commun.* **2005**, *29*, 35–46. [CrossRef]
29. Zhu, F.; Ramadan, G.; Davies, B.; Margolis, D.A.; Keever-Taylor, C.A. Stimulation by means of dendritic cells followed by Epstein-Barr virus-transformed B cells as antigen-presenting cells is more efficient than dendritic cells alone in inducing Aspergillus f16-specific cytotoxic T cell responses. *Clin. Exp. Immunol.* **2008**, *151*, 284–296. [CrossRef]
30. Roy, R.M.; Klein, B.S. Dendritic cells in antifungal immunity and vaccine design. *Cell Host Microbe* **2012**, *11*, 436–446. [CrossRef]
31. Schmidt, S.; Tramsen, L.; Perkhofer, S.; Lass-Florl, C.; Hanisch, M.; Roger, F.; Klingebiel, T.; Koehl, U.; Lehrnbecher, T. Rhizopus oryzae hyphae are damaged by human natural killer (NK) cells, but suppress NK cell mediated immunity. *Immunobiology* **2013**, *218*, 939–944. [CrossRef] [PubMed]
32. Schmidt, S.; Tramsen, L.; Lehrnbecher, T. Natural Killer Cells in Antifungal Immunity. *Front. Immunol.* **2017**, *8*, 1623. [CrossRef] [PubMed]
33. Ziegler, S.; Weiss, E.; Schmitt, A.L.; Schlegel, J.; Burgert, A.; Terpitz, U.; Sauer, M.; Moretta, L.; Sivori, S.; Leonhardt, I.; et al. CD56 Is a Pathogen Recognition Receptor on Human Natural Killer Cells. *Sci. Rep.* **2017**, *7*, 6138. [CrossRef] [PubMed]

34. Schmidt, S.; Tramsen, L.; Hanisch, M.; Latge, J.P.; Huenecke, S.; Koehl, U.; Lehrnbecher, T. Human natural killer cells exhibit direct activity against Aspergillus fumigatus hyphae, but not against resting conidia. *J. Infect. Dis.* **2011**, *203*, 430–435. [CrossRef] [PubMed]

35. Stuehler, C.; Kuenzli, E.; Jaeger, V.K.; Baettig, V.; Ferracin, F.; Rajacic, Z.; Kaiser, D.; Bernardini, C.; Forrer, P.; Weisser, M.; et al. Immune Reconstitution After Allogeneic Hematopoietic Stem Cell Transplantation and Association With Occurrence and Outcome of Invasive Aspergillosis. *J. Infect. Dis.* **2015**, *212*, 959–967. [CrossRef] [PubMed]

36. Park, S.J.; Hughes, M.A.; Burdick, M.; Strieter, R.M.; Mehrad, B. Early NK cell-derived IFN-{gamma} is essential to host defense in neutropenic invasive aspergillosis. *J. Immunol.* **2009**, *182*, 4306–4312. [CrossRef] [PubMed]

37. Benedetto, N.; Sabatini, P.; Sellitto, C.; Romano Carratelli, C. Interleukin-2 and increased natural killer activity in mice experimentally infected with Aspergillus niger. *Microbiologica* **1988**, *11*, 339–345.

38. Tonn, T.; Schwabe, D.; Klingemann, H.G.; Becker, S.; Esser, R.; Koehl, U.; Suttorp, M.; Seifried, E.; Ottmann, O.G.; Bug, G. Treatment of patients with advanced cancer with the natural killer cell line NK-92. *Cytotherapy* **2013**, *15*, 1563–1570. [CrossRef] [PubMed]

39. Arai, S.; Meagher, R.; Swearingen, M.; Myint, H.; Rich, E.; Martinson, J.; Klingemann, H. Infusion of the allogeneic cell line NK-92 in patients with advanced renal cell cancer or melanoma: A phase I trial. *Cytotherapy* **2008**, *10*, 625–632. [CrossRef] [PubMed]

40. Boyiadzis, M.; Agha, M.; Redner, R.L.; Sehgal, A.; Im, A.; Hou, J.Z.; Farah, R.; Dorritie, K.A.; Raptis, A.; Lim, S.H.; et al. Phase 1 clinical trial of adoptive immunotherapy using "off-the-shelf" activated natural killer cells in patients with refractory and relapsed acute myeloid leukemia. *Cytotherapy* **2017**, *19*, 1225–1232. [CrossRef] [PubMed]

41. Klingemann, H.; Grodman, C.; Cutler, E.; Duque, M.; Kadidlo, D.; Klein, A.K.; Sprague, K.A.; Miller, K.B.; Comenzo, R.L.; Kewalramani, T.; et al. Autologous stem cell transplant recipients tolerate haploidentical related-donor natural killer cell-enriched infusions. *Transfusion* **2013**, *53*, 412–418. [CrossRef] [PubMed]

42. Shah, N.; Li, L.; McCarty, J.; Kaur, I.; Yvon, E.; Shaim, H.; Muftuoglu, M.; Liu, E.; Orlowski, R.Z.; Cooper, L.; et al. Phase I study of cord blood-derived natural killer cells combined with autologous stem cell transplantation in multiple myeloma. *Br. J. Haematol.* **2017**, *177*, 457–466. [CrossRef] [PubMed]

43. Dewi, I.; van de Veerdonk, F.; Gresnigt, M. The Multifaceted Role of T-Helper Responses in Host Defense against Aspergillus fumigatus. *J. Fungi* **2017**. [CrossRef] [PubMed]

44. Perruccio, K.; Tosti, A.; Burchielli, E.; Topini, F.; Ruggeri, L.; Carotti, A.; Capanni, M.; Urbani, E.; Mancusi, A.; Aversa, F.; et al. Transferring functional immune responses to pathogens after haploidentical hematopoietic transplantation. *Blood* **2005**, *106*, 4397–4406. [CrossRef]

45. Papadopoulou, A.; Kaloyannidis, P.; Yannaki, E.; Cruz, C.R. Adoptive transfer of Aspergillus-specific T cells as a novel anti-fungal therapy for hematopoietic stem cell transplant recipients: Progress and challenges. *Crit. Rev. Oncol. Hematol.* **2016**, *98*, 62–72. [CrossRef]

46. Tramsen, L.; Koehl, U.; Tonn, T.; Latge, J.P.; Schuster, F.R.; Borkhardt, A.; Uharek, L.; Quaritsch, R.; Beck, O.; Seifried, E.; et al. Clinical-scale generation of human anti-Aspergillus T cells for adoptive immunotherapy. *Bone Marrow Transplant.* **2009**, *43*, 13–19. [CrossRef] [PubMed]

47. Tramsen, L.; Schmidt, S.; Boenig, H.; Latge, J.P.; Lass-Florl, C.; Roeger, F.; Seifried, E.; Klingebiel, T.; Lehrnbecher, T. Clinical-scale generation of multi-specific anti-fungal T cells targeting Candida, Aspergillus and mucormycetes. *Cytotherapy* **2013**, *15*, 344–351. [CrossRef]

48. Khanna, N.; Stuehler, C.; Conrad, B.; Lurati, S.; Krappmann, S.; Einsele, H.; Berges, C.; Topp, M.S. Generation of a multipathogen-specific T-cell product for adoptive immunotherapy based on activation-dependent expression of CD154. *Blood* **2011**, *118*, 1121–1131. [CrossRef]

49. Stuehler, C.; Nowakowska, J.; Bernardini, C.; Topp, M.S.; Battegay, M.; Passweg, J.; Khanna, N. Multispecific Aspergillus T cells selected by CD137 or CD154 induce protective immune responses against the most relevant mold infections. *J. Infect. Dis.* **2015**, *211*, 1251–1261. [CrossRef]

50. Bacher, P.; Jochheim-Richter, A.; Mockel-Tenbrink, N.; Kniemeyer, O.; Wingenfeld, E.; Alex, R.; Ortigao, A.; Karpova, D.; Lehrnbecher, T.; Ullmann, A.J.; et al. Clinical-scale isolation of the total Aspergillus fumigatus-reactive T-helper cell repertoire for adoptive transfer. *Cytotherapy* **2015**, *17*, 1396–1405. [CrossRef]

51. Tramsen, L.; Schmidt, S.; Roeger, F.; Schubert, R.; Salzmann-Manrique, E.; Latge, J.P.; Klingebiel, T.; Lehrnbecher, T. Immunosuppressive compounds exhibit particular effects on functional properties of human anti-Aspergillus Th1 cells. *Infect. Immun.* **2014**, *82*, 2649–2656. [CrossRef] [PubMed]

52. Tzannou, I.; Papadopoulou, A.; Naik, S.; Leung, K.; Martinez, C.A.; Ramos, C.A.; Carrum, G.; Sasa, G.; Lulla, P.; Watanabe, A.; et al. Off-the-Shelf Virus-Specific T Cells to Treat BK Virus, Human Herpesvirus 6, Cytomegalovirus, Epstein-Barr Virus, and Adenovirus Infections After Allogeneic Hematopoietic Stem-Cell Transplantation. *J. Clin. Oncol.* **2017**, *35*, 3547–3557. [CrossRef] [PubMed]

53. Iannitti, R.G.; Carvalho, A.; Romani, L. From memory to antifungal vaccine design. *Trends Immunol.* **2012**, *33*, 467–474. [CrossRef] [PubMed]

54. Cutler, J.E.; Deepe, G.S., Jr.; Klein, B.S. Advances in combating fungal diseases: Vaccines on the threshold. *Nat. Rev. Microbiol.* **2007**, *5*, 13–28. [CrossRef] [PubMed]

55. Sun, Z.; Zhu, P.; Li, L.; Wan, Z.; Zhao, Z.; Li, R. Adoptive immunity mediated by HLA-A*0201 restricted Asp f16 peptides-specific CD8+ T cells against Aspergillus fumigatus infection. *Eur. J. Clin. Microbiol. Infect. Dis. Off. Publ. Eur. Soc. Clin. Microbiol.* **2012**, *31*, 3089–3096. [CrossRef] [PubMed]

56. Maldini, C.R.; Ellis, G.I.; Riley, J.L. CAR T cells for infection, autoimmunity and allotransplantation. *Nat. Rev. Immunol.* **2018**, *18*, 605–616. [CrossRef] [PubMed]

57. Zheng, P.-P.; Kros, J.M.; Li, J. Approved CAR T cell therapies: Ice bucket challenges on glaring safety risks and long-term impacts. *Drug Discov. Today* **2018**, *23*, 1175–1182. [CrossRef]

58. Johnson, L.A.; June, C.H. Driving gene-engineered T cell immunotherapy of cancer. *Cell Res.* **2017**, *27*, 38–58. [CrossRef]

59. Dotti, G.; Gottschalk, S.; Savoldo, B.; Brenner, M.K. Design and development of therapies using chimeric antigen receptor-expressing T cells. *Immunol. Rev.* **2014**, *257*, 107–126. [CrossRef]

60. Kumaresan, P.R.; Silva, T.; Kontoyiannis, D.P. Methods of Controlling Invasive Fungal Infections Using CD8+ T Cells. *Front. Immunol.* **2018**, *8*, 1939. [CrossRef]

61. Bridgeman, J.S.; Hawkins, R.E.; Bagley, S.; Blaylock, M.; Holland, M.; Gilham, D.E. The optimal antigen response of chimeric antigen receptors harboring the CD3zeta transmembrane domain is dependent upon incorporation of the receptor into the endogenous TCR/CD3 complex. *J. Immunol.* **2010**, *184*, 6938–6949. [CrossRef] [PubMed]

62. Abken, H. Costimulation Engages the Gear in Driving CARs. *Immunity* **2016**, *44*, 214–216. [CrossRef] [PubMed]

63. Kumaresan, P.R.; Manuri, P.R.; Albert, N.D.; Maiti, S.; Singh, H.; Mi, T.; Roszik, J.; Rabinovich, B.; Olivares, S.; Krishnamurthy, J.; et al. Bioengineering T cells to target carbohydrate to treat opportunistic fungal infection. *Proc. Natl. Acad. Sci. USA* **2014**, *111*, 10660–10665. [CrossRef] [PubMed]

64. Singh, H.; Manuri, P.R.; Olivares, S.; Dara, N.; Dawson, M.J.; Huls, H.; Hackett, P.B.; Kohn, D.B.; Shpall, E.J.; Champlin, R.E.; et al. Redirecting specificity of T-cell populations for CD19 using the Sleeping Beauty system. *Cancer Res.* **2008**, *68*, 2961–2971. [CrossRef] [PubMed]

65. Wright, C.R.; Ward, A.C.; Russell, A.P. Granulocyte Colony-Stimulating Factor and Its Potential Application for Skeletal Muscle Repair and Regeneration. *Mediat. Inflamm.* **2017**, *2017*, 7517350. [CrossRef] [PubMed]

66. Clark, O.A.; Lyman, G.H.; Castro, A.A.; Clark, L.G.; Djulbegovic, B. Colony-stimulating factors for chemotherapy-induced febrile neutropenia: A meta-analysis of randomized controlled trials. *J. Clin. Oncol.* **2005**, *23*, 4198–4214. [CrossRef] [PubMed]

67. Sionov, E.; Mendlovic, S.; Segal, E. Experimental systemic murine aspergillosis: Treatment with polyene and caspofungin combination and G-CSF. *J. Antimicrob. Chemother.* **2005**, *56*, 594–597. [CrossRef]

68. Liles, W.C.; Huang, J.E.; van Burik, J.A.; Bowden, R.A.; Dale, D.C. Granulocyte colony-stimulating factor administered in vivo augments neutrophil-mediated activity against opportunistic fungal pathogens. *J. Infect. Dis.* **1997**, *175*, 1012–1015. [CrossRef]

69. Smith, T.J.; Khatcheressian, J.; Lyman, G.H.; Ozer, H.; Armitage, J.O.; Balducci, L.; Bennett, C.L.; Cantor, S.B.; Crawford, J.; Cross, S.J.; et al. 2006 update of recommendations for the use of white blood cell growth factors: An evidence-based clinical practice guideline. *J. Clin. Oncol.* **2006**, *24*, 3187–3205. [CrossRef]

70. Ringden, O.; Labopin, M.; Gorin, N.C.; Le Blanc, K.; Rocha, V.; Gluckman, E.; Reiffers, J.; Arcese, W.; Vossen, J.M.; Jouet, J.P.; et al. Treatment with granulocyte colony-stimulating factor after allogeneic bone marrow transplantation for acute leukemia increases the risk of graft-versus-host disease and death: A study

from the Acute Leukemia Working Party of the European Group for Blood and Marrow Transplantation. *J. Clin. Oncol.* **2004**, *22*, 416–423. [CrossRef]

71. Hume, D.A.; MacDonald, K.P. Therapeutic applications of macrophage colony-stimulating factor-1 (CSF-1) and antagonists of CSF-1 receptor (CSF-1R) signaling. *Blood* **2012**, *119*, 1810–1820. [CrossRef] [PubMed]

72. Nemunaitis, J.; Shannon-Dorcy, K.; Appelbaum, F.R.; Meyers, J.; Owens, A.; Day, R.; Ando, D.; O'Neill, C.; Buckner, D.; Singer, J. Long-term follow-up of patients with invasive fungal disease who received adjunctive therapy with recombinant human macrophage colony-stimulating factor. *Blood* **1993**, *82*, 1422–1427. [PubMed]

73. Kandalla, P.K.; Sarrazin, S.; Molawi, K.; Berruyer, C.; Redelberger, D.; Favel, A.; Bordi, C.; de Bentzmann, S.; Sieweke, M.H. M-CSF improves protection against bacterial and fungal infections after hematopoietic stem/progenitor cell transplantation. *J. Exp. Med.* **2016**, *213*, 2269–2279. [CrossRef] [PubMed]

74. Gonzalez, C.E.; Lyman, C.A.; Lee, S.; Del Guercio, C.; Roilides, E.; Bacher, J.; Gehrt, A.; Feuerstein, E.; Tsokos, M.; Walsh, T.J. Recombinant human macrophage colony-stimulating factor augments pulmonary host defences against Aspergillus fumigatus. *Cytokine* **2001**, *15*, 87–95. [CrossRef] [PubMed]

75. Shiomi, A.; Usui, T. Pivotal roles of GM-CSF in autoimmunity and inflammation. *Mediat. Inflamm.* **2015**, *2015*, 568543. [CrossRef] [PubMed]

76. Scriven, J.E.; Tenforde, M.W.; Levitz, S.M.; Jarvis, J.N. Modulating host immune responses to fight invasive fungal infections. *Curr. Opin. Microbiol.* **2017**, *40*, 95–103. [CrossRef] [PubMed]

77. Ruef, C.; Coleman, D.L. Granulocyte-macrophage colony-stimulating factor: Pleiotropic cytokine with potential clinical usefulness. *Rev. Infect. Dis.* **1990**, *12*, 41–62. [CrossRef]

78. Wan, L.; Zhang, Y.; Lai, Y.; Jiang, M.; Song, Y.; Zhou, J.; Zhang, Z.; Duan, X.; Fu, Y.; Liao, L.; et al. Effect of Granulocyte-Macrophage Colony-Stimulating Factor on Prevention and Treatment of Invasive Fungal Disease in Recipients of Allogeneic Stem-Cell Transplantation: A Prospective Multicenter Randomized Phase IV Trial. *J. Clin. Oncol.* **2015**, *33*, 3999–4006. [CrossRef]

79. Safdar, A.; Rodriguez, G.; Zuniga, J.; Al Akhrass, F.; Georgescu, G.; Pande, A. Granulocyte macrophage colony-stimulating factor in 66 patients with myeloid or lymphoid neoplasms and recipients of hematopoietic stem cell transplantation with invasive fungal disease. *Acta Haematol.* **2013**, *129*, 26–34. [CrossRef]

80. Brummer, E.; Maqbool, A.; Stevens, D.A. In vivo GM-CSF prevents dexamethasone suppression of killing of Aspergillus fumigatus conidia by bronchoalveolar macrophages. *J. Leukoc. Biol.* **2001**, *70*, 868–872.

81. Quezada, G.; Koshkina, N.V.; Zweidler-McKay, P.; Zhou, Z.; Kontoyiannis, D.P.; Kleinerman, E.S. Intranasal granulocyte-macrophage colony-stimulating factor reduces the Aspergillus burden in an immunosuppressed murine model of pulmonary aspergillosis. *Antimicrob. Agents Chemother.* **2008**, *52*, 716–718. [CrossRef] [PubMed]

82. Brummer, E.; Kamberi, M.; Stevens, D.A. Regulation by granulocyte-macrophage colony-stimulating factor and/or steroids given in vivo of proinflammatory cytokine and chemokine production by bronchoalveolar macrophages in response to Aspergillus conidia. *J. Infect. Dis.* **2003**, *187*, 705–709. [CrossRef] [PubMed]

83. Cenci, E.; Mencacci, A.; Fe d'Ostiani, C.; Del Sero, G.; Mosci, P.; Montagnoli, C.; Bacci, A.; Romani, L. Cytokine- and T helper-dependent lung mucosal immunity in mice with invasive pulmonary aspergillosis. *J. Infect. Dis.* **1998**, *178*, 1750–1760. [CrossRef] [PubMed]

84. Brieland, J.K.; Jackson, C.; Menzel, F.; Loebenberg, D.; Cacciapuoti, A.; Halpern, J.; Hurst, S.; Muchamuel, T.; Debets, R.; Kastelein, R.; et al. Cytokine networking in lungs of immunocompetent mice in response to inhaled Aspergillus fumigatus. *Infect. Immun.* **2001**, *69*, 1554–1560. [CrossRef] [PubMed]

85. Chai, L.Y.; van de Veerdonk, F.; Marijnissen, R.J.; Cheng, S.C.; Khoo, A.L.; Hectors, M.; Lagrou, K.; Vonk, A.G.; Maertens, J.; Joosten, L.A.; et al. Anti-Aspergillus human host defence relies on type 1 T helper (Th1), rather than type 17 T helper (Th17), cellular immunity. *Immunology* **2010**, *130*, 46–54. [CrossRef] [PubMed]

86. Estrada, C.; Desai, A.G.; Chirch, L.M.; Suh, H.; Seidman, R.; Darras, F.; Nord, E.P. Invasive aspergillosis in a renal transplant recipient successfully treated with interferon-gamma. *Case Rep. Transplant.* **2012**, *2012*, 493758. [CrossRef] [PubMed]

87. Mezidi, M.; Belafia, F.; Nougaret, S.; Pageaux, G.P.; Conseil, M.; Panaro, F.; Boniface, G.; Morquin, D.; Jaber, S.; Jung, B. Interferon gamma in association with immunosuppressive drugs withdrawal and antifungal combination as a rescue therapy for cerebral invasive Aspergillosis in a liver transplant recipient. *Minerva Anestesiol.* **2014**, *80*, 1359–1360. [PubMed]

88. Ellis, M.; Watson, R.; McNabb, A.; Lukic, M.L.; Nork, M. Massive intracerebral aspergillosis responding to combination high dose liposomal amphotericin B and cytokine therapy without surgery. *J. Med. Microbiol.* **2002**, *51*, 70–75. [CrossRef]

89. Kelleher, P.; Goodsall, A.; Mulgirigama, A.; Kunst, H.; Henderson, D.C.; Wilson, R.; Newman-Taylor, A.; Levin, M. Interferon-gamma therapy in two patients with progressive chronic pulmonary aspergillosis. *Eur. Respir. J.* **2006**, *27*, 1307–1310. [CrossRef]

90. The International Chronic Granulomatous Disease Cooperative Study Group. A controlled trial of interferon gamma to prevent infection in chronic granulomatous disease. *N. Engl. J. Med.* **1991**, *324*, 509–516. [CrossRef]

91. Armstrong-James, D.; Teo, I.A.; Shrivastava, S.; Petrou, M.A.; Taube, D.; Dorling, A.; Shaunak, S. Exogenous interferon-gamma immunotherapy for invasive fungal infections in kidney transplant patients. *Am. J. Transplant.* **2010**, *10*, 1796–1803. [CrossRef] [PubMed]

92. Delsing, C.E.; Gresnigt, M.S.; Leentjens, J.; Preijers, F.; Frager, F.A.; Kox, M.; Monneret, G.; Venet, F.; Bleeker-Rovers, C.P.; van de Veerdonk, F.L.; et al. Interferon-gamma as adjunctive immunotherapy for invasive fungal infections: A case series. *BMC Infect. Dis.* **2014**, *14*, 166. [CrossRef] [PubMed]

93. Bandera, A.; Trabattoni, D.; Ferrario, G.; Cesari, M.; Franzetti, F.; Clerici, M.; Gori, A. Interferon-gamma and granulocyte-macrophage colony stimulating factor therapy in three patients with pulmonary aspergillosis. *Infection* **2008**, *36*, 368–373. [CrossRef] [PubMed]

94. Gil-Lamaignere, C.; Winn, R.M.; Simitsopoulou, M.; Maloukou, A.; Walsh, T.J.; Roilides, E. Inteferon gamma and granulocyte-macrophage colony-stimulating factor augment the antifungal activity of human polymorphonuclear leukocytes against Scedosporium spp.: Comparison with Aspergillus spp. *Med. Mycol.* **2005**, *43*, 253–260. [CrossRef] [PubMed]

95. Gaviria, J.M.; van Burik, J.A.; Dale, D.C.; Root, R.K.; Liles, W.C. Comparison of interferon-gamma, granulocyte colony-stimulating factor, and granulocyte-macrophage colony-stimulating factor for priming leukocyte-mediated hyphal damage of opportunistic fungal pathogens. *J. Infect. Dis.* **1999**, *179*, 1038–1041. [CrossRef] [PubMed]

96. Safdar, A.; Rodriguez, G.; Ohmagari, N.; Kontoyiannis, D.P.; Rolston, K.V.; Raad, I.I.; Champlin, R.E. The safety of interferon-gamma-1b therapy for invasive fungal infections after hematopoietic stem cell transplantation. *Cancer* **2005**, *103*, 731–739. [CrossRef]

97. Sainz, J.; Perez, E.; Hassan, L.; Moratalla, A.; Romero, A.; Collado, M.D.; Jurado, M. Variable number of tandem repeats of TNF receptor type 2 promoter as genetic biomarker of susceptibility to develop invasive pulmonary aspergillosis. *Hum. Immunol.* **2007**, *68*, 41–50. [CrossRef]

98. Roilides, E.; Dimitriadou-Georgiadou, A.; Sein, T.; Kadiltsoglou, I.; Walsh, T.J. Tumor necrosis factor alpha enhances antifungal activities of polymorphonuclear and mononuclear phagocytes against Aspergillus fumigatus. *Infect. Immun.* **1998**, *66*, 5999–6003.

99. Nagai, H.; Guo, J.; Choi, H.; Kurup, V. Interferon-gamma and tumor necrosis factor-alpha protect mice from invasive aspergillosis. *J. Infect. Dis.* **1995**, *172*, 1554–1560. [CrossRef]

100. Mehrad, B.; Strieter, R.M.; Standiford, T.J. Role of TNF-alpha in pulmonary host defense in murine invasive aspergillosis. *J. Immunol.* **1999**, *162*, 1633–1640.

101. Roberts, N.J.; Zhou, S.; Diaz, L.A., Jr.; Holdhoff, M. Systemic use of tumor necrosis factor alpha as an anticancer agent. *Oncotarget* **2011**, *2*, 739–751. [CrossRef] [PubMed]

102. Pikman, R.; Ben-Ami, R. Immune modulators as adjuncts for the prevention and treatment of invasive fungal infections. *Immunotherapy* **2012**, *4*, 1869–1882. [CrossRef] [PubMed]

103. Cassone, A.; Casadevall, A. Recent progress in vaccines against fungal diseases. *Curr. Opin. Microbiol.* **2012**, *15*, 427–433. [CrossRef] [PubMed]

104. Ito, J.I.; Lyons, J.M. Vaccination of corticosteroid immunosuppressed mice against invasive pulmonary aspergillosis. *J. Infect. Dis.* **2002**, *186*, 869–871. [CrossRef] [PubMed]

105. Liu, M.; Capilla, J.; Johansen, M.E.; Alvarado, D.; Martinez, M.; Chen, V.; Clemons, K.V.; Stevens, D.A. Saccharomyces as a vaccine against systemic aspergillosis: 'the friend of man' a friend again? *J. Med. Microbiol.* **2011**, *60*, 1423–1432. [CrossRef] [PubMed]

106. Cenci, E.; Mencacci, A.; Bacci, A.; Bistoni, F.; Kurup, V.P.; Romani, L. T cell vaccination in mice with invasive pulmonary aspergillosis. *J. Immunol.* **2000**, *165*, 381–388. [CrossRef] [PubMed]

107. Ito, J.I.; Lyons, J.M.; Hong, T.B.; Tamae, D.; Liu, Y.-K.; Wilczynski, S.P.; Kalkum, M. Vaccinations with recombinant variants of Aspergillus fumigatus allergen Asp f 3 protect mice against invasive aspergillosis. *Infect. Immun.* **2006**, *74*, 5075–5084. [CrossRef]
108. Bozza, S.; Clavaud, C.; Giovannini, G.; Fontaine, T.; Beauvais, A.; Sarfati, J.; D'Angelo, C.; Perruccio, K.; Bonifazi, P.; Zagarella, S.; et al. Immune sensing of Aspergillus fumigatus proteins, glycolipids, and polysaccharides and the impact on Th immunity and vaccination. *J. Immunol.* **2009**, *183*, 2407–2414. [CrossRef]
109. Liu, M.; Machova, E.; Nescakova, Z.; Medovarska, I.; Clemons, K.V.; Martinez, M.; Chen, V.; Bystricky, S.; Stevens, D.A. Vaccination with mannan protects mice against systemic aspergillosis. *Med. Mycol.* **2012**, *50*, 818–828. [CrossRef]
110. Tomblyn, M.; Chiller, T.; Einsele, H.; Gress, R.; Sepkowitz, K.; Storek, J.; Wingard, J.R.; Young, J.A.; Boeckh, M.J. Guidelines for preventing infectious complications among hematopoietic cell transplantation recipients: A global perspective. *Biol. Blood Marrow Transplant.* **2009**, *15*, 1143–1238. [CrossRef]
111. Edwards, J.E., Jr.; Schwartz, M.M.; Schmidt, C.S.; Sobel, J.D.; Nyirjesy, P.; Schodel, F.; Marchus, E.; Lizakowski, M.; DeMontigny, E.A.; Hoeg, J.; et al. A Fungal Immunotherapeutic Vaccine (NDV-3A) for Treatment of Recurrent Vulvovaginal Candidiasis-A Phase 2 Randomized, Double-Blind, Placebo-Controlled Trial. *Clin. Infect. Dis.* **2018**, *66*, 1928–1936. [CrossRef] [PubMed]
112. Casadevall, A.; Pirofski, L.A. Immunoglobulins in defense, pathogenesis, and therapy of fungal diseases. *Cell Host Microbe* **2012**, *11*, 447–456. [CrossRef] [PubMed]
113. Malphettes, M.; Gerard, L.; Galicier, L.; Boutboul, D.; Asli, B.; Szalat, R.; Perlat, A.; Masseau, A.; Schleinitz, N.; Le Guenno, G.; et al. Good syndrome: An adult-onset immunodeficiency remarkable for its high incidence of invasive infections and autoimmune complications. *Clin. Infect. Dis.* **2015**, *61*, e13–e19. [CrossRef] [PubMed]
114. Braem, S.G.; Rooijakkers, S.H.; van Kessel, K.P.; de Cock, H.; Wosten, H.A.; van Strijp, J.A.; Haas, P.J. Effective Neutrophil Phagocytosis of Aspergillus fumigatus Is Mediated by Classical Pathway Complement Activation. *J. Innate Immun.* **2015**, *7*, 364–374. [CrossRef] [PubMed]
115. Torosantucci, A.; Chiani, P.; Bromuro, C.; De Bernardis, F.; Palma, A.S.; Liu, Y.; Mignogna, G.; Maras, B.; Colone, M.; Stringaro, A.; et al. Protection by anti-beta-glucan antibodies is associated with restricted beta-1,3 glucan binding specificity and inhibition of fungal growth and adherence. *PLoS ONE* **2009**, *4*, e5392. [CrossRef] [PubMed]
116. Cenci, E.; Mencacci, A.; Spreca, A.; Montagnoli, C.; Bacci, A.; Perruccio, K.; Velardi, A.; Magliani, W.; Conti, S.; Polonelli, L.; et al. Protection of killer antiidiotypic antibodies against early invasive aspergillosis in a murine model of allogeneic T-cell-depleted bone marrow transplantation. *Infect. Immun.* **2002**, *70*, 2375–2382. [CrossRef] [PubMed]
117. Nosanchuk, J.D.; Dadachova, E. Radioimmunotherapy of fungal diseases: The therapeutic potential of cytocidal radiation delivered by antibody targeting fungal cell surface antigens. *Front. Microbiol.* **2011**, *2*, 283. [CrossRef]
118. Rezvani, K.; Rouce, R.; Liu, E.; Shpall, E. Engineering Natural Killer Cells for Cancer Immunotherapy. *Mol. Ther. J. Am. Soc. Gene Ther.* **2017**, *25*, 1769–1781. [CrossRef]
119. Daver, N.; Kontoyiannis, D.P. Checkpoint inhibitors and aspergillosis in AML: The double hit hypothesis. *Lancet Oncol.* **2017**, *18*, 1571–1573. [CrossRef]
120. Fites, J.S.; Gui, M.; Kernien, J.F.; Negoro, P.; Dagher, Z.; Sykes, D.B.; Nett, J.E.; Mansour, M.K.; Klein, B.S. An unappreciated role for neutrophil-DC hybrids in immunity to invasive fungal infections. *PLoS Pathog.* **2018**, *14*, e1007073. [CrossRef]
121. Ngo, H.X.; Shrestha, S.K.; Garneau-Tsodikova, S. Identification of Ebsulfur Analogues with Broad-Spectrum Antifungal Activity. *ChemMedChem* **2016**, *11*, 1507–1516. [CrossRef] [PubMed]
122. Wiederhold, N.P.; Patterson, T.F.; Srinivasan, A.; Chaturvedi, A.K.; Fothergill, A.W.; Wormley, F.L.; Ramasubramanian, A.K.; Lopez-Ribot, J.L. Repurposing auranofin as an antifungal: In vitro activity against a variety of medically important fungi. *Virulence* **2017**, *8*, 138–142. [CrossRef] [PubMed]

Journal of

Fungi

MDPI

Review

Sporotrichosis In Immunocompromised Hosts

Flavio Queiroz-Telles [1,*], Renata Buccheri [2] and Gil Benard [3]

[1] Department of Public Health, Federal University of Paraná, Curitiba 80060-000, Brazil
[2] Emilio Ribas Institute of Infectious Diseases, São Paulo 05411-000, Brazil; renatabuccheri@gmail.com
[3] Laboratory of Medical Mycology, Department of Dermatology, and Tropical Medicine Institute, University of São Paulo, Sao Paulo 05403-000, Brazil; bengil60@gmail.com
*   Correspondence: queiroz.telles@uol.com.br

Received: 3 December 2018; Accepted: 7 January 2019; Published: 11 January 2019

Abstract: Sporotrichosis is a global implantation or subcutaneous mycosis caused by several members of the genus *Sporothrix*, a thermo-dimorphic fungus. This disease may also depict an endemic profile, especially in tropical to subtropical zones around the world. Interestingly, sporotrichosis is an anthropozoonotic disease that may be transmitted to humans by plants or by animals, especially cats. It may be associated with rather isolated or clustered cases but also with outbreaks in different periods and geographic regions. Usually, sporotrichosis affects immunocompetent hosts, presenting a chronic to subacute evolution course. Less frequently, sporotrichosis may be acquired by inhalation, leading to disseminated clinical forms. Both modes of infection may occur in immunocompromised patients, especially associated with human immunodeficiency virus (HIV) infection, but also diabetes mellitus, chronic alcoholism, steroids, anti-TNF treatment, hematologic cancer and transplanted patients. Similar to other endemic mycoses caused by dimorphic fungi, sporotrichosis in immunocompromised hosts may be associated with rather more severe clinical courses, larger fungal burden and longer periods of systemic antifungal therapy. A prolonged outbreak of cat-transmitted sporotrichosis is in progress in Brazil and potentially crossing the border to neighboring countries. This huge outbreak involves thousands of human and cats, including immunocompromised subjects affected by HIV and FIV (feline immunodeficiency virus), respectively. We reviewed the main epidemiologic, clinical, diagnostic and therapeutic aspects of sporotrichosis in immunocompromised hosts.

Keywords: AIDS; IRIS; cat-transmitted sporotrichosis; immunocompromised hosts; mycoses of implantation; sporotrichosis; *Sporothrix brasiliensis*; *Sporothrix schenckii*; subcutaneous mycoses

---

1. Introduction

Sporotrichosis is a subacute to chronic fungal infection caused by several species of genus *Sporothrix*, a group of thermal dimorphic fungi. Although disease occurs worldwide, most cases are reported in tropical and subtropical zones from Latin America, Africa and Asia [1,2]. Usually, sporotrichosis is an implantation mycosis whose infectious propagules are inoculated from several environmental sources into skin, mucosal or osteoarticular sites [3]. Less frequently, infection may occur by inhalation, resulting in pulmonary disease [4], but in both modalities, immunocompetent and immunocompromised patients can be affected.

In endemic regions, this disease is mainly associated with plant transmission (sapronosis), the main etiologic agents being *S. schenckii* and *S. globosa* [5]. During the last three decades, a new, probable mutant species, *S. brasiliensis*, has emerged in the state of Rio de Janeiro, Brazil. This species is transmitted to humans by infected cats (zoonosis), causing the largest outbreak of sporotrichosis ever reported [6–8]. This epizootic outbreak continues to expand, affecting human and feline patients in several Brazilian regions, possibly reaching neighboring countries [9]. Feline sporotrichosis is unique among infections caused by endemic dimorphic fungi because it is directly transmitted in the yeast phase. The feline lesions typically harbor a high yeast-like fungal burden that can be acquired via cat scratches and bites, by non-traumatic ways, such as a cat's cough or sneezing, and direct contact between patients' integumental barriers and animal secretions [8].

Similar to other endemic mycoses, sporotrichosis in immunocompromised hosts is usually clinically remarkable. The increased clinical severity is related to a decrease of host immune and inflammatory responses, heavy fungal burden, extensive dissemination and higher mortality rates. In addition, in opportunistic sporotrichosis (OS), conventional serology may reveal false negative antibodies levels and long courses of systemic antifungal therapy are usually required.

This mycosis can affect anyone regardless of age, gender or comorbidities, mostly depending on exposure [1]. Human immunodeficiency virus (HIV)/AIDS changes the natural history of sporotrichosis and its opportunistic character depends on the immune status of the host. Comorbidities such as diabetes mellitus, chronic alcoholism, steroid treatment, hematologic cancer and organ transplantation have been sporadically described as risk factors for severe forms of the disease and case reports have focused on unusual manifestations in these scenarios. The aim of this review is to discuss the main epidemiological, clinical, diagnostic and therapeutic aspects of OS, with an emphasis on cat-transmitted sporotrichosis (CTS). In addition, in an attempt to better understand why certain comorbidities may predispose to OS, we performed a critical review of the data on the immune response in sporotrichosis.

2. Epidemiology and Clinical Manifestations

Sapronotic sporotrichosis is mainly related to several types of transcutaneous injuries, occurring in patients in contact with plant material or contaminated soil. Less frequently, animal associated trauma has been associated with *S. schenckii* and *S. globosa* and, to a lesser extent, with *S. pallida* clade (*S. mexicana*, *S. chilensis*, *S. luriei* and *S. pallida*) [5,10]. Zoonotic sporotrichosis is caused by *S. brasiliensis*, and although this expanding and uncontrolled outbreak is apparently limited to Brazil's borders, proven cases have been reported in Argentina and possibly Panama [9,11].

Sporotrichosis is a spectral disease, classified into two categories (cutaneous and extracutaneous sporotrichosis), which comprise four distinct clinical forms: lymphocutaneous (LC), fixed cutaneous (FC), disseminated cutaneous, and extra-cutaneous [12]. LC and FC forms are classical and most common clinical presentations [13,14]. Typically, but not exclusively, disseminated cutaneous and extra-cutaneous, considered severe forms, occur in hosts with depressed cellular immunity [15–19]. Indeed, findings from studies that represent the largest reported outbreaks of this mycosis in regions such as China, Japan, Peru and Brazil indicate the frequency of these severe forms ranges from 1.3% to 9% [13,14,20,21].

The disseminated cutaneous form is a rare variant of sporotrichosis characterized by multiple skin lesions at noncontiguous sites without extracutaneous involvement. It is important to emphasize that in some situations it is difficult to identify whether the clinical presentation is due to dissemination from a single lesion or to multiple inoculations [22]. The extracutaneous or disseminated forms are characterized by the involvement of organs and systems. Skin, eyes, lungs, liver, kidney, heart, central nervous system (CNS) and genitalia have already been described as affected sites. The osteoarticular form may occur by contiguity of the primary lesion or hematogenous spread dissemination from lungs [4,23–26].

Poorly explored until recently, the reemergence of this mycosis in different parts of the world led to a renewed interest in its study, mainly focusing on immunopathogenesis mechanisms and immune response against fungal molecular components; these topics have been reviewed recently [10,27,28]. Nevertheless, our knowledge on the immunopathogenesis of sporotrichosis is still fragmentary.

3. What We (Don´t) Know about the Immune Response in Human Sporotrichosis

In sporotrichosis, exposure does not necessarily result in overt disease since the proportion of those who will develop an illness is smaller than those who control the infection. Although the mechanisms underlying this observation are unknown, some studies suggest that different *Sporothrix* species may present different pathogenicity levels, which may lead to varying degrees in clinical manifestations, with some "susceptible" individuals developing the more benign FC form (which can eventually heal spontaneously), while others evolving to severe disseminated or extracutaneous forms.

Current available data on the immune response to *Sporothrix* spp. (or their components) is predominantly based on in vitro studies and in rodent experimental models, reprising the strategies used in the investigation of the immune reactivity of the other, better studied, endemic deep or subcutaneous mycoses; data gathered directly from human patients is scarce. While those studies differ widely in the species tested, the fungal phase used to infect or to obtain fungal components (conidia vs. yeast vs. germlings), the size and route of inoculation (mostly intraperitoneal and intravenous) and animal models employed (mouse strains, Wistar rats, golden hamster and, more recently, the great wax moth *Galleria mellonella*), the extent to which they contribute to the understanding of the immunopathogenesis of the human disease is still not clear since the pieces do not fit the puzzle. Only recently, mouse models mimicking the human disease (i.e., subcutaneous inoculation) have been explored [29–32].

As a first line of defense against pathogens, innate immunity is considered key to fungal control. Pioneering work by Kajiwara et al. showed that neutrophils and macrophages from mice with chronic granulomatous disease (CGD), who show defective NADPH oxidase complex function and fail to generate microbicidal reactive oxygen species (ROS), were unable to control the growth of *S. schenckii* yeast cells, and those animals developed a disseminated lethal disease upon subcutaneous inoculation, while wild-type counterparts were resistant to systemic infection and survived [32]. Translating those findings, however, to human context seems challenging: while Cunningham et al. observed phagocytosis and intracellular killing of *S. schenckii* by human polymorphonuclear cells in vitro, mediated by the H_2O_2–KI–myeloperoxidase system [33], Schafner et al. found that virulent *S. schenckii* was resistant to killing by neutrophils and H_2O_2 [34].

The controversial role of nitric oxide (NO) in sporotrichosis highlights the complexity of the host's immune response in this mycosis. Experimental data suggest a dual role for NO, supporting both its fungicidal activity against *S. schenckii* in vitro [35] and its association with T cell suppression and poorer outcome in murine models [36]. In patients' biopsies, expression of NO synthase-2 was higher in LC lesions, while FC injuries displayed more intense inflammation, tissue destruction and higher fungal burden [37,38].

Human macrophages were also shown to phagocytose and kill (probably through ROS release) *S. schenckii* conidia and yeast cells [39]. Some studies suggest that melanin expression would protect the isolates from macrophage phagocytosis and oxidative attack [35,40]. However, there are no studies analyzing the in situ expression of melanin by intralesional yeast cells in biopsies. Curiously, in the human monocytic cell line THP-1, engulfment of *S. schenckii* conidia preferentially occurs through mannose receptors while yeast cells internalization relies on complement receptors [39], suggesting the interplay of different receptors in fungi–host interaction.

In parallel to neutrophils and macrophages, it was shown that bone marrow-derived mouse dendritic cells (DC) also participate in the recognition process of fungal components and drive the cellular immune responses [41], regulating the magnitude and balance of Th-1 and Th-17 responses in vitro. The latter were associated with control of the fungal burden in an intraperitoneal infection mouse model [42,43]. Other immune cells, such as mast cells, can also amplify the acute response by releasing mediators (histamine and proinflammatory cytokines that attract neutrophils) that exacerbate the inflammatory process, but with deleterious effects to the host, rather than contributing to control of fungal burden [44,45].

Several studies addressed the recognition of *Sporothrix* spp. and their components by innate immunity receptors (pattern recognition receptors, PRR) and its influence in subsequent cellular immunity. Toll-like receptors (TLR) are conserved membrane-associated proteins that recognize a broad set of microbial components, such as *S. schenckii* lipid antigens, recognized via TLR4 [46], triggering diverse cell responses. TLR2 activation, for example, enhances in vitro phagocytosis of *S. schenckii* yeast cells by mouse macrophages and promotes the release an array of pro- (TNF-α, IL-1β, IL-12) and anti-inflammatory (IL-10) cytokines as well as effector/cytotoxic compounds (e.g., NO) [47,48]. Keratinocytes are also activated through TLR2 and TLR4 to release proinflammatory cytokines when challenged with *S. schenckii* yeast cells [49]. However, it is not yet clear from these studies whether the elicited inflammatory response contributes to enhanced immunopathology or host protection.

Dectin-1 and dectin-2 are important PRRs that trigger Th-17 responses but currently there are only data for participation of dectin-1 in triggering Th-17 responses in an intraperitoneal mouse model of sporotrichosis [50]. Conversely, Zhang et al. showed that both dectin-1 and IL-17 production were dispensable for clearance of *S. schenckii* infection in a rat model [51]. There is also evidence from a mouse model of systemic infection that activation of the inflammasome exerts a transitory protective role, especially due to IL-1, IL-18 and caspase-1 [42,52,53], whose impairment reduced Th-17 and Th-1 mediated inflammatory responses leading to higher susceptibility to *S. schenckii* infection [52]. *S. schenckii* yeast cells can also activate the alternative (antibody independent) complement pathway in vitro but its relevance to in vivo host defenses was not defined [54].

Overall, these studies suggest that *Sporothrix* spp. can be recognized by different innate immunity receptors. Which particular set of these (and their signaling pathways) is involved in human infection, which could also be affected by the different infection routes (percutaneous or inhalatory), remains to be determined. Furthermore, with the identification of new *Sporothrix* spp., the involvement of immune receptors could be species-specific.

In fact, Arrillaga-Moncrieff et al. showed that pathogenicity differs among species: *S. brasiliensis* was the most pathogenic, followed by *S. schenckii*, when compared to *S. albicans*, *S. globosa* and *S. mexicana* [55]. Almeida-Paes et al. suggested those differences might even exist within a single species: *S. brasiliensis* isolates obtained from patients with more severe disease express more putative "virulence" factors, such as urease and melanin, and are able to cause a more disseminated disease [56,57]. In Venezuela, a retrospective study gathered isolates from patients and found that *S. globosa* is isolated mainly from patients with FC sporotrichosis while *S. schenckii* would be related to LC forms [58].

Fibronectin surface adhesins expressed by *S. schenckii* have also been described to increase pathogenicity in C57BL/6 mice. Although analyses of differences according to species were not available at that time, they did not find direct correlation between virulence and the clinical or environmental origin of the isolates: the lowest virulence was observed for an isolate recovered from a patient with meningeal sporotrichosis [59].

Recently, Martinez-Alvarez et al. showed that human peripheral blood mononuclear cells (PBMC) differentially recognize *S. brasiliensis* and *S. schenckii* [60]. The three *S. schenckii* morphologies stimulated higher levels of pro-inflammatory cytokines than *S. brasiliensis*, while the latter stimulated higher IL-10 levels. This finding could help to explain the apparent higher pathogenicity of *S. brasiliensis*. However, as we still do not know the first steps of the infection in humans, the contribution of each morphology to its successful occurrence remains to be established. The authors additionally showed that dectin-1 was a key receptor for cytokine production induced by *S. schenckii*, but was dispensable for *S. brasiliensis* germlings. TLR2 and TLR4 were also involved in sensing of *Sporothrix* cells, with a major role for the former during cytokine production. The mannose receptor had a minor contribution in *S. schenckii* yeast-like cells and germlings recognition, but *S. schenckii* conidia and *S. brasiliensis* yeast-like cells stimulated pro-inflammatory cytokines via this receptor.

Immunochemical studies in the 1970s already suggested that cell wall components elicited immediate and delayed immune responses [61]. Subsequent studies reinforced the important role played by cell-mediated mechanisms (i.e., TCD4+ lymphocytes and activated macrophages) in resistance to intravenous experimental sporotrichosis in athymic nude [62–65] and Swiss mice [66,67]. This research line was resumed more recently in the search for vaccine candidates. Live yeast cells and/or exoantigens were used, and Th-1 and Th-17 responses were, in general, generated [41,43,53], both of which appeared to be required for protection in these models. However, there was also evidence of an important participation of Th-2 responses at later stages and activation of macrophages with anti-inflammatory characteristics (defined as M2 macrophages) such as high levels of IL-10 secretion [47,48,60,68]. It has been suggested that isolates from cutaneous lesions were more potent to activate human monocyte-derived DCs to drive Th-1 responses than isolates from visceral lesions. However, this finding should be regarded with caution since the study was performed before reclassification of the *S. schenckii* complex into several species and thus the isolates' differences could rather reflect different species [69].

Sporothrix spp. components have also been studied with regard to human humoral responses. Sera from extracutaneous or more severe forms of sporotrichosis recognized a wider range of antigens and displayed higher antibody titers than sera from patients with cutaneous/less severe forms of sporotrichosis [56,70,71]. A protective role of antibodies, possibly through facilitation of phagocytosis, has been described in some experimental models [71–74].

Unfortunately, data stemming from patients are scarce and are represented mostly by histopathology studies of biopsies taken from patients, with a limited set of parameters analyzed due to limitations inherent to these methods. Moreover, some studies involved a rather small number of patients. Nonetheless, the cutaneous inflammatory process consists in most cases of a suppurative granulomatous response, with frequent presence of liquefaction and/or necrosis. Of note, the paucity of fungal elements (absent from 65% of the biopsies), associated with better granuloma formation (epithelioid granulomas, higher infiltration of lymphocytes, presence of fibrosis, absence of necrosis) suggests the ability of the human system in partially containing the fungal burden [75,76]. This may help to explain why (a) exposure does not necessarily result in development of illness and (b) some patients self-heal.

Immunohistochemistry studies showed the presence of CD4+ and CD8+ T-cells, CD83+ DC, macrophages and monocytes, and the expression of IFN-γ, but not of iNOS, within granulomas [38,77,78]. Compared to the FC form, LC patients had more intense signs of inflammation (higher infiltration of neutrophils and lymphocytes, and higher expression of nitric oxide synthase 2) and higher fungal burden. IFN-γ expression did not differ but IL-10 was more prominent in LC than FC lesions, consistent with the more intense inflammatory process in the former [77]. Interestingly, these authors also observed a higher ability of PBMC from patients than healthy individuals to release IFN-γ and IL-10 upon in vitro challenge with *S. schenckii* antigen [77]. An early report already noted a trend toward lower T-lymphocyte responsiveness in systemic disease as compared with the LC form [78]. Interestingly, of six systemic sporotrichosis patients, one had bone marrow aplasia and four reported daily consumption of variable amounts of alcohol, while none of the LC patients reported these conditions. Overall, these data reinforce the ability of the human immune response in limiting, at least partially, the disease caused by *S. schenckii*. However, this notion can be challenged by the report of severe extracutaneous sporotrichosis in apparently immunocompetent individuals [79–81]. A summarized, schematic view of the data obtained from experimental studies of the immune response in sporotrichosis is shown in Figure 1.

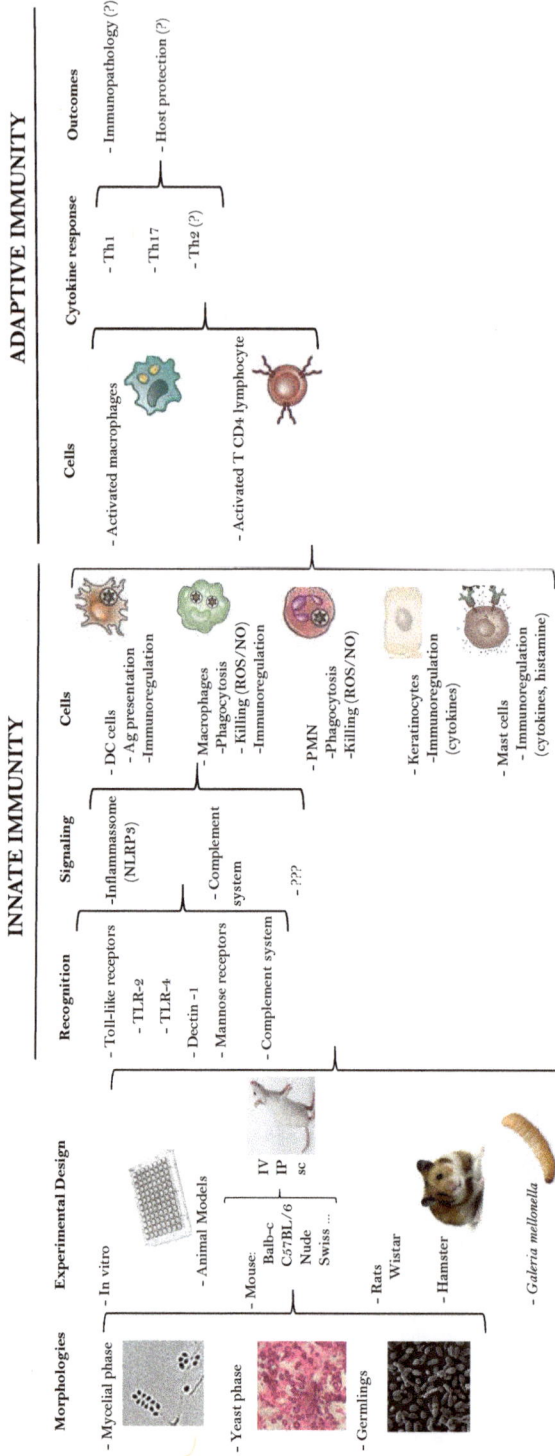

Figure 1. Schematic view of the data obtained from experimental studies of the immune response in sporotrichosis. IV: intravenous; IP: intraperitoneal; sc: subcutaneous; DC: dendritic cells; ROS: reactive oxygen species; NO: nitric oxide; PMN: polymorphonuclear cells; Th: T helper.

4. Sporotrichosis in AIDS Patients

The first case report of OS in HIV-infected individuals dates from 1985 [82]. In the last decade, the reemergence of the disease in Rio de Janeiro, Brazil, was followed by an increase in the number of cases in HIV co-infected patients. Nevertheless, to date, there are no more than 107 cases reported worldwide. Reflecting the rare occurrence of the disease in this group, our knowledge of the clinical features and management principles is based on expert opinions, case studies and retrospective cohort studies [12,83–85]. Still, it is worth noting that HIV/AIDS patients either had disseminated or cutaneous sporotrichosis, or did not become ill after exposure [12].

Data from the largest retrospective cohort study of 3618 cases of sporotrichosis revealed that 1.32% were HIV co-infected. Close to half (44%) were hospitalized over time, much more frequently than HIV negative patients (1%). Although the main cause for hospitalization in both groups was disseminated disease, this corresponded to 90.5% of the hospitalized HIV patients but only 43.2% of HIV negative subjects. In addition, hospitalized patients had a mean CD4 T lymphocyte count of 125 cells/μL and deaths attributed to sporotrichosis occurred 45 times more [83].

Clinical data from a systematic review showed the majority of HIV co-infected patients have cutaneous disease associated with involvement of other organs or systems. Their median CD4 T lymphocyte count was 97 cells/μL. In addition, unusual manifestations cannot be underestimated, with 17% of the cases presenting CNS involvement [84]. CNS disease has already been described in HIV-infected patients without clinical evidence of neurological symptoms [79,85,86]. Ten previous published cases with sufficient clinical details indicate poor prognosis for CNS involvement [85–93]. Thus, investigation of CNS disease in this specific population is strongly recommended [84] in order to provide early diagnosis and aggressive treatment. Most of these patients are males with a mean CD4 count of 101cells/μL. In all cases, skin lesions were present before or associated with the onset of meningeal symptoms. In addition, the concomitant involvement of lungs [85,89,92,93], mucosa [85,86,89], bone [86], kidney [93], testicles, epididymis, bone marrow, lymph nodes and pancreas [90] has also been described.

In the majority of cases, patients present positive cerebrospinal fluid (CSF) cultures on admission or during the follow-up. Only in one case, lumbar puncture was sterile and *Sporothrix* sp. was observed in tissue sections [90]. Most importantly, however, biopsies of skin lesions yielded growth of *Sporothrix* spp. in almost all the previous cases [85–88,91–93], allowing the diagnosis to be made before CSF cultures results were available or became positive. Treatment of sporotrichosis meningoencephalitis is by far the most challenging aspect in the disease management and the poor therapeutic response observed in these few cases is remarkable. Except in one patient [93], amphotericin B formulations were the initial therapeutic choice, although a significant rate of recurrence or relapse of neurological symptoms was observed [85,86,92]. Nine of the patients died; the single patient who survived resolved the infection without sequels [92].

Other unusual manifestations associated with cutaneous lesions were endocarditis, mucosal involvement (ocular and nasal), uveitis, endophthalmitis, and pulmonary and osteoarticular involvement [12,24]. Isolated involvement of the lungs and sinus has also already been described [94–97].

The prevalence of HIV co-infection in patients with less severe forms is not yet well established, because HIV routine testing is not recommended for every patient with sporotrichosis. Usually, only those with severe manifestations or a suspected HIV infection result in a laboratory investigation, which would overestimate the incidence of severe clinical presentations and poor prognosis in this population. Nevertheless, Freitas et al. evaluated the prevalence of HIV co-infection in the Rio de Janeiro epidemic by testing stored blood samples of 850 patients with benign forms of sporotrichosis from 2000 to 2008, and found only one positive result [83]. Moreover, only a few cases of LC and FC forms have been described so far in HIV-infected patients with a mean CD4 T lymphocyte count of 513 cells/μL [12,83,98]. Overall, these data support the notion that HIV co-infection modifies the clinical presentation, severity and outcome of the patients with sporotrichosis, in accordance to their immune status and degree of immunosuppression [83].

5. Sporotrichosis Associated with IRIS

Only six cases of sporotrichosis associated with immune reconstitution inflammatory syndrome (IRIS) have been published to date. There are two cases of paradoxical sporotrichosis meningitis IRIS in patients who exhibited cutaneous lesion and were under itraconazole treatment before the onset of neurological disease. Both patients had confirmation of virologic response to antiretroviral therapy (ART). Despite treatment with amphotericin B at 4–8 weeks, the patients presented recurrence of neurological symptoms during follow-up with itraconazole and/or amphotericin B 2–3 times a week. *Sporothrix* sp. was isolated from CSF at some point [92].

These cases are controversial due to the difficulties in making a definite diagnosis of sporotrichosis IRIS when cultures remain positive, properly excluding other causes of clinical deterioration as therapeutic antifungal failure. Difficulties in defining IRIS still apply to other endemic mycoses such as paracoccidioidomycosis and cryptococcosis [99,100]. Although some debate persists, an apparent consensus for diagnosis of paradoxical IRIS associated with opportunistic mycoses is the worsening or appearance of new clinical and/or radiological manifestations consistent with an inflammatory process occurring during appropriate antifungal therapy with sterile cultures for the initial fungal pathogen within 12 months of ART initiation [100,101].

Another questionable report from Brazil described one patient with disseminated sporotrichosis who was already under treatment for two months with itraconazole when ART started. After six weeks, the patient experienced reactivation of old lesions and development of new cutaneous and mucosal lesions. However, cultures of skin biopsy were positive for *Sporothrix* sp. and the patient recovered well with increased doses of itraconazole combined with amphotericin B [92].

Lyra et al. described two cases of disseminated cutaneous sporotrichosis whose clinical presentations were more consistent with IRIS than progressive fungal infection or failure of treatment. Both patients started antifungal therapy shortly followed by ART. However, after four and five weeks, the patients exhibited paradoxical clinical worsening with recurrence of the lesion as well as development of new lesions along with systemic inflammatory symptoms such as fever and arthralgia. Subsequent mycological examination did not reveal fungal growth. The patients were treated with prednisone, resulting in rapid improvement of arthralgia and fever, followed by resolution of skin lesions [102].

Finally, one case described a patient who had no cutaneous findings before ART, but experienced unmasking of disseminated cutaneous sporotrichosis after five weeks. Cultures of lesion exudates were positive for *Sporothrix* sp. The patient's cat had died of sporotrichosis one month before the patient started ART. He presented complete regression of the lesions after antifungal therapy [92].

Case Presentation

A 59-year-old Brazilian man presented cachexia and disseminated and ulcerated skin lesions with one-year evolution (Figure 2A). Before his illness, he worked as an agriculturist, truck driver and a sewerage system cleaner in his town. During his last professional activity, he was continuously exposed to polluted water. Eight months earlier, the diagnosis of leprosy was made without any microbiological evidence and he was unsuccessfully treated with rifampin, dapsone and clofazimine. Six months ago, HIV infection was detected and lamivudine, tenofovir and efavirenz were added. At admission, he was depressed, febrile and complaining of pain. His body weight was 40 kg, and, besides the cutaneous clinical manifestations, there were no signs of internal organ involvement. The main laboratory findings included anemia with hemoglobin of 9.1 g/dL, leukocytosis (12,100 cell/μL) and protein chain reaction (PCR) of 11mg/L. HIV test was positive with CD4 cell count of 584 cells/mm^3 and viral load of 1558 copies/mL (log 3.1). A skin biopsy depicted a mixed exudative and granulomatous cellular infiltrate with a few round to elongated yeast cells (Figure 2B). The cultures of biopsy fragments yielded a dimorphic fungus phenotypically identified as *Sporothrix* sp., later identified by DNA sequence as *S. schenckii*. The anti-lepromatous therapy was stopped and the patient was treated with itraconazole, 400 mg per day, and cotrimoxazole 360mg/800mg per day for secondary bacterial infection. Because IRIS was suspected, prednisone at the daily dose of 20 mg per day was added and ART was changed to atazanavir/ritonavir due to probable drug-to-drug interactions between itraconazole and the previous antiretrovirals. He improved gradually and corticosteroid and cotrimoxazole were discontinued. After three months of therapy, itraconazole was reduced to 200 mg per day and discontinued after six months. The patient presented complete clinical and mycological responses. (Figure 2C).

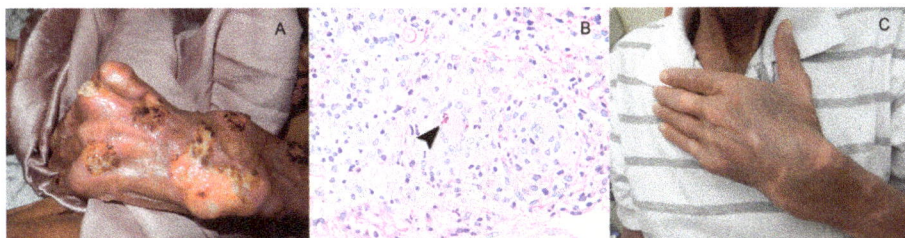

Figure 2. Ulcerated lesions in the hand and fist of a patient with human immunodeficiency virus (HIV) infection ad cutaneous disseminated sporotrichosis and immune reconstitution syndrome (**A**). A skin biopsy (**B**) depicted an exudative and granulomatous infiltrate with cigar shape and round yeast cells (arrow head), imbibed in multinucleate giant cells of the Langerhans type. Periodic Acid-Schiff stain × 400. The patient responded well to long course of continuous itraconazole intercalated with short courses of cotrimoxazole for secondary bacterial infection and prednisone for immune reconstitution inflammatory syndrome (IRIS) control (**C**).

6. Comorbidities as Risk Factors for Sporotrichosis

Diabetes Mellitus and Alcoholism

The main underlying disease reported in sporotrichosis outbreaks is diabetes mellitus, reaching up to 23% of the cases, followed by alcohol consumption, reaching from 5% to 8% [20,22,103]. Despite this observation, in hyperendemic areas of sporotrichosis, little is known about the contribution of these conditions to development of severe forms and there is limited understanding about the immunosuppressive mechanisms involved. To date, no large series or cohort studies described particularities of the clinical presentation in this population and most published reports of sporotrichosis in diabetes and alcoholism have focused on unusual manifestations, which are not necessarily the predominant forms observed in these populations.

Some of the first cases in alcoholic and diabetic patients date back to 1961 and 1970, respectively, and both patients presented with primary pulmonary sporotrichosis [104,105]. A recent systematic analysis of the literature addressing 86 cases of pulmonary sporotrichosis showed that diabetes mellitus was present in six patients and alcohol consumption in 34 cases. Most of these cases (75%) presented with primary pulmonary sporotrichosis. Of note, the cavitary pattern on radiology was the most common finding and 45% of the patients presented extrapulmonary involvement, the skin being the most affected site followed by joint involvement [4].

Putative differences in clinical presentation in diabetic and alcoholic patients occur. One presented disseminated cutaneous lesions and the other particularly severe/destructive localized lesions with a granulomatous aspect, denoting an enhanced pathogenicity in these localized cases. Nevertheless, all patients had a marked improvement with conventional antifungal treatment [106–109]. The most emblematic case of disseminated disease occurred in a diabetic and alcoholic patient who developed a fatal fungemia after 17 days of hospitalization [110]. The other cases corresponded to isolated monoarthritis, endophthalmitis and cutaneous disseminated sporotrichosis [111–113].

A retrospective study of 238 cases in the Peruvian highlands, where *S. schenckii* is hyperendemic, pointed to alcoholism and diabetes mellitus as significant underlying factors. The majority of patients had cutaneous or lymphocutaneous disease; only nine patients presented disseminated cutaneous disease and no cases of extracutaneous involvement were found [13]. In Brazil, clinical data from 178 patients with culture-positive sporotrichosis treated during the period of 1998–2001 showed that 80.9% of the cases presented LC or localized cutaneous forms. Systemic sporotrichosis was not diagnosed even in cases involving alcohol or diabetes as comorbidities, and in 29 patients (16.3%) with skin lesions at multiple locations, this was more likely due to repeated inoculations during persistent contact with sick animals [22].

In another Brazilian case of 24 patients with widespread cutaneous lesions, only two were associated with alcoholism and diabetes, and these conditions may have acted as an immunosuppressive factor favoring either the establishment of the infection and/or its dissemination. None of these presented a history of multiple exposures that could account for the widespread cutaneous lesions. In addition, none of the other patients showed any immunosuppressive condition and were found to be in good general condition, although fever and/or arthralgia were reported in 50% of the cases [103]. Furthermore, Rosa et al. described 304 patients where only four cases with cutaneous disseminated and extracutaneous forms were recognized, but data regarding their comorbidities were not provided [20].

7. Other Immunosuppressive Conditions

The literature reports several cases of cutaneous-disseminated and extra-cutaneous sporotrichosis in patients under immunosuppressive treatments for rheumatologic, autoimmune conditions, solid organ transplantation (SOT), hematologic cancers and primary immunodeficiencies.

Osteo-articular or disseminated sporotrichosis misdiagnosed as rheumatoid arthritis, presumed inflammatory arthritis or sarcoidosis illustrates the role of iatrogenic immunosuppressive regimen in severity and complicated outcome [114–116]. Immunosuppressive therapy included steroids, azathioprine, tocilizumab, tacrolimus and cyclophosphamide—in one case, after almost one year of inappropriate therapy with several immunosuppressives (including prednisolone, tocilizumab, tacrolimus and cyclophosphamide), the patient experienced fungemia and died of respiratory insufficiency due to pulmonary sporotrichosis [114]. Two patients also had delayed diagnosis and progressed to disseminate disease albeit clinical improvement was achieved after antifungal therapy (amphotericin B or itraconazole) was started in parallel to lessening the iatrogenic immunosuppression [115,116].

A retrospective review of 19 cases of sporotrichosis diagnosed at a single service in the United States showed that seven patients were misdiagnosed initially and four received immunosuppressive agents for other diagnoses, such as polyarteritis nodosa, sarcoidosis, pyoderma gangrenosum and vasculitis [117]. One additional patient had received immunosuppressive therapy for a pre-existing polyarthropathy before the development of his cutaneous lesion. In contrast, none of the patients presented extracutaneous disease. The index case was diagnosed as pyoderma gangrenosum for disseminated leg ulcerated lesions and received immunosuppressive treatment with aziatropine, prednisone and cyclosporine with further worsening of the lesions. Treatment required debridement of necrotic tissue, plastic surgery and subsequent staged skin grafting, together with an 18-month course of 600mg/day itraconazole and cessation of immunosuppression. However, no data regarding the response to treatment of these 19 patients were provided.

Similarly, cases of cutaneous disseminated and LC forms in immunosuppressive therapy with tacrolimus, anti-TNF-alpha and prednisone due to lupus nephritis, ankylosing spondylitis and sciatic pain, respectively, have been reported. All these patients had a good clinical response to antifungal therapy (potassium iodide or itraconazole) and discontinuation of immunosuppressive therapy [118–120].

The possible role of immunosuppressive drugs in atypical clinical presentation is reinforced by a systematic analysis of pulmonary sporotrichosis. In this study, of the 86 cases of pulmonary sporotrichosis included, 64 had primary pulmonary disease and 22 also had extra-pulmonary involvement. The only significant difference between the groups that could represent a risk factor for multifocal disease was the increased use of immunosuppressant drugs by the extra-pulmonary group [4].

Sporotrichosis in SOT recipients manifests as more severe disseminated forms than in immunocompetent hosts. Few exceptions respond well to antifungal treatment, being considered uncommon according to a prospective surveillance study of invasive fungal infections conducted in 15 SOT centers in United Sates, which did not identify any case of sporotrichosis [121]. However, in the Rio de Janeiro epidemic, sporotrichosis was retrospectively recognized in one subject, among 42 kidney transplant patients, with extracutaneous disease (LC and bone involvement) [122]. In addition, Caroti et al. followed 774 Italian kidney transplant patients for 18 years and subcutaneous nodules or cutaneous lesions were identified in seven [123]. One patient presented an erythematous papulonodular lesion with positive culture for *S. schenckii*. Despite treatment with fluconazole, seven years after renal transplantation, the patient developed acute osteomyelitis and gangrene in the left foot with ulcers. The patient was treated again with fluconazole together with interruption of the immunosuppressive agent mycophenolate mofetil, presenting gradual regression of the lesions. In India, during a period of two years, 40 renal transplants were performed and pulmonary sporotrichosis was diagnosed in one patient on triple drug immunosuppression [124]. Finally, there are three additional cases in kidney transplant recipients reported in the literature, one case of cutaneous disseminated and two of disseminated disease [23,125]. All patients were taking immunosuppressive agents and were successfully treated with antifungal therapy including amphotericin B deoxycholate, lipid amphotericin B formulations, fluconazole and itraconazole. One unusual case of urinary sporotrichosis after renal transplantation has also been described [25].

Sporotrichosis in SOT other than kidney transplantation is even more rare. Disseminated sporotrichosis with LC, articular and pulmonary involvement was described in a patient 10 years after liver transplantation still on immunosuppressive regimen (tacrolimus and prednisone) [23]. Despite antifungal treatment with itraconazole and reduction of immunosuppressant drugs, after 300 days of follow-up, the patient showed only partial improvement. Pulmonary sporotrichosis in a lung transplant recipient was also reported. On the second day of transplantation, while on induction of immunosuppression with high dose methylprednisolone, tacrolimus, and mycophenolate mofetil, he presented pulmonary diffuse patchy bilateral infiltrates and *S. schenckii* was isolated from bronchoalveolar lavage. Treatment with amphotericin B lipid formulation followed by itraconazole maintenance therapy was successful [126]. Rare cases of sporotrichosis in patients with hematologic cancer have also been described. One patient with multiple myeloma presented disseminated disease, successfully treated with amphotericin B [80]. Two Hodgkin's disease patients, one with fatal meningeal sporotrichosis [127] and the other with a disseminated form refractory to potassium iodide, also responded to a long course of amphotericin B [80]. Two other cases reported patients with hairy cell leukemia, both with disseminated disease, one whose difficult and life-threatening course required liposomal amphotericin B followed by posaconazole (taken indefinitely), and the other exhibited a good responded to itraconazole [128,129].

Finally, primary immunodeficiency has been recognized as a risk factor for severe forms of the disease. A fatal case of disseminated form of sporotrichosis was described in one patient with primary idiopathic CD4 lymphocytopenia [130]. Another unusual case of *S. schenckii* cervical lymphadenitis was identified in a 33-month-old male with X-linked CGD that was successfully treated with surgical excision and voriconazole [131].

Overall, based mostly on published case reports, we suggest that patients on immunosuppressive regimen due to SOT, rheumatologic disease or other comorbidities are at higher risk of more severe clinical presentations of sporotrichosis. However, most cases presented a good outcome when provided with more prolonged and higher doses of antifungal treatment than used in immunocompetent hosts. In this scenario, the drug of choice should be guided by the severity of the disease, with initial therapy with amphotericin B being frequently required [114,116,117,125–128,130,132].

8. Laboratory Diagnosis

The most relevant diagnostic tool for patients with suspicion of sporotrichosis is isolation of etiologic agent from clinical specimens such as secretions, abscess aspirates and biopsied tissue fragments. In extracutaneous clinical forms, synovial fluid, blood, CSF and sputum should be cultivated. Fungal cultures may be obtained in standard media such as Sabouraud dextrose agar with antibiotics, Mycosel, blood agar and brain heart infusion media [133]. After 5 to 10 days, the yeast-like colonies may be observed at 37 °C incubation, although this time may be extended up to 30 days. For phenotypic identification, the micromorphology of mycelial forms must be seen after incubation at room temperature, although it cannot distinguish individual species. Species determination requires molecular methods for definitive identification [10,133].

In contrast to immunocompetent patients, where direct mycologic or histopathologic exams show poor sensitivity, in immunocompromised hosts these methods may depict higher sensitivity, especially in AIDS patients with very low CD4 cell counts [83,84]. In severely immunocompromised AIDS patients, cutaneous and lymphatic lesions may depict a big fungal burden, similar to that observed in cats with *S. brasiliensis* infections [7,134] (Figure 3). Immunocompromised patients with cutaneous disseminated and extracutaneous clinical forms may depict *Sporothrix* spp. yeasts under immunofluorescence, Giemsa, Gram, Grocott-Gomori and periodic acid Schiff (PAS) stains. When observed, yeasts present round to oval and elongated "cigar shape" forms [2,10,135,136]. Non-microbiologic diagnostic tests such as immunoelectrophoresis, immunodiffusion, ELISA and DNA sequencing PCR methods are very important for typical and immune reactive forms but the only commercially available test for immunodiagnostic of sporotrichosis is the latex agglutination technique [10,137,138].

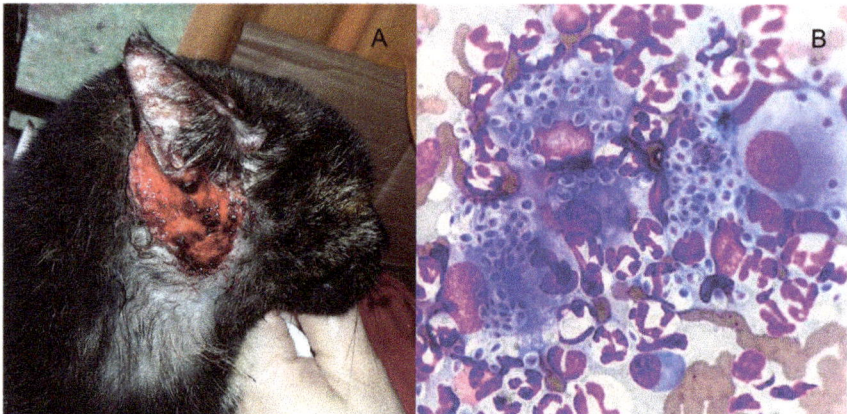

Figure 3. Ulcerated and papular vesicular lesions in the head and ear of a cat with proved *Sporothrix brasiliensis* infection (**A**). Cutaneous feline lesions are highly infective and harbor a great number of yeast cells of the fungus. Feline sporotrichosis can be easily diagnosed by secretion direct exam stained Giemsa × 1000 (**B**).

9. Treatment of Opportunistic Sporotrichosis

Therapy of patients with sporotrichosis associated with impaired host defenses does not differ from treatment modalities applied for immunocompetent individuals, except for inclusion of specific interventions for the underlying conditions leading to opportunistic disease. Although some experimental studies demonstrated different species might show variable in vitro sensitivity to systemic antifungal drugs, no clinical correlation in therapy of human sporotrichosis has been confirmed to date [133,138–141]. For patients with cutaneous or LC forms, itraconazole at the daily dose of 200–400 mg for 3–6 months is the therapy of choice. Exceptionally, if immunosuppression is maintained, a longer period of itraconazole therapy may be required [83,84]. Special attention is recommended for drug interaction between itraconazole and ART drugs such as efavirenz, ritonavir and darunavir [142]. If itraconazole is contraindicated due to intolerance, refractoriness or drug-to-drug interaction, 500 mg of terbinafine twice a day is the second option. Finally, for non-severe forms, super saturated potassium iodine solution, 40–50 drops three times per day, can be tried. Second generation triazoles as posaconazole and isavuconazole have not been evaluated yet.

Although infection with *Sporothrix* spp. is rarely life threatening, all forms of sporotrichosis require some kind of treatment. Unlike HIV/AIDS patients, diabetic and alcoholic subjects do not seem to have a worse prognosis and, in general, all patients show a satisfactory response even if in some cases it is necessary to increase the drug dose or the hospitalization stay due to the comorbidity itself or the severity of the lesions [22].

Regardless of the presence of any comorbidity, in pulmonary, severe or life-threatening disease, amphotericin B should be the initial therapy until the patient shows a favorable response, moving on to itraconazole [140]. In parallel, the management of those risk factors, such as control of chronic alcohol intake and steroid or anti TNF discontinuation should be carried out.

An issue in HIV-patient management concerns the best time to introduce ART. While some authors recommend its initiation should be delayed in patients with CNS disease in order to avoid IRIS [84], currently there is no sufficient evidence to support this recommendation for secondary prophylaxis. Data from AIDS patients and tuberculosis or *Cryptococcus* meningitis suggest that patients should start antifungals before ART [143,144]. Therefore, similar to indications already described in the literature for other opportunistic mycoses, long-term suppressive therapy should be considered in patients with severe forms or CNS infection after at least 1 year of successful treatment and then discontinued in patients with CD4 cells counts $\geq$200 cells/μL and who have undetectable viral loads on ART for >6 months [140,145]. Because the risk of relapse of meningeal sporotrichosis is high, lifelong suppressive therapy seems prudent and recommended in these cases [140].

10. What Can Immunocompromised Patients with Sporotrichosis Teach Us?

Many issues in our understanding of sporotrichosis remain unresolved. First, even though it is expected, a report of specific exposure (i.e., sick cats, gardening, lumbering, farming, hunting, etc.) for contracting this implantation mycosis, such as an epidemiological link, was not evident in most cases compiled in this review, except for CTS. Thus, we hypothesize that the inhalatory route would be the most likely mode of infection. Moreover, in many cases retrieved in this review, sporotrichosis was not suspected initially, delaying diagnosis and appropriate treatment initiation, which might have contributed to a higher severity of the disease.

Second, a yet unknown number of the exposed individuals do not develop disease, suggesting the full list of predisposing factors is completely unknown. Although deficiencies of the host immune response are indisputably a critical factor, reports of systemic or severe extracutaneous disease in individuals without clinical or laboratorial evidence of immunodeficiency [79–81] indicates many factors should be considered in infection pathogenesis.

Not surprisingly, a frequent association between atypical forms of sporotrichosis and HIV/AIDS, transplantation, hematological malignances and iatrogenic immunosuppression for rheumatologic conditions was detected. As seen for other endemic mycoses such as paracoccidioidomycosis,

coccidioidomycosis and histoplasmosis, all these comorbidities, particularly HIV/AIDS, are high-risk factors for disseminated or atypical forms and can even change their natural history [146–148], mostly by promoting a defective T-cell immunity. This was suggested by the experimental works on nude mice reviewed here.

An interesting observation was identification of chronic alcohol abuse as a single predisposing factor to infection. The impact of chronic heavy alcohol consumption on the immune system is complex and time and dose dependent, typically resulting in a subclinical immunosuppression that becomes clinically significant only in the case of a secondary insult [149]. Innate immunity is affected, particularly by inhibiting cellular chemotaxis, phagocytosis (especially for alveolar macrophages) and production of growth factors [150]. Adaptive responses show severe compromise of T-cell function such as lymphopenia, increased cellular differentiation and activation, and reduced migration. The chronic activation of T-cell pool in alcoholic patients would alter its ability to expand and respond to pathogenic challenges or lead to their elimination through increased sensitivity to activation-induced cell death [149,151]. Although the higher risk of infections in alcoholic patients has been related mostly to bacterial (e.g., tuberculosis) and viral infections, our review points to chronic alcohol abuse as a risk factor of atypical, more severe, sporotrichosis. In addition, the higher susceptibility to infections by alcohol abuse may be in part related to behavioral changes that lead to enhanced exposure to pathogens.

Curiously, we only found two reports of primary immunodeficiencies patients with severe sporotrichosis (T CD4 lymphopenia and X-linked CGD). The latter corroborates the finding of increased susceptibility to sporotrichosis in the CGD mice model discussed earlier. Although studies of experimental sporotrichosis suggested that signaling via TLRs and other PRR would be crucial to recognition of the fungus and the mounting of effective immune responses, we did not find cases of OS associated with putative constitutive defects of these pathways. This may be due to the low frequency of these deficiencies in the general population compared with the immunosuppressed conditions aforementioned. We also did not find association between humoral immunodeficiencies and sporotrichosis, despite this subset of immunodeficiency being relatively more common. Thus, although antibodies are protective in some experimental models, they may not play a major role in human sporotrichosis.

Diabetes mellitus is an established risk factor for certain endemic mycoses (e.g., coccidioidomycosis, histoplasmosis, blastomycosis), but surprisingly not in others (e.g., paracoccidioidomycosis), and manifestations were more severe than in non-diabetic patients [152–154]. This was specially related to uncontrolled diabetes (chronic hyperglycemia). Decreased chemotaxis, phagocytic and killing activities of macrophages and neutrophils were described in uncontrolled diabetic patients [155]. However, as for chronic alcohol abuse, the (uncontrolled) diabetes mellitus induced alterations of both innate (neutrophils and macrophages) and adaptive immunity cells (mainly T cells) were related predominantly to enhanced susceptibility to tuberculosis [156–158]. Current evidence suggests underperforming innate immunity followed by a hyper-reactive T-cell-mediated immune response to *Mycobacterium tuberculosis* in patients with tuberculosis disease, but how these altered responses contribute to enhanced susceptibility or more adverse outcomes remains unclear. However, studies of latent tuberculosis individuals showed that diabetes leads to suboptimal induction of protective T-cell responses, thereby providing a possible mechanism for the increased susceptibility to active disease [158]. Conceivably, these alterations would apply to chronic granulomatous infections other than tuberculosis, such as those caused by fungal organisms. Thus, the observation of diabetes as the single underlying condition in an appreciable number of sporotrichosis patients should not be surprising. Further investigation is required to determine which of the immune dysfunctions presented by diabetic patients play a relevant role in the enhanced susceptibility to sporotrichosis.

11. Concluding Remarks and Future Perspectives

The data we gathered neither allow drawing definitive conclusions on several aspects of the opportunistic nature of sporotrichosis nor making a consensus on the management of these cases.

However, they suggest that OS, when not life-threatening, frequently progresses to larger and/or deeper lesions that usually require higher doses of the antifungals and more prolonged courses of therapy. Therapy was generally started with amphotericin B formulations, which were moved to itraconazole after the initial improvement, as judged by the assisting clinician. In general, the patients responded well to treatment, even if slowly. The few fatalities were mainly accounted for by delayed onset of antifungal therapy or by use of immunosuppressors due to misdiagnosis (rheumatologic disease, sarcoidosis, etc.). Potassium iodide was seldom used (mostly in the early case reports), with poorer responses, being then replaced by amphotericin B. Notably, in addition to the skin, the most affected sites were bones, joints, lungs and CNS, with diagnosis based on histopathology/mycological examination plus cultures of specimens such as biopsies, synovial liquid, cerebrospinal fluid, and blood. Unfortunately, non-microbiologic tests, such as antibody/antigen detection, PCR and other molecular methods are not routinely applied for diagnosis. We thus urgently need standardized and commercially available diagnostic tools to discover the deep part of the sporotrichosis iceberg.

Funding: Gil Benard is a senior researcher from Conselho Nacional de Desenvolvimento Científico e Tecnológico and partially supported by Fundação de Amparo à Pesquisa do Estado de São Paulo grant #2016/08739-6.

Acknowledgments: We are grateful to Fabio Seiti for critical reading of the manuscript and English editing and Juliana Ruiz Fernandes for assistance with Figure 1. Figure 3B is a courtesy of Professor Marconi Rodrigues de Faria, Veterinary Catholic School of Curitiba, Paraná, Brazil.

Conflicts of Interest: The authors declare no conflict of interest.

References

1. Chakrabarti, A.; Bonifaz, A.; Gutierrez-Galhardo, M.C.; Mochizuki, T.; Li, S. Global epidemiology of sporotrichosis. *Med. Mycol.* **2015**, *53*, 3–14. [CrossRef] [PubMed]
2. Barros, M.B.; de Almeida Paes, R.; Schubach, A.O. Sporothrix schenckii and Sporotrichosis. *Clin. Microbiol. Rev.* **2011**, *24*, 633–654. [CrossRef] [PubMed]
3. Queiroz-Telles, F.; Nucci, M.; Colombo, A.L.; Tobón, A.; Restrepo, A. Mycoses of implantation in Latin America: An overview of epidemiology, clinical manifestations, diagnosis and treatment. *Med. Mycol.* **2011**, *49*, 225–236. [CrossRef] [PubMed]
4. Aung, A.K.; Teh, B.M.; McGrath, C.; Thompson, P.J. Pulmonary sporotrichosis: Case series and systematic analysis of literature on clinico-radiological patterns and management outcomes. *Med. Mycol.* **2013**, *51*, 534–544. [CrossRef]
5. Rodrigues, A.M.; de Hoog, G.S.; de Camargo, Z.P. Sporothrix Species Causing Outbreaks in Animals and Humans Driven by Animal-Animal Transmission. *PLoS Pathog.* **2016**, *12*, e1005638. [CrossRef] [PubMed]
6. Schubach, A.; Barros, M.B.; Wanke, B. Epidemic sporotrichosis. *Curr. Opin. Infect. Dis.* **2008**, *21*, 129–133. [CrossRef] [PubMed]
7. Gremião, I.D.; Menezes, R.C.; Schubach, T.M.; Figueiredo, A.B.; Cavalcanti, M.C.; Pereira, S.A. Feline sporotrichosis: Epidemiological and clinical aspects. *Med. Mycol.* **2015**, *53*, 15–21. [CrossRef]
8. Queiroz-Telles, F.; Fahal, A.H.; Falci, D.R.; Caceres, D.H.; Chiller, T.; Pasqualotto, A.C. Neglected endemic mycoses. *Lancet Infect. Dis.* **2017**, *17*, e367–e377. [CrossRef]
9. Fernández, N.; Iachini, R.; Farias, L.; Pozzi, N.; Tiraboschi, I. Esporotrichosis: Uma zoonosis em alerta. In Proceedings of the 13th Latin American Forum for Fungal Infections, Cordoba, Argentina, 5–7 November 2015; pp. 10–11.
10. Lopes-Bezerra, L.M.; Mora-Montes, H.M.; Zhang, Y.; Nino-Vega, G.; Rodrigues, A.M.; de Camargo, Z.P.; de Hoog, S. Sporotrichosis between 1898 and 2017: The evolution of knowledge on a changeable disease and on emerging etiological agents. *Med. Mycol.* **2018**, *56*, 126–143. [CrossRef]
11. Rios, M.E.; Suarez, J.; Moreno, J.; Vallee, J.; Moreno, J.P. Zoonotic Sporotrichosis Related to Cat Contact: First Case Report from Panama in Central America. *Cureus* **2018**, *10*, e2906. [CrossRef]
12. Freitas, D.F.; de Siqueira Hoagland, B.; do Valle, A.C.; Fraga, B.B.; de Barros, M.B.; de Oliveira Schubach, A.; de Almeida-Paes, R.; Cuzzi, T.; Rosalino, C.M.; Zancope-Oliveira, R.M.; et al. Sporotrichosis in HIV-infected patients: Report of 21 cases of endemic sporotrichosis in Rio de Janeiro, Brazil. *Med. Mycol.* **2012**, *50*, 170–178. [CrossRef] [PubMed]

13. Pappas, P.G.; Tellez, I.; Deep, A.E.; Nolasco, D.; Holgado, W.; Bustamante, B. Sporotrichosis in Peru: description of an area of hyperendemicity. *Clin. Infect. Dis.* **2000**, *30*, 65–70. [CrossRef] [PubMed]
14. Song, Y.; Li, S.S.; Zhong, S.X.; Liu, Y.Y.; Yao, L.; Huo, S.S. Report of 457 sporotrichosis cases from Jilin province, northeast China, a serious endemic region. *J. Eur. Acad. Dermatol. Venereol.* **2013**, *27*, 313–318. [CrossRef] [PubMed]
15. Yap, F.B. Disseminated cutaneous sporotrichosis in an immunocompetent individual. *Int. J. Infect. Dis.* **2011**, *15*, e727–e729. [CrossRef] [PubMed]
16. Romero-Cabello, R.; Bonifaz, A.; Romero-Feregrino, R.; Sánchez, C.J.; Linares, Y.; Zavala, J.T.; Romero, L.C.; Vega, J.T. Disseminated sporotrichosis. *BMJ Case Rep.* **2011**, *2011*. [CrossRef] [PubMed]
17. Fernandes, B.; Caligiorne, R.B.; Coutinho, D.M.; Gomes, R.R.; Rocha-Silva, F.; Machado, A.S.; Santrer, E.F.R.; Assuncao, C.B.; Guimaraes, C.F.; Laborne, M.S.; et al. A case of disseminated sporotrichosis caused by Sporothrix brasiliensis. *Med. Mycol. Case Rep.* **2018**, *21*, 34–36. [CrossRef] [PubMed]
18. Gandhi, N.; Chander, R.; Jain, A.; Sanke, S.; Garg, T. Atypical Cutaneous Sporotrichosis in an Immunocompetent Adult: Response to Potassium Iodide. *Indian J. Dermatol.* **2016**, *61*, 236. [CrossRef]
19. Hessler, C.; Kauffman, C.A.; Chow, F.C. The Upside of Bias: A Case of Chronic Meningitis Due to Sporothrix Schenckii in an Immunocompetent Host. *Neurohospitalist* **2017**, *7*, 30–34. [CrossRef]
20. da Rosa, A.C.; Scroferneker, M.L.; Vettorato, R.; Gervini, R.L.; Vettorato, G.; Weber, A. Epidemiology of sporotrichosis: A study of 304 cases in Brazil. *J. Am. Acad. Dermatol.* **2005**, *52*, 451–459. [CrossRef]
21. Itoh, M.; Okamoto, S.; Kariya, H. Survey of 200 cases of sporotrichosis. *Dermatologica* **1986**, *172*, 209–213. [CrossRef]
22. Barros, M.B.; Schubach, A.E.O.; do Valle, A.C.; Gutierrez Galhardo, M.C.; Conceição-Silva, F.; Schubach, T.M.; Reis, R.S.; Wanke, B.; Marzochi, K.B.; Conceição, M.J. Cat-transmitted sporotrichosis epidemic in Rio de Janeiro, Brazil: Description of a series of cases. *Clin. Infect. Dis.* **2004**, *38*, 529–535. [CrossRef] [PubMed]
23. da Silva, R.F.; Bonfitto, M.; da Silva Junior, F.I.M.; de Ameida, M.T.G.; da Silva, R.C. Sporotrichosis in a liver transplant patient: A case report and literature review. *Med. Mycol. Case Rep.* **2017**, *17*, 25–27. [CrossRef] [PubMed]
24. Silva-Vergara, M.L.; de Camargo, Z.P.; Silva, P.F.; Abdalla, M.R.; Sgarbieri, R.N.; Rodrigues, A.M.; dos Santos, K.C.; Barata, C.H.; Ferreira-Paim, K. Disseminated Sporothrix brasiliensis infection with endocardial and ocular involvement in an HIV-infected patient. *Am. J. Trop. Med. Hyg.* **2012**, *86*, 477–480. [CrossRef] [PubMed]
25. Agarwal, S.K.; Tiwari, S.C.; Dash, S.C.; Mehta, S.N.; Saxena, S.; Banerjee, U.; Kumar, R.; Bhunyan, U.N. Urinary sporotrichosis in a renal allograft recipient. *Nephron* **1994**, *66*, 485. [CrossRef] [PubMed]
26. Kauffman, C.A. Sporotrichosis. *Clin. Infect. Dis.* **1999**, *29*, 231–236, quiz 237. [CrossRef] [PubMed]
27. Alba-Fierro, C.A.; Pérez-Torres, A.; Toriello, C.; Pulido-Camarillo, E.; López-Romero, E.; Romo-Lozano, Y.; Gutiérrez-Sánchez, G.; Ruiz-Baca, E. Immune Response Induced by an Immunodominant 60 kDa Glycoprotein of the Cell Wall of Sporothrix schenckii in Two Mice Strains with Experimental Sporotrichosis. *J. Immunol. Res.* **2016**, *2016*, 6525831. [CrossRef] [PubMed]
28. Conceição-Silva, F.; Morgado, F.N. Immunopathogenesis of Human Sporotrichosis: What We Already Know. *J. Fungi* **2018**, *4*, 89. [CrossRef]
29. Castro, R.A.; Kubitschek-Barreira, P.H.; Teixeira, P.A.; Sanches, G.F.; Teixeira, M.M.; Quintella, L.P.; Almeida, S.R.; Costa, R.O.; Camargo, Z.P.; Felipe, M.S.; et al. Differences in cell morphometry, cell wall topography and gp70 expression correlate with the virulence of Sporothrix brasiliensis clinical isolates. *PLoS ONE* **2013**, *8*, e75656. [CrossRef]
30. de Almeida, J.R.F.; Jannuzzi, G.P.; Kaihami, G.H.; Breda, L.C.D.; Ferreira, K.S.; de Almeida, S.R. An immunoproteomic approach revealing peptides from Sporothrix brasiliensis that induce a cellular immune response in subcutaneous sporotrichosis. *Sci. Rep.* **2018**, *8*, 4192. [CrossRef]
31. Manente, F.A.; Quinello, C.; Ferreira, L.S.; de Andrade, C.R.; Jellmayer, J.A.; Portuondo, D.L.; Batista-Duharte, A.; Carlos, I.Z. Experimental sporotrichosis in a cyclophosphamide-induced immunosuppressed mice model. *Med. Mycol.* **2018**, *56*, 711–722. [CrossRef]
32. Kajiwara, H.; Saito, M.; Ohga, S.; Uenotsuchi, T.; Yoshida, S. Impaired host defense against Sporothrix schenckii in mice with chronic granulomatous disease. *Infect. Immun.* **2004**, *72*, 5073–5079. [CrossRef] [PubMed]

33. Cunningham, K.M.; Bulmer, G.S.; Rhoades, E.R. Phagocytosis and intracellular fate of Sporothrix schenckii. *J. Infect. Dis.* **1979**, *140*, 815–817. [CrossRef]

34. Schaffner, A.; Davis, C.E.; Schaffner, T.; Markert, M.; Douglas, H.; Braude, A.I. In vitro susceptibility of fungi to killing by neutrophil granulocytes discriminates between primary pathogenicity and opportunism. *J. Clin. Investig.* **1986**, *78*, 511–524. [CrossRef] [PubMed]

35. Fernandes, K.S.; Coelho, A.L.; Lopes Bezerra, L.M.; Barja-Fidalgo, C. Virulence of Sporothrix schenckii conidia and yeast cells, and their susceptibility to nitric oxide. *Immunology* **2000**, *101*, 563–569. [CrossRef] [PubMed]

36. Fernandes, K.S.; Neto, E.H.; Brito, M.M.; Silva, J.S.; Cunha, F.Q.; Barja-Fidalgo, C. Detrimental role of endogenous nitric oxide in host defence against Sporothrix schenckii. *Immunology* **2008**, *123*, 469–479. [CrossRef] [PubMed]

37. Morgado, F.N.; Schubach, A.O.; Barros, M.B.; Conceição-Silva, F. The in situ inflammatory profile of lymphocutaneous and fixed forms of human sporotrichosis. *Med. Mycol.* **2011**, *49*, 612–620. [CrossRef] [PubMed]

38. Morgado, F.N.; de Carvalho, L.M.V.; Leite-Silva, J.; Seba, A.J.; Pimentel, M.I.F.; Fagundes, A.; Madeira, M.F.; Lyra, M.R.; Oliveira, M.M.; Schubach, A.O.; et al. Unbalanced inflammatory reaction could increase tissue destruction and worsen skin infectious diseases—A comparative study of leishmaniasis and sporotrichosis. *Sci. Rep.* **2018**, *8*, 2898. [CrossRef] [PubMed]

39. Guzman-Beltran, S.; Perez-Torres, A.; Coronel-Cruz, C.; Torres-Guerrero, H. Phagocytic receptors on macrophages distinguish between different Sporothrix schenckii morphotypes. *Microbes Infect.* **2012**, *14*, 1093–1101. [CrossRef] [PubMed]

40. Madrid, I.M.; Xavier, M.O.; Mattei, A.S.; Fernandes, C.G.; Guim, T.N.; Santin, R.; Schuch, L.F.; Nobre, M.e.O.; Araújo Meireles, M.C. Role of melanin in the pathogenesis of cutaneous sporotrichosis. *Microbes Infect.* **2010**, *12*, 162–165. [CrossRef] [PubMed]

41. Verdan, F.F.; Faleiros, J.C.; Ferreira, L.S.; Monnazzi, L.G.; Maia, D.C.; Tansine, A.; Placeres, M.C.; Carlos, I.Z.; Santos-Junior, R.R. Dendritic cell are able to differentially recognize Sporothrix schenckii antigens and promote Th1/Th17 response in vitro. *Immunobiology* **2012**, *217*, 788–794. [CrossRef]

42. Goncalves, A.C.; Maia, D.C.; Ferreira, L.S.; Monnazzi, L.G.; Alegranci, P.; Placeres, M.C.; Batista-Duharte, A.; Carlos, I.Z. Involvement of major components from Sporothrix schenckii cell wall in the caspase-1 activation, nitric oxide and cytokines production during experimental sporotrichosis. *Mycopathologia* **2015**, *179*, 21–30. [CrossRef] [PubMed]

43. Ferreira, L.S.; Goncalves, A.C.; Portuondo, D.L.; Maia, D.C.; Placeres, M.C.; Batista-Duharte, A.; Carlos, I.Z. Optimal clearance of Sporothrix schenckii requires an intact Th17 response in a mouse model of systemic infection. *Immunobiology* **2015**, *220*, 985–992. [CrossRef] [PubMed]

44. Romo-Lozano, Y.; Hernandez-Hernandez, F.; Salinas, E. Sporothrix schenckii yeasts induce ERK pathway activation and secretion of IL-6 and TNF-alpha in rat mast cells, but no degranulation. *Med. Mycol.* **2014**, *52*, 862–868. [CrossRef] [PubMed]

45. Romo-Lozano, Y.; Hernandez-Hernandez, F.; Salinas, E. Mast cell activation by conidia of Sporothrix schenckii: Role in the severity of infection. *Scand. J. Immunol.* **2012**, *76*, 11–20. [CrossRef] [PubMed]

46. Sassa, M.F.; Saturi, A.E.; Souza, L.F.; Ribeiro, L.C.; Sgarbi, D.B.; Carlos, I.Z. Response of macrophage Toll-like receptor 4 to a Sporothrix schenckii lipid extract during experimental sporotrichosis. *Immunology* **2009**, *128*, 301–309. [CrossRef]

47. de, C.N.T.; Ferreira, L.S.; Arthur, R.A.; Alegranci, P.; Placeres, M.C.; Spolidorio, L.C.; Carlos, I.Z. Influence of TLR-2 in the immune response in the infection induced by fungus Sporothrix schenckii. *Immunol. Investig.* **2014**, *43*, 370–390. [CrossRef]

48. Negrini Tde, C.; Ferreira, L.S.; Alegranci, P.; Arthur, R.A.; Sundfeld, P.P.; Maia, D.C.; Spolidorio, L.C.; Carlos, I.Z. Role of TLR-2 and fungal surface antigens on innate immune response against Sporothrix schenckii. *Immunol. Investig.* **2013**, *42*, 36–48. [CrossRef] [PubMed]

49. Li, M.; Chen, Q.; Sun, J.; Shen, Y.; Liu, W. Inflammatory response of human keratinocytes triggered by Sporothrix schenckii via Toll-like receptor 2 and 4. *J. Dermatol. Sci.* **2012**, *66*, 80–82. [CrossRef]

50. Jellmayer, J.A.; Ferreira, L.S.; Manente, F.A.; Goncalves, A.C.; Polesi, M.C.; Batista-Duharte, A.; Carlos, I.Z. Dectin-1 expression by macrophages and related antifungal mechanisms in a murine model of Sporothrix schenckii sensu stricto systemic infection. *Microb. Pathog.* **2017**, *110*, 78–84. [CrossRef]

51. Zhang, Z.; Liu, X.; Lv, X.; Lin, J. Variation in genotype and higher virulence of a strain of Sporothrix schenckii causing disseminated cutaneous sporotrichosis. *Mycopathologia* **2011**, *172*, 439–446. [CrossRef]

52. Goncalves, A.C.; Ferreira, L.S.; Manente, F.A.; de Faria, C.; Polesi, M.C.; de Andrade, C.R.; Zamboni, D.S.; Carlos, I.Z. The NLRP3 inflammasome contributes to host protection during Sporothrix schenckii infection. *Immunology* **2017**, *151*, 154–166. [CrossRef] [PubMed]

53. Maia, D.C.; Gonçalves, A.C.; Ferreira, L.S.; Manente, F.A.; Portuondo, D.L.; Vellosa, J.C.; Polesi, M.C.; Batista-Duharte, A.; Carlos, I.Z. Response of Cytokines and Hydrogen Peroxide to Sporothrix schenckii Exoantigen in Systemic Experimental Infection. *Mycopathologia* **2016**, *181*, 207–215. [CrossRef] [PubMed]

54. Scott, E.N.; Muchmore, H.G.; Fine, D.P. Activation of the alternative complement pathway by Sporothrix schenckii. *Infect. Immun.* **1986**, *51*, 6–9. [PubMed]

55. Arrillaga-Moncrieff, I.; Capilla, J.; Mayayo, E.; Marimon, R.; Mariné, M.; Gené, J.; Cano, J.; Guarro, J. Different virulence levels of the species of Sporothrix in a murine model. *Clin. Microbiol. Infect.* **2009**, *15*, 651–655. [CrossRef] [PubMed]

56. Almeida-Paes, R.; Bailao, A.M.; Pizzini, C.V.; Reis, R.S.; Soares, C.M.; Peralta, J.M.; Gutierrez-Galhardo, M.C.; Zancope-Oliveira, R.M. Cell-free antigens of Sporothrix brasiliensis: Antigenic diversity and application in an immunoblot assay. *Mycoses* **2012**, *55*, 467–475. [CrossRef] [PubMed]

57. Almeida-Paes, R.; de Oliveira, L.C.; Oliveira, M.M.; Gutierrez-Galhardo, M.C.; Nosanchuk, J.D.; Zancope-Oliveira, R.M. Phenotypic characteristics associated with virulence of clinical isolates from the Sporothrix complex. *Biomed. Res. Int.* **2015**, *2015*, 212308. [CrossRef] [PubMed]

58. Camacho, E.; León-Navarro, I.; Rodríguez-Brito, S.; Mendoza, M.; Niño-Vega, G.A. Molecular epidemiology of human sporotrichosis in Venezuela reveals high frequency of Sporothrix globosa. *BMC Infect. Dis.* **2015**, *15*, 94. [CrossRef]

59. Teixeira, P.A.; de Castro, R.A.; Nascimento, R.C.; Tronchin, G.; Torres, A.P.; Lazéra, M.; de Almeida, S.R.; Bouchara, J.P.; Loureiro y Penha, C.V.; Lopes-Bezerra, L.M. Cell surface expression of adhesins for fibronectin correlates with virulence in Sporothrix schenckii. *Microbiology* **2009**, *155*, 3730–3738. [CrossRef]

60. Martínez-Álvarez, J.A.; Pérez-García, L.A.; Mellado-Mojica, E.; López, M.G.; Martínez-Duncker, I.; Lópes-Bezerra, L.M.; Mora-Montes, H.M. Sporothrix schenckii sensu stricto and Sporothrix brasiliensis Are Differentially Recognized by Human Peripheral Blood Mononuclear Cells. *Front. Microbiol.* **2017**, *8*, 843. [CrossRef]

61. Shimonaka, H.; Noguchi, T.; Kawai, K.; Hasegawa, I.; Nozawa, Y.; Ito, Y. Immunochemical studies on the human pathogen Sporothrix schenckii: Effects of chemical and enzymatic modification of the antigenic compounds upon immediate and delayed reactions. *Infect. Immun.* **1975**, *11*, 1187–1194.

62. Shiraishi, A.; Nakagaki, K.; Arai, T. Experimental sporotrichosis in congenitally athymic (nude) mice. *J. Reticuloendothel. Soc.* **1979**, *26*, 333–336.

63. Shiraishi, A.; Nakagaki, K.; Arai, T. Role of cell-mediated immunity in the resistance to experimental sporotrichosis in mice. *Mycopathologia* **1992**, *120*, 15–21. [CrossRef] [PubMed]

64. Tachibana, T.; Matsuyama, T.; Mitsuyama, M. Involvement of CD4+ T cells and macrophages in acquired protection against infection with Sporothrix schenckii in mice. *Med. Mycol.* **1999**, *37*, 397–404. [CrossRef] [PubMed]

65. Dickerson, C.L.; Taylor, R.L.; Drutz, D.J. Susceptibility of congenitally athymic (nude) mice to sporotrichosis. *Infect. Immun.* **1983**, *40*, 417–420. [PubMed]

66. Carlos, I.Z.; Sgarbi, D.B.; Angluster, J.; Alviano, C.S.; Silva, C.L. Detection of cellular immunity with the soluble antigen of the fungus Sporothrix schenckii in the systemic form of the disease. *Mycopathologia* **1992**, *117*, 139–144. [CrossRef] [PubMed]

67. Carlos, I.Z.; Sgarbi, D.B.; Placeres, M.C. Host organism defense by a peptide-polysaccharide extracted from the fungus Sporothrix schenckii. *Mycopathologia* **1998**, *144*, 9–14. [CrossRef] [PubMed]

68. Alegranci, P.; de Abreu Ribeiro, L.C.; Ferreira, L.S.; Negrini Tde, C.; Maia, D.C.; Tansini, A.; Goncalves, A.C.; Placeres, M.C.; Carlos, I.Z. The predominance of alternatively activated macrophages following challenge with cell wall peptide-polysaccharide after prior infection with Sporothrix schenckii. *Mycopathologia* **2013**, *176*, 57–65. [CrossRef] [PubMed]

69. Uenotsuchi, T.; Takeuchi, S.; Matsuda, T.; Urabe, K.; Koga, T.; Uchi, H.; Nakahara, T.; Fukagawa, S.; Kawasaki, M.; Kajiwara, H.; et al. Differential induction of Th1-prone immunity by human dendritic cells

activated with Sporothrix schenckii of cutaneous and visceral origins to determine their different virulence. *Int. Immunol.* **2006**, *18*, 1637–1646. [CrossRef]

70.   Almeida-Paes, R.; Pimenta, M.A.; Pizzini, C.V.; Monteiro, P.C.; Peralta, J.M.; Nosanchuk, J.D.; Zancopé-Oliveira, R.M. Use of mycelial-phase Sporothrix schenckii exoantigens in an enzyme-linked immunosorbent assay for diagnosis of sporotrichosis by antibody detection. *Clin. Vaccine Immunol.* **2007**, *14*, 244–249. [CrossRef]

71.   Portuondo, D.L.; Batista-Duharte, A.; Ferreira, L.S.; Martínez, D.T.; Polesi, M.C.; Duarte, R.A.; de Paula E Silva, A.C.; Marcos, C.M.; Almeida, A.M.; Carlos, I.Z. A cell wall protein-based vaccine candidate induce protective immune response against Sporothrix schenckii infection. *Immunobiology* **2016**, *221*, 300–309. [CrossRef]

72.   Nascimento, R.C.; Espíndola, N.M.; Castro, R.A.; Teixeira, P.A.; Loureiro y Penha, C.V.; Lopes-Bezerra, L.M.; Almeida, S.R. Passive immunization with monoclonal antibody against a 70-kDa putative adhesin of Sporothrix schenckii induces protection in murine sporotrichosis. *Eur. J. Immunol.* **2008**, *38*, 3080–3089. [CrossRef] [PubMed]

73.   Franco Dde, L.; Nascimento, R.C.; Ferreira, K.S.; Almeida, S.R. Antibodies Against Sporothrix schenckii Enhance TNF-alpha Production and Killing by Macrophages. *Scand. J. Immunol.* **2012**, *75*, 142–146. [CrossRef] [PubMed]

74.   de Almeida, J.R.; Kaihami, G.H.; Jannuzzi, G.P.; de Almeida, S.R. Therapeutic vaccine using a monoclonal antibody against a 70-kDa glycoprotein in mice infected with highly virulent Sporothrix schenckii and Sporothrix brasiliensis. *Med. Mycol.* **2015**, *53*, 42–50. [CrossRef] [PubMed]

75.   Zhang, Y.Q.; Xu, X.G.; Zhang, M.; Jiang, P.; Zhou, X.Y.; Li, Z.Z.; Zhang, M.F. Sporotrichosis: Clinical and histopathological manifestations. *Am. J. Dermatopathol.* **2011**, *33*, 296–302. [CrossRef] [PubMed]

76.   Quintella, L.P.; Passos, S.R.; do Vale, A.C.; Galhardo, M.C.; Barros, M.B.; Cuzzi, T.; Reis, R.O.S.; de Carvalho, M.H.; Zappa, M.B.; Schubach, A.E.O. Histopathology of cutaneous sporotrichosis in Rio de Janeiro: A series of 119 consecutive cases. *J. Cutan. Pathol.* **2011**, *38*, 25–32. [CrossRef]

77.   Morgado, F.N.; Schubach, A.O.; Pimentel, M.I.; Lyra, M.R.; Vasconcellos, É.; Valete-Rosalino, C.M.; Conceição-Silva, F. Is There Any Difference between the In Situ and Systemic IL-10 and IFN-γ Production when Clinical Forms of Cutaneous Sporotrichosis Are Compared? *PLoS ONE* **2016**, *11*, e0162764. [CrossRef]

78.   Plouffe, J.F.; Silva, J.; Fekety, R.; Reinhalter, E.; Browne, R. Cell-mediated immune responses in sporotrichosis. *J. Infect. Dis.* **1979**, *139*, 152–157. [CrossRef]

79.   Mialski, R.; de Oliveira, J.N., Jr.; da Silva, L.H.; Kono, A.; Pinheiro, R.L.; Teixeira, M.J.; Gomes, R.R.; de Queiroz-Telles, F.; Pinto, F.G.; Benard, G. Chronic Meningitis and Hydrocephalus due to Sporothrix brasiliensis in Immunocompetent Adults: A Challenging Entity. *Open Forum Infect. Dis.* **2018**, *5*, ofy081. [CrossRef]

80.   Lynch, P.J.; Voorhees, J.J.; Harrell, E.R. Systemic sporotrichosis. *Ann. Intern. Med.* **1970**, *73*, 23–30. [CrossRef]

81.   Almeida-Paes, R.; de Oliveira, M.M.; Freitas, D.F.; do Valle, A.C.; Zancopé-Oliveira, R.M.; Gutierrez-Galhardo, M.C. Sporotrichosis in Rio de Janeiro, Brazil: Sporothrix brasiliensis is associated with atypical clinical presentations. *PLoS Negl. Trop. Dis.* **2014**, *8*, e3094. [CrossRef]

82.   Lipstein-Kresch, E.; Isenberg, H.D.; Singer, C.; Cooke, O.; Greenwald, R.A. Disseminated Sporothrix schenckii infection with arthritis in a patient with acquired immunodeficiency syndrome. *J. Rheumatol.* **1985**, *12*, 805–808. [PubMed]

83.   Freitas, D.F.; Valle, A.C.; da Silva, M.B.; Campos, D.P.; Lyra, M.R.; de Souza, R.V.; Veloso, V.G.; Zancopé-Oliveira, R.M.; Bastos, F.I.; Galhardo, M.C. Sporotrichosis: An emerging neglected opportunistic infection in HIV-infected patients in Rio de Janeiro, Brazil. *PLoS Negl. Trop. Dis.* **2014**, *8*, e3110. [CrossRef]

84.   Moreira, J.A.; Freitas, D.F.; Lamas, C.C. The impact of sporotrichosis in HIV-infected patients: A systematic review. *Infection* **2015**, *43*, 267–276. [CrossRef] [PubMed]

85.   Donabedian, H.; O'Donnell, E.; Olszewski, C.; MacArthur, R.D.; Budd, N. Disseminated cutaneous and meningeal sporotrichosis in an AIDS patient. *Diagn. Microbiol. Infect. Dis.* **1994**, *18*, 111–115. [CrossRef]

86.   Paixão, A.G.; Galhardo, M.C.G.; Almeida-Paes, R.; Nunes, E.P.; Gonçalves, M.L.C.; Chequer, G.L.; Lamas, C.D.C. The difficult management of disseminated Sporothrix brasiliensis in a patient with advanced AIDS. *AIDS Res. Ther.* **2015**, *12*, 16. [CrossRef] [PubMed]

87.   Penn, C.C.; Goldstein, E.; Bartholomew, W.R. Sporothrix schenckii meningitis in a patient with AIDS. *Clin. Infect. Dis.* **1992**, *15*, 741–743. [CrossRef] [PubMed]

88. Hardman, S.; Stephenson, I.; Jenkins, D.R.; Wiselka, M.J.; Johnson, E.M. Disseminated Sporothix schenckii in a patient with AIDS. *J. Infect.* **2005**, *51*, e73–e77. [CrossRef]
89. Dong, J.A.; Chren, M.M.; Elewski, B.E. Bonsai tree: Risk factor for disseminated sporotrichosis. *J. Am. Acad. Dermatol.* **1995**, *33*, 839–840. [CrossRef]
90. Silva-Vergara, M.L.; Maneira, F.R.; De Oliveira, R.M.; Santos, C.T.; Etchebehere, R.M.; Adad, S.J. Multifocal sporotrichosis with meningeal involvement in a patient with AIDS. *Med. Mycol.* **2005**, *43*, 187–190. [CrossRef] [PubMed]
91. Vilela, R.; Souza, G.F.; Fernandes Cota, G.; Mendoza, L. Cutaneous and meningeal sporotrichosis in a HIV patient. *Rev. Iberoam. Micol.* **2007**, *24*, 161–163. [CrossRef]
92. Galhardo, M.C.; Silva, M.T.; Lima, M.A.; Nunes, E.P.; Schettini, L.E.; de Freitas, R.F.; Paes Rde, A.; Neves Ede, S.; do Valle, A.C. Sporothrix schenckii meningitis in AIDS during immune reconstitution syndrome. *J. Neurol. Neurosurg. Psychiatry* **2010**, *81*, 696–699. [CrossRef] [PubMed]
93. Rotz, L.D.; Slater, L.N.; Wack, M.F.; Boyd, A.L.; Nan, S.E.; Greenfield, R.A. Disseminated Sporotrichosis with Meningitis in a Patient with AIDS. *Infect. Dis. Clin. Pract.* **1996**, *5*, 566–568. [CrossRef]
94. Callens, S.F.; Kitetele, F.; Lukun, P.; Lelo, P.; Van Rie, A.; Behets, F.; Colebunders, R. Pulmonary Sporothrix schenckii infection in a HIV positive child. *J. Trop. Pediatr.* **2006**, *52*, 144–146. [CrossRef] [PubMed]
95. Losman, J.A.; Cavanaugh, K. Cases from the Osler Medical Service at Johns Hopkins University. Diagnosis: *P. carinii* pneumonia and primary pulmonary sporotrichosis. *Am. J. Med.* **2004**, *117*, 353–356. [CrossRef] [PubMed]
96. Gori, S.; Lupetti, A.; Moscato, G.; Parenti, M.; Lofaro, A. Pulmonary sporotrichosis with hyphae in a human immunodeficiency virus-infected patient. A case report. *Acta Cytol.* **1997**, *41*, 519–521. [CrossRef] [PubMed]
97. Morgan, M.; Reves, R. Invasive sinusitis due to Sporothrix schenckii in a patient with AIDS. *Clin. Infect. Dis.* **1996**, *23*, 1319–1320. [CrossRef] [PubMed]
98. Bustamante, B.; Lama, J.R.; Mosquera, C.; Soto, L. Sporotrichosis in Human Immunodeficiency Virus Infected Peruvian Patients: Two Case Reports and Literature Review. *Infect. Dis. Clin. Pract.* **2009**, *17*, 78–83. [CrossRef]
99. Buccheri, R.; Benard, G. Opinion: Paracoccidioidomycosis and HIV Immune Recovery Inflammatory Syndrome. *Mycopathologia* **2018**, *183*, 495–498. [CrossRef]
100. Haddow, L.J.; Colebunders, R.; Meintjes, G.; Lawn, S.D.; Elliott, J.H.; Manabe, Y.C.; Bohjanen, P.R.; Sungkanuparph, S.; Easterbrook, P.J.; French, M.A.; et al. Cryptococcal immune reconstitution inflammatory syndrome in HIV-1-infected individuals: Proposed clinical case definitions. *Lancet Infect. Dis.* **2010**, *10*, 791–802. [CrossRef]
101. Singh, N.; Perfect, J.R. Immune reconstitution syndrome associated with opportunistic mycoses. *Lancet Infect. Dis.* **2007**, *7*, 395–401. [CrossRef]
102. Lyra, M.R.; Nascimento, M.L.; Varon, A.G.; Pimentel, M.I.; Antonio, L.E.F.; Saheki, M.N.; Bedoya-Pacheco, S.J.; Valle, A.C. Immune reconstitution inflammatory syndrome in HIV and sporotrichosis coinfection: report of two cases and review of the literature. *Rev. Soc. Bras. Med. Trop.* **2014**, *47*, 806–809. [CrossRef] [PubMed]
103. de Lima Barros, M.B.; de Oliveira Schubach, A.; Galhardo, M.C.; Schubach, T.M.; dos Reis, R.S.; Conceição, M.J.; do Valle, A.C. Sporotrichosis with widespread cutaneous lesions: Report of 24 cases related to transmission by domestic cats in Rio de Janeiro, Brazil. *Int. J. Dermatol.* **2003**, *42*, 677–681. [CrossRef] [PubMed]
104. Scott, S.M.; Peasley, E.D.; Crymes, T.P. Pulmonary sporotrichosis. Report of two cases with cavitation. *N. Engl. J. Med.* **1961**, *265*, 453–457. [CrossRef] [PubMed]
105. Mohr, J.A.; Patterson, C.D.; Eaton, B.G.; Rhoades, E.R.; Nichols, N.B. Primary pulmonary sporotrichosis. *Am. Rev. Respir. Dis.* **1972**, *106*, 260–264. [CrossRef] [PubMed]
106. Zhang, Y.; Hagen, F.; Wan, Z.; Liu, Y.; Wang, Q.; de Hoog, G.S.; Li, R.; Zhang, J. Two cases of sporotrichosis of the right upper extremity in right-handed patients with diabetes mellitus. *Rev. Iberoam. Micol.* **2016**, *33*, 38–42. [CrossRef] [PubMed]
107. Mohamad, N.; Badrin, S.; Wan Abdullah, W.N.H. A Diabetic Elderly Man with Finger Ulcer. *Korean J. Fam. Med.* **2018**, *39*, 126–129. [CrossRef] [PubMed]
108. Ramirez-Soto, M.; Lizarraga-Trujillo, J. Granulomatous sporotrichosis: Report of two unusual cases. *Rev. Chil. Infectol.* **2013**, *30*, 548–553. [CrossRef]

109. Nassif, P.W.; Granado, I.R.; Ferraz, J.S.; Souza, R.; Nassif, A.E. Atypical presentation of cutaneous sporotrichosis in an alcoholic patient. *Dermatol. Online J.* **2012**, *18*, 12.

110. Castrejon, O.V.; Robles, M.; Zubieta Arroyo, O.E. Fatal fungaemia due to Sporothrix schenckii. *Mycoses* **1995**, *38*, 373–376. [CrossRef]

111. Solorzano, S.; Ramirez, R.; Cabada, M.M.; Montoya, M.; Cazorla, E. Disseminated cutaneous sporotrichosis with joint involvement in a woman with type 2 diabetes. *Rev. Peru. Med. Exp. Salud Publica* **2015**, *32*, 187–190. [CrossRef]

112. Agger, W.A.; Caplan, R.H.; Maki, D.G. Ocular sporotrichosis mimicking mucormycosis in a diabetic. *Ann. Ophthalmol.* **1978**, *10*, 767–771. [PubMed]

113. Benvegnu, A.M.; Stramari, J.; Dallazem, L.N.D.; Chemello, R.M.L.; Beber, A.A.C. Disseminated cutaneous sporotrichosis in patient with alcoholism. *Rev. Soc. Bras. Med. Trop.* **2017**, *50*, 871–873. [CrossRef] [PubMed]

114. Yamaguchi, T.; Ito, S.; Takano, Y.; Umeda, N.; Goto, M.; Horikoshi, M.; Hayashi, T.; Goto, D.; Matsumoto, I.; Sumida, T. A case of disseminated sporotrichosis treated with prednisolone, immunosuppressants, and tocilizumab under the diagnosis of rheumatoid arthritis. *Intern. Med.* **2012**, *51*, 2035–2039. [CrossRef]

115. Yang, D.J.; Krishnan, R.S.; Guillen, D.R.; Schmiege, L.M.; Leis, P.F.; Hsu, S. Disseminated sporotrichosis mimicking sarcoidosis. *Int. J. Dermatol.* **2006**, *45*, 450–453. [CrossRef] [PubMed]

116. Mauermann, M.L.; Klein, C.J.; Orenstein, R.; Dyck, P.J. Disseminated sporotrichosis presenting with granulomatous inflammatory multiple mononeuropathies. *Muscle Nerve* **2007**, *36*, 866–872. [CrossRef] [PubMed]

117. Byrd, D.R.; El-Azhary, R.A.; Gibson, L.E.; Roberts, G.D. Sporotrichosis masquerading as pyoderma gangrenosum: Case report and review of 19 cases of sporotrichosis. *J. Eur. Acad. Dermatol. Venereol.* **2001**, *15*, 581–584. [CrossRef] [PubMed]

118. Ursini, F.; Russo, E.; Leporini, C.; Calabria, M.; Bruno, C.; Tripolino, C.; Naty, S.; Grembiale, R.D. Lymphocutaneous Sporotrichosis during Treatment with Anti-TNF-Alpha Monotherapy. *Case Rep. Rheumatol.* **2015**, *2015*, 614504. [CrossRef]

119. Tochigi, M.; Ochiai, T.; Mekata, C.; Nishiyama, H.; Anzawa, K.; Kawasaki, M. Sporotrichosis of the face by autoinoculation in a patient undergoing tacrolimus treatment. *J. Dermatol.* **2012**, *39*, 796–798. [CrossRef]

120. Severo, L.C.; Festugato, M.; Bernardi, C.; Londero, A.T. Widespread cutaneous lesions due to Sporothrix schenckii in a patient under a long-term steroids therapy. *Rev. Inst. Med. Trop. Sao Paulo* **1999**, *41*, 59–62. [CrossRef]

121. Pappas, P.G.; Alexander, B.D.; Andes, D.R.; Hadley, S.; Kauffman, C.A.; Freifeld, A.; Anaissie, E.J.; Brumble, L.M.; Herwaldt, L.; Ito, J.; et al. Invasive fungal infections among organ transplant recipients: Results of the Transplant-Associated Infection Surveillance Network (TRANSNET). *Clin. Infect. Dis.* **2010**, *50*, 1101–1111. [CrossRef]

122. Guimaraes, L.F.; Halpern, M.; de Lemos, A.S.; de Gouvea, E.F.; Goncalves, R.T.; da Rosa Santos, M.A.; Nucci, M.; Santoro-Lopes, G. Invasive Fungal Disease in Renal Transplant Recipients at a Brazilian Center: Local Epidemiology Matters. *Transplant. Proc.* **2016**, *48*, 2306–2309. [CrossRef] [PubMed]

123. Caroti, L.; Zanazzi, M.; Rogasi, P.; Fantoni, E.; Farsetti, S.; Rosso, G.; Bertoni, E.; Salvadori, M. Subcutaneous nodules and infectious complications in renal allograft recipients. *Transplant. Proc.* **2010**, *42*, 1146–1147. [CrossRef] [PubMed]

124. Rao, K.H.; Jha, R.; Narayan, G.; Sinha, S. Opportunistic infections following renal transplantation. *Indian J. Med. Microbiol.* **2002**, *20*, 47–49.

125. Gewehr, P.; Jung, B.; Aquino, V.; Manfro, R.C.; Spuldaro, F.; Rosa, R.G.; Goldani, L.Z. Sporotrichosis in renal transplant patients. *Can. J. Infect. Dis. Med. Microbiol.* **2013**, *24*, e47–e49. [CrossRef] [PubMed]

126. Bahr, N.C.; Janssen, K.; Billings, J.; Loor, G.; Green, J.S. Respiratory Failure due to Possible Donor-Derived Sporothrix schenckii Infection in a Lung Transplant Recipient. *Case Rep. Infect. Dis.* **2015**, *2015*, 925718. [CrossRef] [PubMed]

127. Ewing, G.E.; Bosl, G.J.; Peterson, P.K. Sporothrix schenckii meningitis in a farmer with Hodgkin's disease. *Am. J. Med.* **1980**, *68*, 455–457. [CrossRef]

128. Bunce, P.E.; Yang, L.; Chun, S.; Zhang, S.X.; Trinkaus, M.A.; Matukas, L.M. Disseminated sporotrichosis in a patient with hairy cell leukemia treated with amphotericin B and posaconazole. *Med. Mycol.* **2012**, *50*, 197–201. [CrossRef]

129. Kumar, S.; Kumar, D.; Gourley, W.K.; Alperin, J.B. Sporotrichosis as a presenting manifestation of hairy cell leukemia. *Am. J. Hematol.* **1994**, *46*, 134–137. [CrossRef] [PubMed]

130. Yagnik, K.J.; Skelton, W.P.T.; Olson, A.; Trillo, C.A.; Lascano, J. A rare case of disseminated Sporothrix schenckii with bone marrow involvement in a patient with idiopathic CD4 lymphocytopenia. *IDCases* **2017**, *9*, 70–72. [CrossRef]

131. Trotter, J.R.; Sriaroon, P.; Berman, D.; Petrovic, A.; Leiding, J.W. Sporothrix schenckii lymphadenitis in a male with X-linked chronic granulomatous disease. *J. Clin. Immunol.* **2014**, *34*, 49–52. [CrossRef] [PubMed]

132. Gullberg, R.M.; Quintanilla, A.; Levin, M.L.; Williams, J.; Phair, J.P. Sporotrichosis: Recurrent cutaneous, articular, and central nervous system infection in a renal transplant recipient. *Rev. Infect. Dis.* **1987**, *9*, 369–375. [CrossRef] [PubMed]

133. Orofino-Costa, R.; Macedo, P.M.; Rodrigues, A.M.; Bernardes-Engemann, A.R. Sporotrichosis: An update on epidemiology, etiopathogenesis, laboratory and clinical therapeutics. *An. Bras. Dermatol.* **2017**, *92*, 606–620. [CrossRef] [PubMed]

134. Sanchotene, K.O.; Madrid, I.M.; Klafke, G.B.; Bergamashi, M.; Della Terra, P.P.; Rodrigues, A.M.; de Camargo, Z.P.; Xavier, M.O. Sporothrix brasiliensis outbreaks and the rapid emergence of feline sporotrichosis. *Mycoses* **2015**, *58*, 652–658. [CrossRef] [PubMed]

135. Bonifaz, A.; Tirado-Sánchez, A. Cutaneous Disseminated and Extracutaneous Sporotrichosis: Current Status of a Complex Disease. *J. Fungi* **2017**, *3*, 6. [CrossRef] [PubMed]

136. Tirado-Sanchez, A.; Bonifaz, A. Nodular Lymphangitis (Sporotrichoid Lymphocutaneous Infections). Clues to Differential Diagnosis. *J. Fungi* **2018**, *4*, 56. [CrossRef] [PubMed]

137. Bernardes-Engemann, A.R.; de Lima Barros, M.; Zeitune, T.; Russi, D.C.; Orofino-Costa, R.; Lopes-Bezerra, L.M. Validation of a serodiagnostic test for sporotrichosis: a follow-up study of patients related to the Rio de Janeiro zoonotic outbreak. *Med. Mycol.* **2015**, *53*, 28–33. [CrossRef] [PubMed]

138. Ottonelli Stopiglia, C.D.; Magagnin, C.M.; Castrillon, M.R.; Mendes, S.D.; Heidrich, D.; Valente, P.; Scroferneker, M.L. Antifungal susceptibilities and identification of species of the Sporothrix schenckii complex isolated in Brazil. *Med. Mycol.* **2014**, *52*, 56–64. [CrossRef]

139. Brilhante, R.S.; Rodrigues, A.M.; Sidrim, J.J.; Rocha, M.F.; Pereira, S.A.; Gremiao, I.D.; Schubach, T.M.; de Camargo, Z.P. In vitro susceptibility of antifungal drugs against Sporothrix brasiliensis recovered from cats with sporotrichosis in Brazil. *Med. Mycol.* **2016**, *54*, 275–279. [CrossRef]

140. Kauffman, C.A.; Bustamante, B.; Chapman, S.W.; Pappas, P.G. Infectious Diseases Society of America. Clinical practice guidelines for the management of sporotrichosis: 2007 update by the Infectious Diseases Society of America. *Clin. Infect. Dis.* **2007**, *45*, 1255–1265. [CrossRef]

141. de Lima Barros, M.B.; Schubach, A.O.; de Vasconcellos Carvalhaes de Oliveira, R.; Martins, E.B.; Teixeira, J.L.; Wanke, B. Treatment of cutaneous sporotrichosis with itraconazole—Study of 645 patients. *Clin. Infect. Dis.* **2011**, *52*, e200–e206. [CrossRef]

142. Shikanai-Yasuda, M.A.; Mendes, R.P.; Colombo, A.L.; Queiroz-Telles, F.; Kono, A.S.G.; Paniago, A.M.M.; Nathan, A.; Valle, A.; Bagagli, E.; Benard, G.; et al. Brazilian guidelines for the clinical management of paracoccidioidomycosis. *Rev. Soc. Bras. Med. Trop.* **2017**, *50*, 715–740. [CrossRef] [PubMed]

143. Torok, M.E.; Yen, N.T.; Chau, T.T.; Mai, N.T.; Phu, N.H.; Mai, P.P.; Dung, N.T.; Chau, N.V.; Bang, N.D.; Tien, N.A.; et al. Timing of initiation of antiretroviral therapy in human immunodeficiency virus (HIV)—Associated tuberculous meningitis. *Clin. Infect. Dis.* **2011**, *52*, 1374–1383. [CrossRef] [PubMed]

144. Boulware, D.R.; Meya, D.B.; Muzoora, C.; Rolfes, M.A.; Huppler Hullsiek, K.; Musubire, A.; Taseera, K.; Nabeta, H.W.; Schutz, C.; Williams, D.A.; et al. Timing of antiretroviral therapy after diagnosis of cryptococcal meningitis. *N. Engl. J. Med.* **2014**, *370*, 2487–2498. [CrossRef] [PubMed]

145. Kaplan, J.E.; Benson, C.; Holmes, K.K.; Brooks, J.T.; Pau, A.; Masur, H.; CDC; National Institutes of Health; HIV Medicine Association of the Infectious Diseases Society of America. Guidelines for prevention and treatment of opportunistic infections in HIV-infected adults and adolescents: Recommendations from CDC, the National Institutes of Health, and the HIV Medicine Association of the Infectious Diseases Society of America. *MMWR Recomm. Rep.* **2009**, *58*, 1–207, quiz CE201-204. [PubMed]

146. Benard, G.; Duarte, A.J. Paracoccidioidomycosis: A model for evaluation of the effects of human immunodeficiency virus infection on the natural history of endemic tropical diseases. *Clin. Infect. Dis.* **2000**, *31*, 1032–1039. [CrossRef] [PubMed]

147. Hajjeh, R.A. Disseminated histoplasmosis in persons infected with human immunodeficiency virus. *Clin. Infect. Dis.* **1995**, *21* (Suppl. 1), S108–S110. [CrossRef] [PubMed]
148. Galgiani, J.N. Coccidioidomycosis: Changes in clinical expression, serological diagnosis, and therapeutic options. *Clin. Infect. Dis.* **1992**, *14* (Suppl. 1), S100–S105. [CrossRef]
149. Szabo, G.; Saha, B. Alcohol's Effect on Host Defense. *Alcohol Res.* **2015**, *37*, 159–170.
150. Barr, T.; Helms, C.; Grant, K.; Messaoudi, I. Opposing effects of alcohol on the immune system. *Prog. Neuropsychopharmacol. Biol. Psychiatry* **2016**, *65*, 242–251. [CrossRef]
151. Pasala, S.; Barr, T.; Messaoudi, I. Impact of Alcohol Abuse on the Adaptive Immune System. *Alcohol Res.* **2015**, *37*, 185–197.
152. Pan, B.; Chen, M.; Pan, W.; Liao, W. Histoplasmosis: A new endemic fungal infection in China? Review and analysis of cases. *Mycoses* **2013**, *56*, 212–221. [CrossRef]
153. Santelli, A.C.; Blair, J.E.; Roust, L.R. Coccidioidomycosis in patients with diabetes mellitus. *Am. J. Med.* **2006**, *119*, 964–969. [CrossRef] [PubMed]
154. Lemos, L.B.; Baliga, M.; Guo, M. Blastomycosis: The great pretender can also be an opportunist. Initial clinical diagnosis and underlying diseases in 123 patients. *Ann. Diagn. Pathol.* **2002**, *6*, 194–203. [CrossRef] [PubMed]
155. Geerlings, S.E.; Hoepelman, A.I. Immune dysfunction in patients with diabetes mellitus (DM). *FEMS Immunol. Med. Microbiol.* **1999**, *26*, 259–265. [CrossRef] [PubMed]
156. Restrepo, B.I.; Schlesinger, L.S. Host-pathogen interactions in tuberculosis patients with type 2 diabetes mellitus. *Tuberculosis* **2013**, *93*, S10–S14. [CrossRef]
157. Martinez, N.; Kornfeld, H. Diabetes and immunity to tuberculosis. *Eur. J. Immunol.* **2014**, *44*, 617–626. [CrossRef]
158. Kumar Nathella, P.; Babu, S. Influence of diabetes mellitus on immunity to human tuberculosis. *Immunology* **2017**, *152*, 13–24. [CrossRef]

Journal of
Fungi

MDPI

Review

Paracoccidioidomycosis in Immunocompromised Patients: A Literature Review

João N. de Almeida Jr. [1], Paula M. Peçanha-Pietrobom [2] and Arnaldo L. Colombo [2,*]

[1] Central Laboratory Division, Hospital das Clínicas da Faculdade de Medicina da Universidade de São Paulo, CEP 05403-000 São Paulo, Brazil; jnaj99@gmail.com
[2] Department of Medicine, Division of Infectious Diseases, Escola Paulista de Medicina, Universidade Federal de São Paulo, CEP 04039-032 São Paulo, Brazil; paulampecanha@gmail.com
* Correspondence: arnaldolcolombo@gmail.com; Tel.: +55-11-5576-4848; Fax: +55-11-5081-3240

Received: 14 November 2018; Accepted: 21 December 2018; Published: 26 December 2018

Abstract: *Paracoccidioidomycosis* (PCM) is an endemic mycosis found in Latin America that causes systemic disease mostly in immunocompetent hosts. A small percentage of PCM occurs in immunocompromised patients where low clinical suspicion of the infection, late diagnosis, and uncertainties about its management are factors that negatively impact their outcomes. We conducted a literature review searching reports on PCM associated to HIV, cancer, maligned hemopathies, solid organ transplantation, and immunotherapies, in order to check for peculiarities in terms of natural history and challenges in the clinical management of PCM in this population. HIV patients with PCM usually had low T CD4$^+$ cell counts, pulmonary and lymph nodes involvement, and a poorer prognosis ($\approx$50% mortality). Most of the patients with PCM and cancer had carcinoma of the respiratory tract. Among maligned hemopathies, PCM was more often related to lymphoma. In general, PCM prognosis in patients with malignant diseases was related to the cancer stage. PCM in transplant recipients was mostly associated with the late phase of kidney transplantation, with a high mortality rate (44%). Despite being uncommon, reactivation of latent PCM may take place in the setting of immunocompromised patients exhibiting clinical particularities and it carries higher mortality rates than normal hosts.

Keywords: paracoccidioidomycosis; HIV; cancer; lymphoma; kidney transplant; TNF inhibitors; literature review

1. Introduction

Paracoccidioidomycosis (PCM) is a systemic endemic mycosis caused by *Paracoccidioides brasiliensis* and *Paracoccidioides lutzii*, exhibiting geographic distribution restricted to Latin America, mainly Brazil, Argentina, Colombia and Venezuela [1].

The real burden of the disease is still not defined but incidence rates ranging from 0.7–3 and 9–40 cases/100,000 inhabitants have been reported in endemic and hyperendemic areas of Latin America, respectively [2,3]. The vast majority of cases are reported in normal hosts and continuous exposure to infecting propagules in rural areas is considered to be the main risk factor for this condition [1].

PCM may present in two clinical forms: (i) an acute/subacute form, usually reported in children (or adults < 30years old), with high fever, disseminated lymphadenopathy, hepatosplenomegaly, skin lesions with limited or absent lung involvement and eventually bone lesions, among other symptoms; (ii) a chronic form that represents >80% of all cases reported, presenting with exuberant lung involvement, skin, mucocutaneous lesions or both, and eventually, lesions of the central nervous system and other organs [4,5]. In both clinical forms, adrenal involvement may take place. The polarization between clinical forms is related to the pattern of the adaptive T cell immune response, with a Th2 and

Th9 response leading to an uncontrolled inflammatory process in the acute form, and a deficient Th1 mixed with a Th17 immune response leading to the chronic form of the disease [6,7]. Mortality rates of PCM in normal hosts is usually <5%, but sequelae are frequently documented and include chronic respiratory failure and Addison disease [5,8]. Diagnosis is mainly obtained by conventional methods, including direct examination, culture, histopathology, and detection of specific anti-*Paracoccidioides* antibodies (immunodiffusion or counterimmunoelectrophoresis). PCR-based methods and assays for specific antigen detection were developed by reference labs but are not available in the vast majority of the medical centers in Latin America [9].

Antifungal treatment of mild and moderate cases usually relies on itraconazole or the combination of sulfamethoxazole-trimethoprim. Severe and disseminated infections may require the use of amphotericin B formulations followed by consolidation therapy with itraconazole or sulfamethoxazole-trimethoprim. Patients are usually treated for 12–24 months, depending on clinical presentation [5].

All the aforementioned knowledge applies to PCM in the normal host, and data regarding PCM and immunocompromised patients are scarce and limited. Lack of its clinical suspicion, late diagnosis, and uncertainties about its management are factors that may negatively impact the outcomes of PCM in this population. The present paper describes the peculiarities in terms of natural history and challenges in the clinical management of PCM in patients with HIV, cancer, malignant hemopathies, solid organ transplantation, and immunobiological drugs.

2. Material and Methods

2.1. Search Strategy

We searched the Pubmed database for reports of PCM in immunocompromised patients that were published in the last 30 years. We made all efforts to identify papers addressing epidemiology, fungal diagnosis and antifungal therapy in five different scenarios: HIV, cancer, hematologic patients, solid organ transplant, and related to use of immunobiological agents such as TNF inhibitors and anti-CD20 blockers. Search terms included various combinations of the terms "paracoccidioidomycosis" or "*Paracoccidioides*" with one of the following: "HIV", "AIDS", "cancer", "leukemia", "lymphoma", "myelodysplastic syndrome", "aplastic anemia", "stem cell transplantation", "kidney transplantation", "liver transplantation", "heart transplantation", "lung transplantation"; "TNF inhibitor"; "infliximab"; "etanercept"; "adalimumab"; "anti-CD20"; "rituximab"; "natalizumab". We reviewed all articles retrieved from these search terms and relevant references cited in those articles. There were no language restrictions and only cases with proven or probable PCM were included, following criteria defined elsewhere [5].

Data on age, sex, underlying diseases, immunological status, risk factors (e.g., rural work, tobacco or alcohol consumption), immunosuppressive therapy, laboratorial diagnosis, details of antifungal induction, maintenance therapies, as well as relapse and outcome were recorded. The outcome was considered to be favorable if the patient met the cure criteria according to the Brazilian Guidelines for the Clinical Management of PCM [5]. Recurrence of clinical signs and symptoms, or radiology findings in the presence of any laboratory results suggesting active PCM, were used to define a relapse episode.

2.2. Statistical Analysis

Comparisons between groups were performed using Fisher's exact or chi-square tests when appropriate for the categorical variables (SPSS v.25, IBM, Armonk, NY, USA). p values of <0.05 were considered to be statistically significant.

3. Results and Discussion

3.1. Paracoccidioidomycosis and HIV Infection

The first two cases of PCM associated with HIV infection were reported in 1989 [10]. Since then, PCM/HIV coinfection occurrence has been reported as small case-series in endemic areas in Brazil [11–13], and in isolated cases reports in Colombia and Argentina [14,15]. Two retrospective case-control studies have been conducted up until the present date. In the first study published in 2009, Morejón, Machado and Martinez reported 53 cases of PCM and HIV coinfection in Brazil [16]. In the second controlled study, Almeida et al. 2016, reported thirty-one HIV-infected patients with PCM between 1993 and 2014 [17]. After compiling the data from these two case-control studies [16,17], two case-series reports [13,18], and 30 single case reports [10,14,19–35], we retrieved 136 cases of PCM and HIV coinfection reported in the last 30 years. They aged between 13 and 59 years, with a mean age of 35.9 years. Twenty-six (19%) were female, and only two were in the adult PCM form (6%). The higher proportion of females in the casuistic of immunocompromised patients with PCM compared to the usual gender distribution observed in normal adult hosts (over 6%) suggests that the hormonal protection described for normal hosts is mitigated in the setting of immunosuppression [5]. Most patients worked in the urban area (68%), which is different from the usual epidemiology of the disease. Thus, an ancient exposure and activation of a latent infection might have occurred on these coinfected patients.

In 56 cases, data regarding the awareness of the HIV status at the PCM diagnosis were provided, and only 31 (55%) patients were known to be HIV-infected at that time. Over 80% of the PCM-HIV coinfected patients for whom CD4+ cell counts were available had <200 cells/mm^3. This finding suggests that *Paracoccidioides* spp. may take advantage of the T-cell immunosuppression related to AIDS to shift from quiescent infection to systemic disease. Fever, generalized lymphadenopathy, splenomegaly, and skin lesions, which are generally reported in the acute form of the PCM disease, were more common in PCM-HIV coinfected patients than in the immunocompetent group (see Table 1, Figures 1 and 2).

Table 1. Comparison of clinical, laboratory and outcome data from adult patients with paracoccidioidomycosis coinfected and non-coinfected with HIV virus.

	PCM and HIV (%)	PCM (%)	*p* Value
Clinical Data			
Fever	82.7	35.4 **	<0.001
Lymphadenopathy	72.9 #	50.6 *	<0.001
Splenomegaly	22.6	4.7 *	<0.001
Skin Lesions	58.9	29.6 *	<0.001
Pneumopathy	70.3	63.8 *	0.15
Oral mucosa	29	50 *	<0.001
Laboratory data (positivity rates)			
Direct microscopy	57.4	44 **	0.052
Culture	42.2	25.3 *	<0.001
Histopathology	94.5	64.7 *	<0.001
Serology	74.6	97.2 *	<0.001
Outcomes			
Relapse rates	11	8.2 **	0.48
Mortality rates	35	7.9 **	<0.001

* Data extracted from the references: [2,4]; ** data extracted from the references [16,17]. # PCM coinfected patients usually present multiple lymph nodes involvement, in different anatomic sites.

Pulmonary disease was reported in most of the coinfected population despite the age of the patient, characterizing a mixed clinical form (juvenile–adult) exhibiting simultaneous lung involvement and disseminated infection of the reticuloendothelial system. This is a clinical presentation suggestive of an opportunistic manifestation of PCM in this specific population. This mixed clinical form, already reported by other authors, may be a consequence of the inefficient immune control of the primary lung foci followed by lymphohematogenous dissemination [36,37].

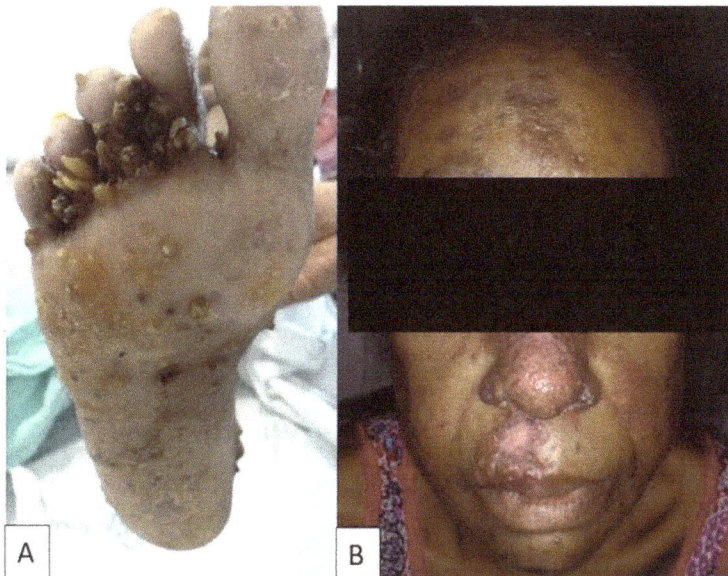

Figure 1. Skin involvement in PCM-HIV coinfection. (**A**) Verrucous lesions on the foot caused by hematogenous dissemination. (**B**) Papulonodular ulcerative lesions caused by hematogenous dissemination. Illustration provided by Prof. Paulo Mendes Peçanha from Infectious Disease Unit, Universidade Federal do Espírito Santo.

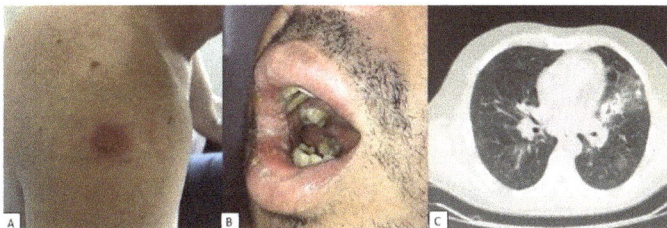

Figure 2. Clinical presentation of PCM-HIV coinfection. (**A**) Ulcerative lesions on the arm. (**B**) Moriform ulcerative perioral involvement. (**C**) Diffuse ground-glass opacities and consolidation in left superior lobe on chest computed tomography. Illustration provided by Dr. Adriana Maria Porro from the Dermatology Department, Escola Paulista de Medicina-Universidade Federal de São Paulo.

Coinfected patients presented more often with positive histopathology and culture results when compared to HIV-negative patients with PCM (see Table 1). These data reflect the higher fungal burden in the coinfected patients. In contrast, the detection of anti-*Paracoccidioides* serum antibody by quantitative double immunodiffusion test or counterimmunoelectrophoresis had lower positive rates in the PCM-HIV population (74.6% vs. 97.2%, $p < 0.001$). Therefore, in patients with AIDS,

Paracoccidioides sp. negative serological results do not exclude the PCM diagnosis, which should be investigated with alternative microbiological tests and histopathological examination.

Two case reports described PCM in patients receiving cotrimoxazole prophylaxis [24,38]. Similarly, Morejón et al. described 10 out 25 coinfected patients that were diagnosed with PCM under cotrimoxazole prophylaxis. Consequently, PCM should not be ruled out in patients under trimethoprim–sulfamethoxazole prophylaxis.

The overall mortality rate was higher in the coinfected population (35% vs. 7.9%, $p < 0.001$, see Table 1), mainly as a consequence of the severe immunodepression seen in most of the patients. Information about the presence of any other (non-PCM) concomitant opportunistic infections was mentioned by the authors in only 65 (47.7%) of 136 patients. Of note, only 19 cases (29%) of the 65 patients where this information was available had other severe infections concomitantly with PCM. This means that the outcome of PCM in this population was not impacted by other severe conditions in 71% of the cases. Fungemia by *Paracoccidioides* seems to be a predictor of poor prognosis since all three patients with positive blood cultures died. The choice of the initial antifungal therapy did not influence the outcome. Amphotericin B deoxycholate (AMB) was used as primary therapy in 19 cases, and 9 patients died (47.3%). Of note, all these nine patients with a fatal outcome had concomitant severe opportunist infections. Among 18 cases in which cotrimoxazole or itraconazole was used as initial therapy, seven died (38.8%, $p = 0.7$). Otherwise, one could suggest that there was one imbalance in the clinical severity of patients exposed to amphotericin B and other drugs that may certainly influence the expected mortality in both groups.

3.2. Paracoccidioidomycosis and Solid Organ Malignancies

The relationship between PCM and solid cancer was first described more than 80 years ago [39]. Concomitant PCM and solid cancer are reported in 0.16 to 11% of the cases, according to some cohorts [39–41]. Most of the solid cancers are carcinomas (>80%) from the respiratory or digestive tract [42], and are related to the chronic form of PCM and its risk factors, such as male sex, rural workers with a history of smoking and alcohol intake [1,42]. Likewise, among the 36 cases that fulfilled the inclusion criteria (31 from two case series [42,43] and five from isolated case reports [44–48]), 26 (72%) were related to carcinomas of the upper and lower respiratory ($n = 16$, 44%) and digestive tracts, mainly the oropharynx and esophagus ($n = 7$, 19%). Carcinoma was diagnosed at the same anatomical site of the fungal lesion in twenty-one cases (58%). In thirteen cases (36%) PCM appeared before cancer, and in 19 patients (53%) cancer was diagnosed simultaneously with PCM. These findings have led researchers to raise the hypothesis that PCM may be an additive factor for cancer development due to chronic antigenic stimulation of the pathogenic yeasts on the epithelial cells [43].

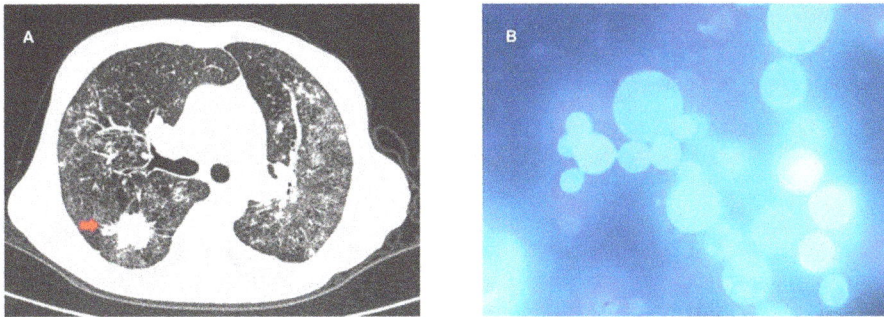

Figure 3. (**A**) Chest computed tomography scan showing spiculated nodule from adenocarcinoma (red arrow) and bilateral diffuse bronchoalveolar infiltrates from PCM. (**B**) Bronchoalveolar lavage with yeast cells with multiple daughter buds (Calcofluor white, 1000×). Figure **A** provided by Dr. Drielle Peixoto Bittencourt from Hospital do Cancer, Universidade de São Paulo; Figure **B** provided by the authors.

So far, we were not able to identify any particularity in terms of clinical presentation of PCM in patients with solid cancer. Serology was useful for PCM diagnosis in this population in only 50% of cases [43]. Figure 3 illustrates a case of concomitant pulmonary adenocarcinoma and PCM. In a retrospective analysis of 25 cases of PCM and cancer, Rodrigues et al. reported a mortality rate of 16%, a rate apparently higher than that associated to normal hosts [43,49].

3.3. Paracoccidioidomycosis and Hematologic Malignancies

Hematologic malignancies are rarely reported in patients with PCM, with an estimated prevalence of 0–3% [42]. A dozen cases were reported in the last 30 years, most of them were B cell lymphomas (either Hodgkin or non-Hodgkin) that were diagnosed after the PCM disease (1–8 years) [41–43,50]. Resende et al. reported four detailed cases of PCM and lymphoma, all of them had lymph nodes with PCM yeasts found in histopathology and positive serology [50]. The patients were treated for a long time, from 2 to 10 years, and two of them showed PCM recurrence. Two patients died due to complications related to the lymphoma [50]. The authors suggested that the chronic PCM antigenic stimulation may have had a role in the development of B cell lymphoma [50], but further investigation is necessary to confirm this hypothesis. The limited casuistic precludes any conclusion in terms of putative peculiarities of natural history or diagnostic tools for PCM in this specific setting.

3.4. Paracoccidioidomycosis and Solid Organ Transplant

Among the different solid organ transplant modalities, chronic PCM has been described predominantly in kidney transplant recipients. Nine cases of PCM in kidney transplant patients and a single episode in a liver transplant recipient fulfilled the inclusion criteria [51–55]. We excluded from the present series one episode of PCM mentioned in a report of a lung transplant recipient without any further details [56,57]. The patients had a median age of 55 years; three of them were rural workers before the transplantation, and three (60%) out of five cases with gender description were male. Seven cases developed PCM after one year of transplantation (range 1–14 years), none of them were having cotrimoxazole prophylaxis at diagnosis, and symptoms of PCM appeared 2 to 6 months before diagnosis. One case lacked a description of the time elapsed between transplant and PCM diagnosis [51]. Three (33%) of those reported cases had skin lesions, either combined with oral mucosal and lung infiltrates (two cases), or with lymphadenopathy (one case). One illustrative case is presented in Figure 4.

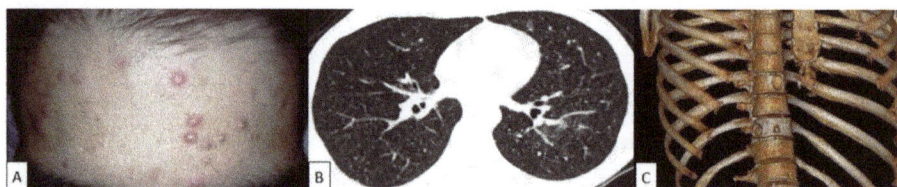

Figure 4. Disseminated paracoccidioidomycosis in a kidney transplant recipient. (**A**) Vesiculopapular lesions on the face. (**B**) Chest computerized tomography showing a miliary nodular pattern. (**C**) Chest computerized tomography and 3D reconstruction of the thoracic bone structure showing osteolytic vertebral lesions. Illustration provided by Dr. Daniel Wagner from Hospital do Rim, Universidade Federal de São Paulo.

Three reports provided information regarding immunosuppressive therapy, that consisted of corticosteroids and the combination of other immunosuppressants, such as cyclosporine and azathioprine [52], mycophenolate mofetil (MMF) and tacrolimus [53], and mesalazine and tacrolimus [54]. The diagnosis was mainly achieved through direct examination or histopathology of clinical samples. All patients required hospitalization, four were initially treated with AMB formulations, three out nine patients (33%) died due to the mycosis. Of note, a 57-year-old female patient developed acute respiratory failure due to PCM only two days after the kidney transplantation [55]. This particular patient lived in a rural area and was diagnosed with a solitary pulmonary nodule before the transplant, which was considered to be latent tuberculosis. The pre-transplant immunosuppression therapy consisted of anti-thymocyte globulin (ATG) and MMF. The patient was successfully treated with liposomal AMB (1 mg/Kg) for 14 days followed by itraconazole (200mg/day) for one year. ID serology tested in the acute phase of the disease and during clinical follow-up were both negative.

One case of severe disseminated PCM in a 3-year-old girl was reported 24 months after liver transplantation due to congenital biliary atresia [56]. She was initially treated with cotrimoxazole (200/40 mg, q12h), and due to a poor clinical response, AMB was also prescribed. Initial CIE serology was positive (titer 1/64), and after six months of cotrimoxazole (100/20 mg q12h) maintenance therapy, the patient was considered cured [56].

Prophylaxis with cotrimoxazole after transplantation explains the rarity of PCM in this scenario, as this drug is active against *P. brasiliensis*. Most of the reported cases occur after the first year of transplantation when immunosuppression is tapered and cotrimoxazole prophylaxis is no longer necessary. However, despite being rare, PCM in kidney transplant patients seems to have a poor prognosis, possibly related to its low clinical suspicion, difficult and late diagnosis (negative serological tests), and immunosuppression. Indeed, it has been demonstrated that immunosuppressed kidney transplant recipients have a persistent poor Th1 immune response to *Paracoccidioides* antigen gp43 [58].

Due to the rarity of PCM in organ transplant recipients, there is no formal recommendation for living donors and recipients that have lived in endemic areas to be routinely screened for PCM latent infection before transplant. Serological tests have poor value for the clinical management of PCM disease in kidney transplant recipients, and AMB lipid formulations have to be considered as initial therapy in severe cases [59].

3.5. Paracoccidioidomycosis and Immunotherapies

Only three cases of PCM disease related to immunobiological agents have been reported so far [60,61]. A 60-year-old man with rheumatoid arthritis and on immunosuppressive therapy, including methotrexate, leflunomide and adalimumab (40mg every 15 days), for 3 years, was hospitalized to investigate a chronic hip pain that was further diagnosed as an osteosarcoma. Chest CT-scan revealed an excavated pulmonary lesion on the lower left lobe and biopsy revealed granuloma and yeasts compatible with *P. brasiliensis*. Despite AMB induction therapy and adalimumab interruption, the patient developed sepsis after a hip surgery for tumor resection and died [60]. A 47-year old man with psoriatic spondyloarthritis and on infliximab therapy for 18 months developed fever and respiratory symptoms. CT-scan revealed a left inferior pulmonary lobe nodule and mediastinal lymphadenopathy. Lung biopsy histological analysis diagnosed PCM. Investigation of anti-*Paracoccidioides* antibodies by the immunodiffusion technique was negative. Infliximab was suspended and the patient was successfully treated with cotrimoxazole for 29 months. A 46-year old man with multiple sclerosis developed pulmonary PCM after 15 months of natalizumab therapy [62]. The diagnosis was made by lung biopsy and the patient was treated with itraconazole and natalizumab was discontinued.

In the USA, TNF inhibitors, mainly infliximab, have been related to histoplasmosis, with a median of 15 months after initiation of the immunobiological drug [63]. Likewise, the reported cases of PCM in patients undergoing TNF inhibitor therapy occurred more than one year after the introduction of the medication. These cases highlight the need to include PCM in the list of opportunistic infections in patients under long-term immunotherapy from endemic areas. Moreover, negative serology should not exclude the diagnosis of PCM in these patients and infection must be managed with interruption of the TNF blocker agent and prolonged antifungal therapy.

4. Conclusions

Despite being uncommon, PCM may be reported in patients with immunosuppression, including AIDS, cancer, patients with solid organ transplantations, or on immunobiological therapy. The vast majority of PCM in immunosuppressed patients has been reported in HIV patients, where this disease may exhibit simultaneously the clinical characteristics of chronic (lung involvement) and acute forms (generalized lymph adenomegaly and hepatosplenomegaly) of the disease. In this population, more disease relapses and higher mortality rates are reported than in non-immunocompromised hosts.

Lymphoma was the most common hematologic malignancy reported with PCM, and few cases of PCM have been reported in the late period after kidney transplantation. We were not able to identify any change in terms of the natural history of PCM documented in patients with cancer or solid organ transplant recipients.

Diagnosis of PCM may be challenging in immunosuppressed patients where serology usually has a low sensitivity and PCR-based methods or assays for antigen detection are not available in the vast majority of routine laboratories. Consequently, invasive procedures (e.g., biopsy) may be required to confirm the diagnosis [17,62]. Of note, in endemic areas, paracoccidioidomycosis should be included in the differential diagnosis of any patient with a disease associated with T-cell immunodeficiency who presents with pulmonary infiltrates with nodules, cavitation or chronic alveolar consolidation, as well as skin or mucocutaneous lesions with a chronic evolution.

Regarding specific therapy of patients with severe clinical forms of PCM, amphotericin B should be promptly initiated, followed by 12 to 24 months of treatment with itraconazole or sulfamethoxazole-trimethoprim. Treatment duration relies on the severity of clinical presentation, sites of infection, restoration of the host immune response, as well as the clinical and laboratorial response to therapy [5].

Finally, immunocompromised patients who travel to endemic areas of PCM should be counseled before traveling to avoid high-risk exposures. Once the transplant recipient returns from an endemic area, the clinician must instigate a complete diagnostic investigation if any sign or symptom of PCM appears, in order to treat it rapidly and to mitigate against morbidity and a poor outcome [59].

Author Contributions: J.N.d.A.J. and P.M.P.-P. reviewed the literature and were responsible for writing the first draft of the manuscript. A.L.C. designed the study strategy, wrote the introduction and conclusions, as well as reviewed the whole manuscript.

Funding: Authors has not received funding to make this manuscript.

Acknowledgments: A.L.C. received grants from Conselho Nacional de Desenvolvimento Científico e Tecnológico, Brazil, (CNPq, Grant 307510/2015-8). We thank Paulo Mendes Peçanha from Infectious Disease Unit, Universidade Federal do Espírito Santo; Adriana Maria Porro from the Dermatology Department, Universidade Federal de São Paulo; Daniel Wagner from Hospital do Rim, Universidade Federal de São Paulo; Drielle Peixoto Bittencourt from Hospital do Cancer, University of São Paulo; that kindly provided us the photos.

Conflicts of Interest: Arnaldo Lopes Colombo has received educational grants from Biotoscana, Gilead, MSD, Pfizer, and United Medical and research grants from Astellas and Pfizer. All other authors: none to declare.

References

1. Martinez, R. New Trends in Paracoccidioidomycosis Epidemiology. *J. Fungi* **2017**, *3*, 1. [CrossRef] [PubMed]
2. Bellissimo-Rodrigues, F.; Machado, A.A.; Martinez, R. Paracoccidioidomycosis epidemiological features of a 1,000-cases series from a hyperendemic area on the southeast of Brazil. *Am. J. Trop. Med. Hyg.* **2011**, *85*, 546–550. [CrossRef] [PubMed]
3. Vieira, G.D.; Alves, T.D.; Lima, S.M.; Camargo, L.M.; Sousa, C.M. Paracoccidioidomycosis in a western Brazilian Amazon State: Clinical-epidemiologic profile and spatial distribution of the disease. *Rev. Soc. Bras. Med. Trop.* **2014**, *47*, 63–68. [CrossRef] [PubMed]
4. Bellissimo-Rodrigues, F.; Bollela, V.R.; Da Fonseca, B.A.L.; Martinez, R. Endemic paracoccidioidomycosis: Relationship between clinical presentation and patients' demographic features. *Med. Mycol.* **2013**, *51*, 313–318. [CrossRef] [PubMed]
5. Shikanai-Yasuda, M.A.; Mendes, R.P.; Colombo, A.L.; de Queiroz-Telles, F.; Kono, A.S.G.; Paniago, A.M.M.; Nathan, A.; do Valle, A.C.F.; Bagagli, E.; Benard, G.; et al. Brazilian guidelines for the clinical management of paracoccidioidomycosis. *Rev. Soc. Brasil. Med. Trop.* **2017**, *50*, 715–740. [CrossRef] [PubMed]
6. De Castro, L.F.; Ferreira, M.C.; da Silva, R.M.; Blotta, M.H.; Longhi, L.N.A.; Mamoni, R.L. Characterization of the immune response in human paracoccidioidomycosis. *J. Infect.* **2013**, *67*, 470–485. [CrossRef] [PubMed]
7. De Oliveira, H.C.; Assato, P.A.; Marcos, C.M.; Scorzoni, L.; de Paula, E.; Silva, A.C.A.; Da Silva, J.D.F.; Singulani, J.D.; Alarcon, K.M.; Fusco-Almeida, A.M.; et al. Paracoccidioides-host Interaction: An Overview on Recent Advances in the Paracoccidioidomycosis. *Front. Microbiol.* **2015**, *6*, 1319. [CrossRef] [PubMed]
8. Colombo, A.L.; Tobón, A.; Restrepo, A.; Queiroz-Telles, F.; Nucci, M. Epidemiology of endemic systemic fungal infections in Latin America. *Med. Mycol.* **2011**, *49*, 785–798. [CrossRef] [PubMed]
9. Queiroz-Telles, F.; Fahal, A.H.; Falci, D.R.; Caceres, D.H.; Chiller, T.; Pasqualotto, A.C. Neglected endemic mycoses. *Lancet Infect. Dis.* **2017**, *17*, e367–e377. [CrossRef]
10. Pedro, R.D.; Aoki, F.H.; Boccato, R.S.; Branchini, M.L.; Gonçales Júnior, F.L.; Papaiordanou, P.M.; Ramos, M.D. Paracoccidioidomicose e infecção pelo virus da imunodeficiência humana. *Rev. Inst. Med. Trop São Paulo* **1989**, *31*, 119–125. [CrossRef]
11. Marques, S.A.; Conterno, L.O.; Sgarbi, L.P.; Villagra, A.M.; Sabongi, V.P.; Bagatin, E.; Gonçalves, V.L. Paracoccidioidomycosis associated with acquired immunodeficiency syndrome. Report of seven cases. *Rev. Inst. Med. Trop. Sao Paulo* **1995**, *37*, 261–265. [CrossRef]
12. Silva-Vergara, M.L.; Teixeira, A.C.; Curi, V.G.M.; Costa Júnior, J.C.; Vanunce, R.; Carmo, W.M.; Silva, M.R. Paracoccidioidomycosis associated with human immunodeficiency virus infection. Report of 10 cases. *Med. Mycol.* **2003**, *41*, 259–263. [CrossRef] [PubMed]
13. Paniago, A.M.M.; de Freitas, A.C.C.; Aguiar, E.S.A.; Aguiar, J.I.A.; da Cunha, R.V.; Castro, A.R.C.M.; Wanke, B. Paracoccidioidomycosis in patients with human immunodeficiency virus: Review of 12 cases observed in an endemic region in Brazil. *J. Infect.* **2005**, *51*, 248–252. [CrossRef] [PubMed]
14. Tobon, A.M.; Orozco, B.; Estrada, S.; Jaramillo, E.; de Bedout, C.; Arango, M.; Restrepo, A. Paracoccidioidomycosis and AIDS: Report of the first two Colombian cases. *Rev. Inst. Med. Trop. Sao Paulo* **1998**, *40*, 377–381. [CrossRef] [PubMed]
15. Corti, M.; Trione, N.; Risso, D.; Soto, I.; Villafañe, M.F.; Yampolsky, C.; Negroni, R. Disseminated paracoccidioidomycosis with a single brainstem lesion. A case report and literature review. *Neuroradiol. J.* **2010**, *23*, 454–458. [CrossRef] [PubMed]
16. Morejón, K.M.L.; Machado, A.A.; Martinez, R. Paracoccidioidomycosis in patients infected with and not infected with human immunodeficiency virus: A case-control study. *Am. J. Trop. Med. Hyg.* **2009**, *80*, 359–366. [CrossRef] [PubMed]

17. De Almeida, F.A.; Neves, F.F.; Mora, D.J.; Reis, T.A.D.; Sotini, D.M.; Ribeiro, B.D.M.; Andrade-Silva, L.E.; Nascentes, G.N.; Ferreira-Paim, K.; Silva-Vergara, M.L. Paracoccidioidomycosis in Brazilian Patients with and Without Human Immunodeficiency Virus Infection. *Am. J. Trop. Med. Hyg.* **2017**, *96*, 368–372. [CrossRef] [PubMed]

18. Macedo, P.M.; Almeida-Paes, R.; Almeida, M.D.; Coelho, R.A.; Andrade, H.B.; Ferreira, A.B.T.B.C.; Zancopé-Oliveira, R.M.; Valle, A.C.F. Paracoccidioidomycosis due to Paracoccidioides brasiliensis S1 plus HIV co-infection. *Mem. Inst. Oswaldo Cruz* **2018**, *113*, 167–172. [CrossRef]

19. Goldani, L.Z.; Martinez, R.; Landell, G.A.; Machado, A.A.; Coutinho, V. Paracoccidioidomycosis in a patient with acquired immunodeficiency syndrome. *Mycopathologia* **1989**, *105*, 71–74. [CrossRef]

20. Bakos, L.; Kronfeld, M.; Hampe, S.; Castro, I.; Zampese, M. Disseminated paracoccidioidomycosis with skin lesions in a patient with acquired immunodeficiency syndrome. *J. Am. Acad. Dermatol.* **1989**, *20*, 854–855. [CrossRef]

21. Benard, G.; Bueno, J.P.; Yamashiro-Kanashiro, E.H.; Shikanai-Yasuda, M.A.; Del Negro, G.M.; Melo, N.T.; Sato, M.N.; Amato Neto, V.; Shiroma, M.; Duarte, A.J. Paracoccidioidomycosis in a patient with HIV infection: Immunological study. *Trans. R. Soc. Trop. Med. Hyg.* **1990**, *84*, 151–152. [CrossRef]

22. De Lima, M.A.; Silva-Vergara, M.L.; Demachki, S.; dos Santos, J.A. Paracoccidioidomycosis in a patient with human immunodeficiency virus infection. A necropsy case. *Rev. Soc. Bras. Med. Trop.* **1995**, *28*, 279–284. [PubMed]

23. Silletti, R.P.; Glezerov, V.; Schwartz, I.S. Pulmonary paracoccidioidomycosis misdiagnosed as Pneumocystis pneumonia in an immunocompromised host. *J. Clin. Microbiol.* **1996**, *34*, 2328–2330. [PubMed]

24. Dos Santos, J.W.; Costa, J.M.; Cechella, M.; Michel, G.T.; de Figueiredo, C.W.; Londero, A.T. An unusual presentation of paracoccidioidomycosis in an AIDS patient: A case report. *Mycopathologia* **1998**, *142*, 139–142. [CrossRef] [PubMed]

25. Nobre, V.; Braga, E.; Rayes, A.; Serufo, J.C.; Godoy, P.; Nunes, N.; Antunes, C.M.; Lambertucci, J.R. Opportunistic infections in patients with AIDS admitted to an university hospital of the Southeast of Brazil. *Rev. Inst. Med. Trop. Sao Paulo* **2003**, *45*, 69–74. [CrossRef] [PubMed]

26. Corti, M.; Villafañe, M.F.; Negroni, R.; Palmieri, O. Disseminated paracoccidioidomycosis with pleuritis in an AIDS patient. *Rev. Inst. Med. Trop. Sao Paulo* **2004**, *46*, 47–50. [CrossRef] [PubMed]

27. Caseiro, M.M.; Etzel, A.; Soares, M.C.B.; Costa, S.O.P. Septicemia caused by Paracoccidioides brasiliensis (Lutz, 1908) as the cause of death of an AIDS patient from Santos, São Paulo State, Brazil—A nonendemic area. *Rev. Inst. Med. Trop. Sao Paulo* **2005**, *47*, 209–211. [CrossRef] [PubMed]

28. Godoy, P.; Lelis, S.S.R.; Resende, U.M. Paracoccidioidomycosis and acquired immunodeficiency syndrome: Report of necropsy. *Rev. Soc. Bras. Med. Trop.* **2006**, *39*, 79–81. [CrossRef]

29. Castro, G.; Martinez, R. Images in clinical medicine. Disseminated paracoccidioidomycosis and coinfection with HIV. *N. Engl. J. Med.* **2006**, *355*, 2677. [CrossRef]

30. Brunaldi, M.O.; Rezende, R.E.F.; Zucoloto, S.; Garcia, S.B.; Módena, J.L.P.; Machado, A.A. Co-infection with paracoccidioidomycosis and human immunodeficiency virus: Report of a case with esophageal involvement. *Am. J. Trop. Med. Hyg.* **2010**, *82*, 1099–1101. [CrossRef]

31. De Freitas, R.S.; Dantas, K.C.; Garcia, R.S.P.; Magri, M.M.C.; de Andrade, H.F. Paracoccidioides brasiliensis causing a rib lesion in an adult AIDS patient. *Hum. Pathol.* **2010**, *41*, 1350–1354. [CrossRef] [PubMed]

32. Nunura, R.J.; Salazar, M.D.; Vásquez, L.T.; Endo, G.S.; Rodríguez, F.A.; Zerpa, L.R. Paracoccidioidomicosis and multidrug-resistant tuberculosis (TBC-MDR) in patient coinfected with HIV and hepatitis C. *Rev. Chilena Infectol.* **2010**, *27*, 551–555.

33. Lambertucci, J.R.; Vale, T.C.; Voieta, I. Images in infectious diseases. Concomitant progressive multifocal leukoencephalopathy and disseminated paracoccidioidomycosis in an AIDS patient. *Rev. Soc. Bras. Med. Trop.* **2010**, *43*, 758. [CrossRef] [PubMed]

34. Nogueira, L.M.C.; Santos, M.; Ferreira, L.C.; Talhari, C.; Rodrigues, R.R.; Talhari, S. AIDS-associated paracoccidioidomycosis in a patient with a CD4+ T-cell count of 4 cells/mm^3. *Anais Bras. Dermatol.* **2011**, *86*, S129–S132. [CrossRef]

35. Silva-Vergara, M.L.; Rocha, I.H.; Vasconcelos, R.R.; Maltos, A.L.; Neves, F.; Teixeira, L.; Mora, D.J. Central nervous system paracoccidioidomycosis in an AIDS patient: Case report. *Mycopathologia* **2014**, *177*, 137–141. [CrossRef] [PubMed]

36. Cimerman, S.; Bacha, H.A.; Ladeira, M.C.; Silveira, O.S.; Colombo, A.L. Paracoccidioidomycosis in a boy infected with HIV. *Mycoses* **1997**, *40*, 343–344. [CrossRef] [PubMed]

37. Benard, G.; Duarte, A.J. Paracoccidioidomycosis: A model for evaluation of the effects of human immunodeficiency virus infection on the natural history of endemic tropical diseases. *Clin. Infect. Dis.* **2000**, *31*, 1032–1039. [CrossRef] [PubMed]

38. Hadad, D.J.; Pieres, M.; Petry, T.C.; Orozco, S.F.; Melhem, M.; Paes, R.A.; Gianini, M.J. Paracoccidioides brasiliensis (Lutz, 1908) isolated by hemoculture in a patient with the acquired immunodeficiency syndrome (AIDS). *Rev. Inst. Med. Trop. Sao Paulo* **1992**, *34*, 565–567. [CrossRef]

39. Rabello Filho, E. Lupus eritematoso disseminado, blastomicose e epitelioma do lábio superior. *Anais Bras. Derm. Sif.* **1933**, *8*, 38–39.

40. Padilha-Gonçalves, A. Adenopatia na Micose de Lutz. Thesis, Escola de Medicina e Cirurgia do Rio de Janeiro, Rio de Janeiro, Brazil, 1971; p. 234.

41. Da Conceição, Y.T.M. Frequência da Associação em Estudo de Necropsias. Master's Thesis, Faculdade Medicina Universidade de São Paulo, São Paulo, Brazil, 1996; p. 102.

42. Shikanai-Yasuda, M.A.; Conceição, Y.M.T.; Kono, A.; Rivitti, E.; Campos, A.F.; Campos, S.V. Neoplasia and paracoccidioidomycosis. *Mycopathologia* **2008**, *165*, 303–312. [CrossRef]

43. Rodrigues, G.D.; Severo, C.B.; Oliveira, F.D.; Moreira, J.D.; Prolla, J.C.; Severo, L.C. Association between paracoccidioidomycosis and cancer. *J. Bras. Pneumol.* **2010**, *36*, 356–362. [CrossRef]

44. Maymó Argañaraz, M.; Luque, A.G.; Tosello, M.E.A.; Perez, J. Paracoccidioidomycosis and larynx carcinoma. *Mycoses* **2003**, *46*, 229–232. [CrossRef] [PubMed]

45. Marques, S.A.; Lastória, J.C.; Marques, M.E.A. Paracoccidioidomicose em paciente com carcinoma do colo uterino. *Anais Bras. Dermatol.* **2011**, *86*, 587–588. [CrossRef]

46. Porro, A.M.; Rotta, O. Cutaneous and pulmonary paracoccidioidomycosis in a patient with a malignant visceral tumor. *Anais Bras. Dermatol.* **2011**, *86*, 1220–1221. [CrossRef]

47. Azevedo, R.S.; Gouvêa, A.F.; Lopes, M.A.; Corrêa, M.B.; Jorge, J. Synchronous oral paracoccidioidomycosis and oral squamous cell carcinomas with submandibular enlargement. *Med. Mycol.* **2011**, *49*, 84–89. [CrossRef] [PubMed]

48. Tubino, P.V.A.; Sarmento, B.J.; dos Santos, V.M.; Borges, E.R.; da Silva, L.E.C.; Lima, R. Synchronous oral paracoccidioidomycosis and esophageal carcinoma. *Mycopathologia* **2012**, *174*, 157–161. [CrossRef] [PubMed]

49. Peçanha, P.M.; Ferreira, M.E.; Peçanha, M.A.; Schmidt, E.B.; de Araújo, M.L.; Zanotti, R.L.; Potratz, F.F.; Nunes, N.E.; Ferreira, C.U.; Delmaestro, D.; et al. Paracoccidioidomycosis: Epidemiological and Clinical Aspects in 546 Cases Studied in the State of Espírito Santo, Brazil. *Am. J. Trop. Med. Hyg.* **2017**, *97*, 836–844. [CrossRef]

50. Ruiz e Resende, L.S.; Yasuda, A.G.; Mendes, R.P.; Marques, S.A.; Niéro-Melo, L.; Defaveri, J.; Domingues, M.A.C. Paracoccidioidomycosis in patients with lymphoma and review of published literature. *Mycopathologia* **2015**, *179*, 285–291. [CrossRef]

51. Reis, M.A.; Costa, R.S.; Ferraz, A.S. Causes of death in renal transplant recipients: A study of 102 autopsies from 1968 to 1991. *J. R. Soc. Med.* **1995**, *88*, 24–27.

52. Zavascki, A.P.; Bienardt, J.C.; Severo, L.C. Paracoccidioidomycosis in organ transplant recipient: Case report. *Rev. Inst. Med. Trop. Sao Paulo* **2004**, *46*, 279–281. [CrossRef]

53. Pontes, A.M.; Borborema, J.; Correia, C.R.B.; de Almeida, W.L.; Maciel, R.F. A rare paracoccidioidomycosis diagnosis in a kidney transplant receptor: Case report. *Transplant. Proc.* **2015**, *47*, 1048–1050. [CrossRef] [PubMed]

54. Góes, H.F.; Durães, S.M.B.; Lima, C.; Souza, M.B.; Vilar, E.A.G.; Dalston, M.O. Paracoccidioidomycosis in a Renal Transplant Recipient. *Rev. Inst. Med. Trop. Sao Paulo* **2016**, *58*, 12. [CrossRef] [PubMed]

55. Radisic, M.V.; Linares, L.; Afeltra, J.; Pujato, N.; Vitale, R.G.; Bravo, M.; Dotta, A.C.; Casadei, D.H. Acute pulmonary involvement by paracoccidiodomycosis disease immediately after kidney transplantation: Case report and literature review. *Transpl. Infect. Dis.* **2017**, *19*, e12655. [CrossRef] [PubMed]

56. Lima, T.C.; Bezerra, R.O.F.; Siqueira, L.T.; Menezes, M.R.; Leite, C.; Porta, G.; Cerri, G.G. Paracoccidioidomycosis in a liver transplant recipient. *Rev. Soc. Bras. Med. Trop.* **2017**, *50*, 138–140. [CrossRef] [PubMed]

57. Campos, S.; Caramori, M.; Teixeira, R.; Afonso, J.; Carraro, R.; Strabelli, T.; Samano, M.; Pêgo-Fernandes, P.; Jatene, F. Bacterial and Fungal Pneumonias After Lung Transplantation. *Transpl. Proc.* **2008**, *40*, 822–824. [CrossRef] [PubMed]

58. Batista, M.V.; Sato, P.K.; Pierrotti, L.C.; de Paula, F.J.; Ferreira, G.F.; Ribeiro-David, D.S.; Nahas, W.C.; Duarte, M.I.S.; Shikanai-Yasuda, M.A. Recipient of kidney from donor with asymptomatic infection by Paracoccidioides brasiliensis. *Med. Mycol.* **2012**, *50*, 187–192. [CrossRef] [PubMed]

59. Abdala, E.; Miller, R.; Pasqualotto, A.C.; Muñoz, P.; Colombo, A.L.; Cuenca-Estrella, M. Endemic Fungal Infection Recommendations for Solid-Organ Transplant Recipients and Donors. *Transplantation* **2018**, *102*, S52–S59. [CrossRef] [PubMed]

60. Woyciechowsky, T.G.; Dalcin, D.C.; dos Santos, J.W.A.; Michel, G.T. Paracoccidioidomycosis induced by immunosuppressive drugs in a patient with rheumatoid arthritis and bone sarcoma: Case report and review of the literature. *Mycopathologia* **2011**, *172*, 77–81. [CrossRef]

61. Covre, L.C.P.; Hombre, P.M.; Falqueto, A.; Peçanha, P.M.; Valim, V. Pulmonary paracoccidioidomycosis: A case report of reactivation in a patient receiving biological therapy. *Rev. Soc. Bras. Med. Trop.* **2018**, *51*, 249–252. [CrossRef]

62. Almeida, K.J.; Barreto-Soares, R.V.; Campos-Sousa, R.N.; Campos-Sousa, M.G.; Bor-Seng-Shu, E. Pulmonary paracoccidioidomycosis associated with the use of natalizumab in multiple sclerosis. *Mult. Scler.* **2018**, *24*, 1002–1004. [CrossRef]

63. Vergidis, P.; Avery, R.K.; Wheat, L.J.; Dotson, J.L.; Assi, M.A.; Antoun, S.A.; Hamoud, K.A.; Burdette, S.D.; Freifeld, A.G.; McKinsey, D.S.; et al. Histoplasmosis complicating tumor necrosis factor-α blocker therapy: A retrospective analysis of 98 cases. *Clin. Infect. Dis.* **2015**, *61*, 409–417. [CrossRef] [PubMed]

Journal of
Fungi

MDPI

Review

Understanding Pathogenesis and Care Challenges of Immune Reconstitution Inflammatory Syndrome in Fungal Infections

Sarah Dellière [1], Romain Guery [1], Sophie Candon [2], Blandine Rammaert [3], Claire Aguilar [1], Fanny Lanternier [1,4], Lucienne Chatenoud [2] and Olivier Lortholary [1,4,*]

[1] Medical School, Paris-Descartes University, APHP, Necker-Enfants Malades Hospital, Infectious Disease Center Necker-Pasteur, IHU Imagine, 75015 Paris, France; sarah.delliere@gmail.com (S.D.); romain.guery@aphp.fr (R.G.); ag_claire@yahoo.fr (C.A.); fanny.lanternier@aphp.fr (F.L.)
[2] Medical School, Paris-Descartes University, INSERM U1151-CNRS UMR 8253APHP, Necker-Enfants Malades Hospital, APHP, Clinical Immunology, 75015 Paris, France; sophie.candon@inserm.fr (S.C.); lucienne.chatenoud@inserm.fr (L.C.)
[3] Medical School, Poitiers University, Poitiers, France; Poitiers University Hospital, Infectious Disease Unit, Poitiers, France; INSERM U1070, 86022 Poitiers, France; blandine.rammaert.paltrie@univ-poitiers.fr
[4] Pasteur Institute, Molecular Mycology Unit, National Reference Center for Invasive Fungal Disease and Antifungals, CNRS UMR 2000, 75015 Paris, France
* Correspondence: olivier.lortholary@aphp.fr

Received: 19 November 2018; Accepted: 15 December 2018; Published: 17 December 2018

Abstract: Immune deficiency of diverse etiology, including human immunodeficiency virus (HIV), antineoplastic agents, immunosuppressive agents used in solid organ recipients, immunomodulatory therapy, and other biologics, all promote invasive fungal infections. Subsequent voluntary or unintended immune recovery may induce an exaggerated inflammatory response defining immune reconstitution inflammatory syndrome (IRIS), which causes significant mortality and morbidity. Fungal-associated IRIS raises several diagnostic and management issues. Mostly studied with *Cryptococcus*, it has also been described with other major fungi implicated in human invasive fungal infections, such as *Pneumocystis*, *Aspergillus*, *Candida*, and *Histoplasma*. Furthermore, the understanding of IRIS pathogenesis remains in its infancy. This review summarizes current knowledge regarding the clinical characteristics of IRIS depending on fungal species and existing strategies to predict, prevent, and treat IRIS in this patient population, and tries to propose a common immunological background to fungal IRIS.

Keywords: invasive fungal infections; mycoses; immune reconstitution inflammatory syndrome; fungal immunity

1. Introduction

The increasing prevalence of acquired immunodeficiency, subsequent to the human immunodeficiency virus (HIV) pandemic and medical advances, such as organ transplant, stem cell transplant, intensive anti-neoplastic chemotherapy, or immunomodulatory biological agents, has tremendously raised the prevalence of opportunistic infectious diseases, including fungal ones [1]. Further progress, such as anti-retroviral therapy (ART) in HIV patients, has managed to restore immunity, therefore shedding light on a new syndrome: immune reconstitution inflammatory syndrome (IRIS). IRIS is now known to occur during the course of various invasive fungal diseases (IFD). It can be defined as a clinical worsening, or the new presentation, of infectious disease after reversal of immune deficiency. This reversal can be driven by the introduction of ART in HIV patients, neutrophil recovery after chemotherapy and/or stem cell transplant, inadequate balancing of immunosuppressive therapy after solid organ transplantation (SOT) [2,3], and even by post-partum immunological changes after pregnancy [4]. IRIS is triggered by the recovery of immune cells, resulting in a "cytokine storm" and an exaggerated host inflammatory response. IRIS has been best described in HIV-infected patients as a syndrome occurring in the first 6 months of ART and associated with a wide range of opportunistic pathogens, such as JC virus, cytomegalovirus, non-tuberculous mycobacteria, *Mycobacterium tuberculosis*, cryptococci, and *Histoplasma* species [5]. IRIS is commonly divided into two clinical pictures. "Paradoxical" IRIS refers to a primarily diagnosed and treated infectious disease with a secondary inflammatory increase occurring during antimicrobial treatment and immunodeficiency reversal [6]. "Unmasking" IRIS refers to disease symptoms that first appear after immune recovery [6]. When occurring during fungal infections, these entities have been extensively studied in the context of cryptococcosis. Cryptococcal IRIS develops in 8-49% of patients with known cryptococcal disease before ART [6]. The panel and clinical presentation of IFD responsible for IRIS after immune recovery varies with the underlying primary or acquired immunodeficiency. It depends on multifactorial conditions, including the nature of the immune defect, host genetics, and fungal pathogenesis and exposure. Our understanding of IRIS's pathogenesis remains poor. It is a true diagnosis challenge to distinguish IRIS from sole fungal infection or treatment failure due to a similar clinical presentation. Misdiagnosis and the resulting ineffective treatment with antifungals instead of anti-inflammatory drugs may result in the disease having a fatal course [7]. Furthermore, the therapeutics attempted for IRIS include many anti-inflammatory agents and biologic immunomodulators; however, these remain poorly codified in guidelines. In this review, we will summarize current knowledge on the risk factors and the clinical and biological manifestations of IRIS associated with various fungal infections. Current knowledge and cues to understanding the immunopathogenesis of IRIS will be detailed. Finally, IRIS management, including therapeutics and prevention strategies, will be discussed and research priorities highlighted.

2. Fungal-Pathogen-Associated IRIS Characteristics

2.1. Cryptococcus

The yeast *Cryptococcus* is the most frequently described genus in IRIS [6]. The difference between paradoxical and unmasking IRIS is also best described and defined for this pathogen [6]. In cases where cryptococcal disease was not recognized at ART initiation, it may be difficult to differentiate between IRIS (caused by the restoration of immune functions then called unmasking IRIS) and the progression of a disease in the context of persisting immunodeficiency. Therefore, Haddow et al. proposed a panel of criteria to support this controversial entity [6]. These criteria mainly include unusual, exaggerated, and heightened inflammatory features. The epidemiology of cryptococcal IRIS may describe various incidence rates. In a review, including 12 studies and 598 patients with diagnosed cryptococcal disease before ART initiation, paradoxical IRIS developed in 8–49% of patients [6]. Interestingly, it appears that the incidence is lower in high-income countries (i.e., 8% in France, 13% in Thailand) while the highest incidence is found in low-income countries (i.e., 49% in Uganda). Furthermore, it seems that cryptococcal IRIS is seen less frequently in the most recent studies [6,8]. The increase in access to improved antifungal therapies may explain such figures. The incidence of unmasking IRIS is much lower, ranging from 0 to 7%; however, a case definition has not been uniformly addressed. Cryptococcal paradoxical IRIS has been described in HIV, SOT (mostly in kidney and liver transplant recipients) [3,9,10], and in early post-partum after pregnancy [4]. IRIS may occur a few days or several months after ART initiation in HIV-infected patients. In SOT recipients, a mean occurrence of 6 weeks after the introduction of antifungal treatment has been described and is triggered by a reduction in immunosuppressive treatment [11]. In those cases, patients may experience allograft loss temporally related to the onset of IRIS through Th1 upregulation. IRIS seems to occur more frequently in patients receiving a combination of tacrolimus, mycophenolate mofetil, and prednisone than in patients receiving another immunosuppressive regimen [2,12]. Clinical manifestations are described in

Table 1. IRIS non-specific symptoms may occur in organs that were not apparently initially infected by the fungus, resulting from exuberant tissue inflammation [13]. Diagnosis remains clinical and requires exclusion of other diagnoses, including worsening or relapse of infection, other opportunistic infections, tumors, and drug-related adverse effects. Cerebrospinal fluid (CSF) culture is typically sterile in IRIS; however, if IRIS arises shortly after antifungal treatment, culture may remain positive. In this case, comparing the fungal burden to CSF at the initial diagnosis of cryptococcosis can help to differentiate IRIS from relapse; however, this technique is not part of routine management. Moreover, it has been suggested that monitoring (1-3)-β-D-glucan (BDG) in CSF could be helpful [14]. Indeed, it has been shown in a recent Ugandan and South African cohort of HIV-infected patients with cryptococcal meningitis that BDG measured in CSF could contribute to the differentiation of fungal progression (i.e., positive BDG) from cryptococcal paradoxical IRIS (i.e., negative BDG) [14]. Furthermore, a PCR-based assay might also be useful: the FilmArray system was evaluated in 39 HIV-infected patients from Uganda with suspected cryptococcal meningitis and was able to detect *Cryptococcus* with 100% sensitivity and to distinguish relapse from IRIS in a limited number of patients [15]. IRIS-like syndrome has been described in immunocompetent patients. Cases have been described with both *Cryptococcus neoformans* and *Cryptococcus gattii* [16–18]. They report an extended overwhelming inflammatory immune response despite CSF culture sterilization with a poor prognosis and subsequent neurological sequelae. In one case, thalidomide was successfully used to decrease inflammation when corticosteroids were inefficient [17]. It is possible that antifungal therapy reduces the burden of *Cryptococcus*, thereby facilitating the reversion of a Th2 response to a Th1 response. However, all of these cases failed to study if the strains may belong to a hypervirulent clade and if a host immune polymorphism could have led to this specific clinical presentation.

2.2. Candida

Chronic disseminated candidiasis (CDC), also called hepato-splenic candidiasis, has been suspected to be a form of candidiasis-related IRIS [19]. This clinical entity develops in patients who recently experienced profound and prolonged neutropenia, especially at neutrophil recovery after chemotherapy for acute leukemia. Before the introduction of posaconazole as the primary antifungal prophylaxis, its incidence ranged from 3 to 29% in this population [19], and a diagnosis is usually obtained within 2 weeks following immune recovery but can sometimes be diagnosed as late as 165 days thereafter [19]. Symptoms, which are described in Table 1, usually persist despite antifungal therapy. MRI has a much better sensitivity than ultrasonography and computed tomography (CT) to detect micro-abscesses, which are most frequently localized in the liver and the spleen and result from an exuberant inflammatory response [20]. Positron emission tomography (PET) has been increasingly used and shows promising results in CDC diagnosis [21,22]. At a histopathological level, epithelioid granulomas and micro-abscesses are encountered in most cases [19]. Blood cultures are negative in more than 80% of cases, and microscopy shows the presence of the yeast in less than 50% of cases [19,23]. Biomarkers could be useful in diagnosing CDC. BDG are usually highly increased (>500 ng/L); however, serum detection of mannan antigens and antibodies (Platelia®Candida) appears to be more sensitive to detect the disease and monitor treatment efficacy [24]. Ongoing pathophysiological and imaging studies from our group aim to better decipher the CDC entity. To our knowledge, no form of paradoxical candida-related IRIS after ART initiation in HIV-infected patient has ever been reported.

2.3. Aspergillus

Aspergillus-related IRIS also occurs during neutrophil recovery, especially after a course of chemotherapy to treat acute leukemia, or after stem cell recipients [25]. The mean time to clinical and radiological findings of IRIS from an absolute neutrophil count >100/µL and >500/µL was 3.5 days and 2 days, respectively, in a cohort of 19 patients [26]. A few cases of *Aspergillus*-related IRIS have also been described in lung transplant recipients [27]. The clinical manifestations, which are summarized in Table 1, include non-specific worsening, or new onset, of hypoxia, chest pain, cough, dyspnea, and hemoptysis [26]. A CT scan can show a size increase in pulmonary infiltrates, a pleural effusion, nodular lesions, and/or cavitation of a pre-existing lesion [26]. As with all fungi-related IRIS, it remains a diagnosis of exclusion. However, a decrease in serum galactomannan together with a perceived clinical and radiological worsening is good supporting evidence [28]. Overall, *Aspergillus*-related IRIS in neutropenic patients appears to have a good prognosis [28].

2.4. Histoplasma

The dimorphic yeast *Histoplasma capsulatum* is represented by two species found in endemic regions: *Histoplasma capsulatum* variety *capsulatum* is mostly encountered in North and South America as well as a few regions in Africa and Asia, while *capsulatum* var. *duboisii* is found in Africa only. Immunocompetent patients infected with *Histoplasma capsulatum* are mostly asymptomatic (>90%), while in immunocompromised patients, a reactivation of infection is common and may lead to a disseminated disease associated with a poor prognosis. *Histoplasma*-related IRIS may present as "unmasking" or "paradoxical". Most cases have been described with *H. capsulatum* var. *capsulatum* in HIV patients [29–31], solid organ transplant recipients [32], and patients receiving TNFα inhibitors [33]. To our knowledge, only one case was described with *H. capsulatum* var. *duboisii* in a HIV patient [34]. The HIV patient population has been the most studied. The incidence of histoplasmosis appears to be higher in patients treated with anti-TNF monoclonal antibodies (e.g., infliximab) than in patients receiving soluble TNF-α receptors (e.g., etanercept) [33]. Overall, IRIS occurs in 9.2% of this patient population [33]. The median time of onset of IRIS symptoms from TNF-α inhibitors discontinuation was 6 weeks (1–45) in 9 patients [33]. Clinical presentations are described in Table 1.

2.5. Pneumocystis jiroveci

Pneumocystis jiroveci is a unique fungal organism and was only recently reclassified from a protozoan to an ascomycetous fungus after an analysis of the ribosomal DNA (rDNA) subunit [35]. It is a common opportunistic fungal pathogen and causes pneumonitis in immunocompromised patients. While *P. jirovecii* transmission occurs via person-to-person contact during the first years of life and is controlled by the immune system, it may rapidly multiply in the lungs of immunocompromised patients and lead to severe hypoxia and death [36]. Cases of *Pneumocystis*-associated IRIS have been described mostly in HIV-infected patients and patients receiving high-dose corticosteroids secondarily tapered [37–39]. In an analysis of 15 reports, time to IRIS symptom onset following ART initiation was 15 days (3–301 days) [40]. *P. jiroveci* pneumonia (PJP)-IRIS presents as a recurrence of fever, dyspnea, cough, and night sweat in patients treated for PJP. When performed, a CT scan often shows a recurrence of ground glass opacities; however, many cases also report atypical radiologic manifestations, including nodules, consolidations, and organizing pneumonia [37,41,42]. No adequate diagnostic criteria have been described for this entity. Similarly to other fungi-associated IRIS, microbiological tests (a direct examination of the broncho-alveolar lavage, PCR) are more likely to be negative. Nowadays, a PJP diagnosis often relies on quantitative PCR performed on the bronchoalveolar lavage or induced-sputum [43]. However, the lack of consensus on molecular threshold values for fungal load makes the PCR results difficult to interpret, even more so in the case of IRIS. Furthermore, the lack of a culture technique for *P. jiroveci* creates an additional challenge to distinguish relapse from IRIS. When performed, especially in organizing pneumonia forms, biopsies also showed granulomatous inflammation [37,41,42,44]. These histopathological results have also been described in liver and renal transplant recipients; however, whether a decrease in immunosuppression was responsible or not for IRIS was unclear [45,46].

2.6. Other Fungi

Fungi-related IRIS has also been described during infections with *Talaromyces marneffei* (ex-*Penicillium marneffei*) [47–51], *Coccidioides* spp. [52], *Paracoccidioides* spp. [53], *Sporothrix schenckii* [54], *Fusarium* spp. [55], or the newly described *Emergomyces africanus* [56]. To the best of our knowledge, no case of IRIS has been described with any of the Mucorales and phaeohyphomycetes. All IRIS cases related to these more unusual fungi occurred in HIV-infected patients after introduction of ART except for the case involving *Fusarium*, which occurred after neutrophil recovery in a patient treated for acute myeloid leukemia.

3. Is There a Common Background for Fungal IRIS?

Finding a common pathophysiological explanation for all fungi-associated IRIS may appear to be impossible in the light of the diversity of organisms, clinical presentation, and immunological mechanisms underlying a sudden increase in immune response. Although all data come from an isolated clinical picture and a limited series, the histopathological hallmarks when a biopsy is performed seem to consistently involve numerous non-growing fungi, necrosis, macrophages, and, more specifically, granuloma (Table 1). However, objective histopathological data are lacking concerning *Aspergillus*-associated IRIS [26,28]. This granulomatous reaction is described as clusters of epithelioid macrophages, which are sometimes vacuolated and always CD68-positive [57,58]. A lymphocytic infiltrate can also be observed, predominantly at the periphery of granulomatous lesions. Hence, a common immunological explanation might link the underlying mechanisms of fungal IRIS in these very different clinical pictures.

An inadequate balance between pro-inflammatory Th1 response and anti-inflammatory Th2 response was commonly admitted to be the origin of IRIS. With the recent discovery of Th17 and regulatory T cell (Treg) responses, this model evolved to an inbalance between pro-inflammatory Th1/Th17 and anti-inflammatory Th2/Treg axes [3,59–61]. Cytokines driving the differentiation of naïve Th0 cells into Th1, Th2, Th17, and Treg stimulated by cytokines are summarized in Figure 1. Specific transcription factor pathways are detailed for each subset (Figure 1) [3]. IRIS is now believed to arise from an unregulated Th1/Th17 leading to increased production of interferon-γ (IFNγ) [3].

Table 1. Clinical presentation and characteristics of fungi-associated immune reconstitution inflammatory syndrome (IRIS) by fungal species.

Pathogen	Patient Background	Symptoms	Diagnostic Test	Histopathology	Reference
Cryptococcus species	HIV SOT Pregnancy	Headaches, seizures, neurological deficits Lymphadenopathy, pneumonitis, chorioretinitis, skin and soft tissues lesions	Diagnosis of exclusion * No consensual diagnostic test Sterile CSF culture BDG- CSF PCR-	Granulomatous lesions	[7,14,15]
Candida species	Neutropenia (acute leukemia, lymphoma, stem cell transplant)	Fever, abdominal pain, liver and spleen enlargement MRI: mm-sized abscesses in liver, spleen, kidney, and/or lungs	Diagnosis of exclusion * ↑ liver enzymes ↑ BDG Mannan/anti-mannan antibody detection	Epithelioid granuloma, necrosis with minimal inflammatory reaction, micro-abscesses with major inflammatory reaction	[19,22,24]
Aspergillus species	Neutropenia (Stem cell transplant and acute leukemia)	Hypoxia, chest pain, dyspnea, hemoptysis CT scan: ↑ pulmonary infiltrates	Diagnosis of exclusion * ↑ galactomannan	Insufficiently studied	[25,26,28]
Histoplasma capsulatum	HIV SOT TNF-α receptor inhibitors	Hemoptysis, dyspnea, lymphadenopathy, skin nodules CT scan: pulmonary bilateral nodules and ground-glass opacities	Diagnosis of exclusion * ↑ Histoplasma serum antigen Sterile culture	Well-formed granulomatous inflammation	[29–33]
Pneumocystis jirovecii	HIV Corticosteroid-treated patients	Fever, cough, dyspnea, night sweat	Diagnosis of exclusion *	Organizing pneumonia: organizing granulation tissue	[39–42,44]

* Exclusion of other diagnosis: microbial progression, other opportunistic infection, tumors, a drug-related adverse effect. SOT, Solid Organ Transplant; BDG, (1-3)-β-D-glucan; CSF, cerebral spinal fluid; HIV, human immunodeficiency virus.

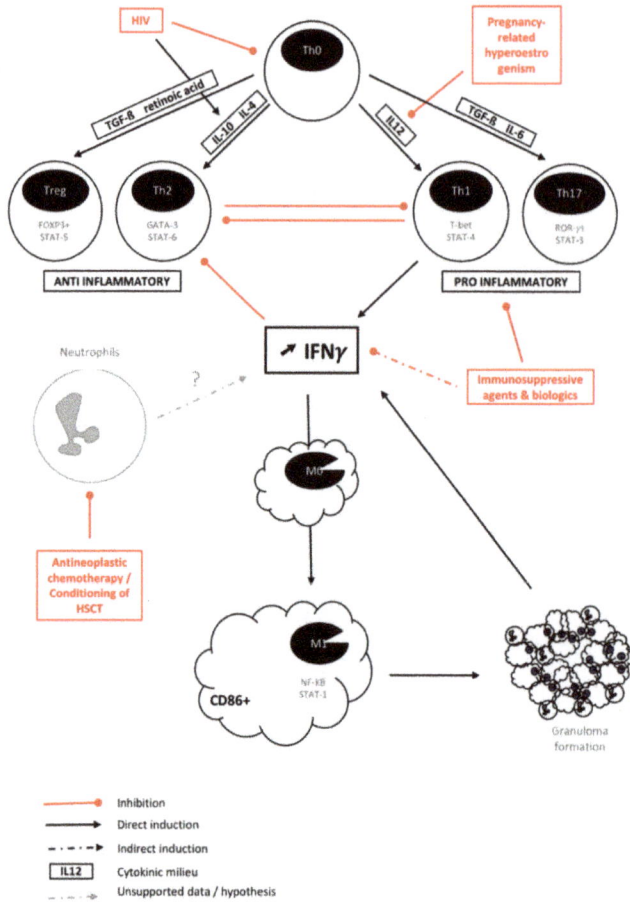

Figure 1. Proposed immune pathophysiology of IRIS. Precursor T helper cells (Th0) differentiate into Th1, Th17, or Th2 cells or Tregs according to cytokines produced in the surrounding milieu through induction of specific transcription factor expression: FOXP3/STAT-5, GATA-3/STAT-6, T-bet/STAT-4, and ROR-γt/STAT-3 for Treg, Th2, Th1, and Th17, respectively [60,61]. Th1 and Th17 cells are pro-inflammatory cells and produce IFN-γ driving macrophage differentiation into M1 macrophages and their activation. M1 macrophages promote granuloma formation and subsequently produce more IFNγ, thus creating an amplification loop leading to an inflammation burst [62]. HIV drives depletion of all Th cell subsets and favors an anti-inflammatory response in remaining cells through a change in cytokine balance [63,64]. Pregnancy hormones inhibit differentiation of naive Th0 cells into Th1 cells, thus promoting a Th2 environment [65]. Biologic agents, such as infliximab, inhibit the pro-inflammatory cytokine TNFα [66]. Immunosuppressive therapeutics inhibit Th1/Th17 response and allow graft acceptance in solid organ transplant [3]. Neutropenia resulting from chemotherapy or conditioning for hematopoeitic stem cell therapy leads to a decrease in IFNγ production through cytokinic interaction with T-cell production; however, data are lacking [67]. Many feedback loops exist between these different players depending on cytokine production. Sudden dysregulation at any level may create an exaggerated inflammatory response with an IFN-γ-unregulated increase leading to IRIS. FOXP3, fox head box P3; IFN, interferon; IL, interleukin; PMN, polymorphonuclear; TGF, transforming growth factor; ROR, retinoid orphan receptor; STAT, signal transducer and activation of transcription; HSCT, Hematopoietic Stem Cell Therapy.

Overall, this exaggerated inflammatory response translates in radiological findings as edema and abnormal contrast medium uptake into pulmonary and brain lesions [58]. Cells of the innate immune system, such as monocytes, macrophages, and neutrophils, are of increasing interest in IRIS pathophysiology, since granuloma appears to be frequently found in IRIS lesions [19,41,47]. Indeed, granuloma is the histopathology hallmark in chronic disseminated candidiasis [19] and is commonly found in other fungi-related IRIS [37,41,48,49,56]. An extended description of granuloma in fungi-related IRIS is unavailable, and should be better studied in particular through immune staining to understand which cell types are implicated and their degree of differentiation/activation. Granulomas may display very distinct features, may be activated or latent, and their cell-type composition may vary according to the situation and pathogen [68]. In a simplistic approach, excess IFNγ produced by Th1 cells, neutrophils, or activated macrophages will elicit the differentiation of monocytes towards macrophages and activate their phagocytic activity as well as stimulate granuloma formation. IFNγ favors the classical activation of macrophages into M1-phenotype macrophages [62]. These macrophages secrete large amounts of pro-inflammatory cytokines, such as IL-1β, TNFα, IL-12, IL-18, and IL-23, driving in return a Th1/Th17 cell inflammatory response. Phenotypically, M1 macrophages express high levels of histocompatibility complex class II and the CD86 marker. M1 macrophages are implicated in initiating and sustaining inflammation, and can, therefore, be detrimental. Furthermore, M1 polarization appears to be predominant in granuloma formation [69]. Conversely, M2-polarized or alternately activated macrophages produce IL-4, IL-13, and IL-10 and are prone to immune regulation and tissue remodeling [70]. A further analysis would provide significant data to help us understand fungi-associated IRIS.

The immune pathophysiology appears to vary according to the underlying type of immunosuppression before immune recovery. HIV, in addition to depleting the total pool of CD4+ T-cells by apoptosis, including Th0 cells, induces preferential death of Th1 cells and differentiation of the remaining Th0 into Th2 cells through a critical change in cytokine balance [63,64]. Introduction of ART leads to an initial redistribution of CD4+ T-cells to the blood rather than proliferation. After 4–6 weeks, the production of naive CD4+ T-cells and memory T-cells coincides with the mean onset of paradoxical IRIS [71]. A concomitant infection responsible for increased circulating pro-inflammatory cytokines may also contribute to dominant Th1 differentiation and IRIS. In addition, HIV induces Tregs to migrate and accumulate in peripheral and mucosal lymphoid tissues [72]. Montes et al. have described an initial redistribution of Tregs to the blood at ART initiation [73]. This tissular depletion in Tregs may also participate in a locally uncontrolled inflammatory response. Moreover, the functional capacities of Tregs appear to be reduced [74].

After solid organ transplantation, graft survival relies on the inhibition of alloreactive Th1/Th17 responses by immunosuppressive drugs (i.e., mycophenolate mofetil, calcineurin inhibitors, corticosteroids ...) through different mechanisms [3]. Calcineurin inhibitors, especially tacrolimus, strongly suppress the Th1 response while rapamycin promotes Treg survival and function and suppresses the differentiation of Th17 cells [3]. Corticosteroids decrease Th1 responses but also expand Th2 cells and Tregs [3]. Post-transplant IRIS is subsequent to a decrease in immunosuppression due to drug–drug interactions or an intentional modification in drug dosage in the context of an ongoing infection, therefore increasing Th1/Th17 responses. However, a recent study showed that only a discontinuation of calcineurin inhibitors influenced the development of IRIS in cryptococcosis by a 5-fold higher risk [11]. The excessive Th1 response observed in IRIS may also lead to collateral damage to the graft. Indeed, in a prospective study involving 54 renal allograft recipients with cryptococcosis, 5.5% of patients presented with IRIS and the renal graft was lost due to chronic rejection in 66% of patients with IRIS as compared to 5.9% of IRIS-free patients [12].

Immunomodulatory biologic agents are increasingly being used to treat chronic autoimmune and inflammatory conditions, such as rheumatoid arthritis and inflammatory colitis. Tumor necrosis factor (TNF)-α, a cytokine mainly produced by activated macrophages, is involved in systemic inflammation and plays a key role in the recruitment of immune cells and granuloma formation. Histoplasmosis is the most prevalent IFD in patients undergoing TNFα inhibitors therapy, followed by candidiasis and aspergillosis [66]. Most mycoses described in those patients were associated with the use of the monoclonal antibodies infliximab/adalimumab rather than the soluble TNFα receptor etanercept, reflecting the difference in the mechanism of action between those drugs [66]. Discontinuation of TNFα inhibitors results in IRIS in 9.2% of patients with histoplasmosis [33]. Alemtuzumab, a humanized CD52 monoclonal antibody that depletes T and B cell populations, has also been involved in fungi-related IRIS with *C. neoformans* and *P. jiroveci* upon its withdrawal [75,76].

A prompt recovery of neutrophils after a stem-cell transplant or a chemotherapy cycle has been linked to a severe pulmonary complication subsequent to inflammation during invasive pulmonary aspergillosis [25]. Likewise, chronic disseminated candidiasis manifests during neutrophil recovery [19]. Despite the lack of studies and data concerning this IRIS-like syndrome following the fast expansion of neutrophils after a stem-cell transplantation, it could be expected that a direct or indirect increase in IFNγ plays a role in pathogenesis. Indeed, a subset of neutrophils, Gr-1+/CD11b+ cells, has been shown in mice to produce IFNγ and mediate early graft loss or take part in severe renal ischemia reperfusion injury [67,77]. However, production of IFNγ by human neutrophils is controversial [78]. They may, nonetheless, participate through indirect mechanisms, especially through a crosstalk with Th17 cells, in cytokine production to increase IFNγ levels [79]. More data are needed to understand IRIS following neutrophil recovery, and its pathogenesis is currently being studied in the CANPHARI study (NCT01916057) headed by our group.

During pregnancy, mechanisms of fetal tolerance lead to downregulation of the Th1/Th17 axis. Interestingly, regulation of M2 macrophage polarization is required for successful pregnancy, and is sustained by pregnancy hormones, such as estrogen and HCG [65]. A shift in cytokine pattern is observed during the post-partum period that may be associated with pathological inflammatory syndrome and has been documented 3–6 weeks after delivery [4,80].

In addition to this immune dysregulation, commonly used antifungals are known to bear some immunomodulatory properties that could contribute to IRIS. These data only supported by in vitro studies, and the few in vivo studies in mice are to be taken with caution. Amphotericin B deoxycholate upregulates Th1 response through toll-like receptor 2 (TLR-2)-mediated transcription and inflammatory cytokine transcription [81,82]. Conversely, the lipid formulation of polyene has no effect on, or even downregulates, the inflammatory response directly due to the intrinsic properties of the liposomes [82]. However, it may confer an increase in the phagocytic properties of macrophages and neutrophils against numerous fungi [82–85]. Echinocandins unmask β-glucan from the fungal cell wall, eliciting pro-inflammatory cytokine release upon its recognition through pattern recognition receptors (PRR). Azoles seem to be the least active with respect to modulation of the host's immune system [82]. However, fluconazole and voriconazole have been shown to enhance phagocytic pro-inflammatory activity through a TLR2 interaction [82]. Overall, the initial choice of antifungal could influence the subsequent inflammatory response; however, their immunomodulatory effects have not been studied in the context of IRIS.

Lastly, three more elements, often omitted and far less studied, may be discussed to explain IRIS's pathophysiology: fungal strain immunomodulatory characteristics, host immune system features, and autoimmunity disorders. Firstly, *Cryptococcus* itself inhibits the Th1 response while inducing a Th2 response that compromises the host's resistance in mice [86]. Furthermore, the cryptococcal genotype has been shown to influence the immune response and human clinical outcome after meningitis [87]. In addition, Desnos-Ollivier et al. showed that mixed strain infections are seen in 20% of patients and could drive overstimulation of the immune system [88]. Similarly, infection by *P. jiroveci* often results from mixed strain genotypes [89]. However, the impact of strain genotype in IRIS has not been studied. Secondly, on the host side, human immune system genetics is a growing field of interest and may be involved in IRIS development. Polymorphisms in cytokine genes may play a role in host susceptibility to IRIS [90]. For instance, a non-synonymous polymorphism in IL-23R is associated with a reduced risk of schistosomiasis-associated IRIS in a Kenyan population [91]. Similarly, a common haplotype of the IL-4 promoter was over-represented in patients with CDC [92]. Also, a single nucleotide polymorphism in the promoter of leukotriene A4 hydroxylase (LTA4H) regulating the balance between anti-inflammatory lipoxins and the pro-inflammatory LTB4 is responsible for a higher incidence of severe tuberculosis-associated IRIS [93]. One could easily hypothesize that polymorphisms of the human IFNγ receptor may lead to different susceptibility patterns to IRIS. Third, opportunistic infection results in tissue damage and the epitope spreading that is known to predispose to autoimmunity [71]. Furthermore, various autoimmune disorders have been discovered in HIV-infected patients after ART initiation that may play a role in the excessive inflammatory response [71]. Nevertheless, these hypotheses require further investigations, which are currently ongoing through a collaborative work about chronic disseminated candidiasis between the group of Lausanne and ours.

Overall, any rapid immune recovery can lead to IRIS, driven by multifactorial factors, such as patient response to drugs (ART, antifungals, ...), host immune genetics, and the microbial strain. Fungal pathogen immunogenicity may trigger an overshoot of IFNγ production either through an enhancement of the Th1/Th17 axis and/or rapid neutrophil reconstitution leading to overproduction/activation of CD68+ macrophages and promoting granuloma formation and inflammatory disease.

4. A Limited Therapeutic Arsenal Against IRIS

The ideal management of fungi-associated IRIS in general remains unknown. No recognized clinical guidelines have been dedicated to this challenging subject matter except those from IDSA for the management of IRIS in cryptococcal disease [7]. They do not recommend a specific treatment for minor IRIS manifestations that usually resolve within a few days or weeks. However, for complications involving central nervous system (CNS) inflammation associated with increased intracranial pressure, they advise corticosteroids (0.5–1 mg/kg/day of prednisone equivalent) and dexamethasone at higher doses for more severe CNS signs and symptoms. No data support recommendations for the length of treatment; however, a 2–6-week course with close monitoring and concomitant antifungal treatment is widely accepted [7].

Corticosteroids are the only drug class that has been recognized for the treatment of IRIS so far [94]. They show an anti-inflammatory effect on most immune cells by altering the transcription of inflammatory mediators, interfering with the nuclear factor-ĸB, and directly enhancing the effect of the anti-inflammatory proteins [95]. An evaluation of the efficacy of corticosteroids compared to a placebo in a randomized controlled trial has only been done in TB-associated IRIS, and showed a reduced length of hospitalization and surgical procedure [96,97]. Regarding fungal infections, only case reports and small series account for the benefit of corticosteroids in fungi-associated IRIS, especially in patients with impending respiratory failure after neutrophil recovery and aspergillosis [26,98], chronic disseminated candidiasis [57], and cryptococcosis [13,58,99]. Also, despite the lack of formal diagnostic criteria for IRIS, old studies support the adjunction of corticosteroids to prevent early deterioration in patients with moderately severe *P. jiroveci* pneumonia and HIV [100]. In cryptococcal meningitis, adjunction of dexamethasone at baseline did not reduce mortality among patients with HIV, nor was it associated with a reduced incidence of IRIS in a large cohort of 451 patients [8]. On the contrary, it was correlated with more adverse effects and disability than a placebo [8]. Indeed, complications associated with the use of corticosteroids must be considered. Their non-specific immunosuppressive effect can lead to subsequent infectious complications, such as herpes virus reactivation, strongyloides hyperinfection, worsening of chronic hepatitis B, and Kaposi sarcoma progression [94,101,102]. In addition, corticosteroids can cause metabolic complications, such as dysglycemia, hypertension, and cushingoïd features, or be responsible for other adverse effects, such as worsening psychiatric disorders and drug interactions [94].

TNF-α is a pro-inflammatory cytokine required for macrophage activation and granuloma formation. The anti-TNF-α antibodies infliximab and adalimumab have been used in several case reports to reduce inflammation in IRIS. Etanercept is a soluble TNFα receptor; however, it has never been reported to be effective in IRIS treatment. Infliximab has primarily been used [103,104]; however, adalimumab has shown the ability to treat corticosteroid- and/or infliximab-resistant IRIS associated with TB and Cryptococcus [105–107]. Although the optimal length of treatment with such therapy has not been determined yet, the successful management of IRIS has been reported to require several months of treatment. Overall, adalimumab should probably be preferred when considering anti-TNF-α for treating IRIS, since more fungal complications have been reported with infliximab [66].

Thalidomide also acts as an immunomodulatory drug inhibiting TNFα synthesis among other cytokines (i.e. IFNγ, IL-10, and IL-12 and cyclooxygenase 2 (COX-2)) [17]. In several cases, it has been shown to be an interesting molecule for corticosteroid-dependent IRIS in cryptococcal meningitis that allows for steroid tapering [17,58,108]. The treatment duration in these studies ranged from 4 weeks to 14 months with no relapse [19].

Non-steroidal anti-inflammatory drugs (NSAIDs) inhibit cyclooxygenase, an enzyme required for prostaglandin synthesis and inflammation mediation. Many cases report the use of NSAIDs in TB-IRIS [94], and some clinical guidelines recommend their use for mild IRIS related to mycobacterial infection [109]. Nonetheless, no clinical trial supports these recommendations and data are lacking regarding fungi-associated IRIS. Moreover, nephrotoxicity is a concern with long-term use and concomitant administration of other nephrotoxic drugs, such as amphotericin B or calcineurin inhibitors.

Statins inhibit 3-hydroxy-3-methyl-glutary-CoA (HMG-CoA) reductase. It has been hypothesized that, considering the close homology between fungal and human HMG-CoA reductase, statins may have a potential antifungal effect. However, a meta-analysis of five retrospective studies showed no positive effects during fungal infections [110]. Nevertheless, statins display immunomodulatory activity promoting the Th2/Treg axis through various mechanisms [111]. Subsequently, many authors speculate that they could play a role in the management of IRIS; however, data is still lacking [111,112].

Finally, intravenous immunoglobulins have been successfully used in virus-related IRIS [113,114]; however, they have never been tried in the fungal context.

To conclude, the best-studied and most-used therapy for IRIS is corticosteroids despite their several drawbacks. Anti-TNFα, especially adalimumab, as well as thalidomide, appear to be promising in treating fungi-associated IRIS or corticosteroid-dependent IRIS. However, more studies are required. Symptomatic treatment alone, including analgesia and anti-epileptic treatment, is often sufficient to manage symptoms and should not be underestimated.

5. Predict and Prevent: The Cornerstone of IRIS Management Today

Since treating IRIS remains uncertain and challenging with no guidelines to rely on, patient care has been focusing on identifying risk factors and developing preventive strategies.

5.1. Prediction with Diagnostic Markers

The identification of risk factors mostly depends on biological markers assessed in diagnostic labs and a few clinical risk factors, especially information concerning patient treatment. Once again, cryptococcal-related paradoxical IRIS in HIV-infected patients has been the most studied situation. Therefore, our current knowledge on predictive markers is limited to these circumstances in predictive markers as well. Risk factors for cryptococcal-associated IRIS can be divided into three categories: (i) host-related factors, (ii) pathogen-related factors, and (iii) treatment-related factors. Host-related factors include various measurable immunological blood and cerebrospinal fluid (CSF) parameters. A lower pre-ART CD4 count has been demonstrated many times in bacterial- or viral-associated IRIS as well [115–118]. However, no cut-off has been established. Other blood parameters reflecting a lack of an immune response toward cryptococcosis at diagnosis appear to be relevant to predict the occurrence of IRIS. A lower plasma total IgM, a specific anti-fungal IgM (glucuronoxylomannan-IgM and β-glucan-binding IgM), and a specific IgG prior to initiation of ART were observed in patients who developed IRIS [119]. A lack of pro-inflammatory cytokines in a serum, such as TNFα, IFNγ, granulocyte-colony stimulating factor (G-CSF), and granulocyte-macrophage CSF (GM-CSF), predicted future IRIS [120]. An increase in Th2 response reflected by the IL-4 level was also associated with IRIS [120]. The use of a modified IFNγ release assay of whole blood stimulated with a cryptococcal mannoprotein has confirmed that lower IFNγ responses before ART initiation are associated with a higher risk to develop IRIS [117]. Similarly, a lack of an immune response in CSF with a decrease in the leucocytes count to ≤25 cells/μL and a reduced level of IFNγ, IL-6, IL-8, and TNF-α were associated with the development of IRIS [118,121]. In these circumstances, a global CSF protein level ≤50 mg/dL was also an independent risk factor [121]. A higher CSF ratio of CCL2/CXCL10 and CCL3/CXCL10 were also found in patients who subsequently developed IRIS [122]. CCL2 and CCL3 are chemokines known to attract monocytes, macrophages, neutrophils, and T-cells, whereas CXCL10 is only chemotactic to CXCR3+ lymphocytes (Th1 cells) [122]. Thus, an increase in the former chemokines may promote the infiltration of macrophages and neutrophils into the CSF and be responsible for IRIS.

Pathogen-related risk factors correspond to the fungal burden, which can be assessed by a serum cryptococcal antigen (CrAg) titer, the colony forming unit (CFU)/mL of CSF, and the presence of fungemia [58,118,123]. Patients with IRIS had a 4-fold higher median CrAg level pre-ART [120]. In addition, patients with a negative cryptococcal culture from a CSF sample pre-ART initiation experienced fewer CNS deterioration symptoms and a lower IRIS rate than patients with a positive culture [118].

Treatment-related risk factors include a shorter duration of antifungal treatment prior to starting ART and/or a rapid suppression of HIV viral load. Indeed, a decrease in HIV viral RNA to >2.5 log at the time of IRIS compared with RNA levels before the initiation of ART was associated with subsequent IRIS [116]. In addition, a rapid immunologic response to ART reflected by a more important rise in CD4 cells over a 6-month period was associated with IRIS [124,125]. Furthermore, some ART regimens, especially those using a boosted protease inhibitor, were a risk factor for developing IRIS [116]. Boosted protease inhibitors appear to have direct immunomodulatory effects, including anti-apoptotic effects and an increase in pro-inflammatory cytokines [116]. Recent European studies found that the use of integrase inhibitors, especially dolutegravir, increases the risk of IRIS by an odds ratio of 1.96–3.25 [126–128].

Regarding *Aspergillus*-related IRIS, the use of a colony-stimulating factor appears to be associated with the occurrence of IRIS in patients with invasive pulmonary aspergillosis with neutropenia [28]. Similarly, a personal case describes a severe exacerbation of CDC after G-CSF administration [129]. Concerning other fungal pathogens, no factors have been studied to predict IRIS to our knowledge. Yet, in *Aspergillus*-related IRIS, one can expect higher galactomannan titers by analogy to CrAg titers. Similarly, β-D-glucans may be higher during fungemia before chronic disseminated candidiasis. However, these statements remain hypotheses and require proof.

To conclude, no markers are yet consensual among the community, and more studies are needed to include one or several of them with proper cut-offs in standard patient care guidelines. A selection of a few of these markers, based on ease of use in the laboratory, reproducibility, price, and effectiveness to predict IRIS, should provide a strong algorithm and robust tool for stratifying patients with high, moderate, and low risk to develop IRIS.

5.2. Prevention by Delaying and/or Tapering Immune Restoration

IRIS depends on the critical time point when the immune system is restored. It seems that the shorter this period is, the more likely the occurrence of IRIS [25,116]. In many situations, this period cannot be controlled and only monitored to identify patients at risk to develop IRIS secondarily (i.e., neutrophil recovery, unbalanced immunosuppressive treatment in SOT). However, in HIV-infected patients, immune recovery is elicited by ART and can be adjusted. Two pioneer studies on IRIS in cryptococcal meningitis showed that initiation of ART closer to the diagnosis of the fungal disease was associated with subsequent development of IRIS [58,125]. This suggests that the inflammatory response is likely higher when the fungal burden or its remnants (i.e., antigen titers) is still substantial. Interestingly, the timing of ART appears to be more essential in *Cryptococcus*-IRIS than in TB-IRIS, in which early ART increased survival [130]. Nonetheless, this was not the case when TB meningitis was present, highlighting that CNS involvement in IRIS is the most deleterious form, and requires extra caution and specific guidelines [131].

The timing of ART initiation has been studied in four trials involving HIV-infected patients [132–135]. In the oldest one, all opportunistic infections combined, ART initiation after 2 weeks was associated with a reduced likelihood of progression or death compared to ART initiation after 6–7 weeks [132]. Opportunistic infections were mostly fungal, including 63% of *Pneumocystis* pneumonia, 12% of cryptococcal meningitis, and 4% histoplasmosis. No subanalysis was made in those groups. Surprisingly, IRIS was uncommon (7%) and was not more prevalent in the early or delayed therapy group [132]. This may be related to the smaller incidence of IRIS in *Pneumocystis* pneumonia, where corticosteroids are frequently used in severe cases. The second study focused on cryptococcal meningitis in a cohort of 54 patients in Zimbabwe, in which ART was initiated at 72-h after diagnosis versus 10 weeks later [133]. The 3-year mortality rate was significantly higher in the early ART group (88% versus 54%; p <0.006), and could be attributed to IRIS according to the authors [133]. Similarly, the third study concerned a small cohort of patients from Botswana (n = 27) with cryptococcal meningitis [134]. Initiation of ART within 7 days following diagnosis of fungal disease, as compared to 28 days after, was associated with a significantly increased risk of IRIS; however, there was no difference in mortality [134]. Lastly, Boulware et al. conducted an open-label randomized trial in Uganda and South Africa that enrolled HIV patients diagnosed with cryptococcal meningitis [135]. Early ART was given between 1 and 2 weeks after diagnosis, while deferred ART was given after 5 weeks. The 6-month mortality rate was significantly higher in the early ART arm (45% versus 30%; *p* = 0.03), which prematurely ended the study. The rate of IRIS was increased in the early ART arm, but not significantly different from the delayed one (20% versus 13%; *p* = 0.32). No other cause (i.e., antifungal toxicity) could explain the difference [135]. Scriven et al. attempted to explain the difference in mortality by exploring CSF macrophage activation, and found an increase of activation markers (CD206+, CD163+) on monocytes and macrophages in the early ART arm versus the delayed ART arm [136]. More data are required to determine the implications of recent ART initiation for the immune system; however, these results point to the possible involvement of innate immune response mechanisms [136].

While the IDSA guidelines had previously recommended the introduction of ART 2–10 weeks after diagnosis [7], this gap has narrowed to 4–6 weeks in newer recommendations taking into account these studies [137,138]. Though delaying ART is recommended, predictive factors should not be underestimated. Achieving a negative CSF culture prior to starting ART might be a better target to aim for to reduce IRIS risk than considering a consensual time limit, since the immune response may differ among patients. Regarding other forms of IRIS, no studies have been done concerning preventive strategies. Close monitoring of inflammation and clinical worsening is recommended to enable early care in those specific cases.

6. Conclusions/Perspective

IRIS is certainly underdiagnosed and many times considered as a failure of antifungal treatment. No consensual diagnostic test is used and the diagnosis remains clinical. As far as we know, the three following criteria need to be satisfied: (1) the new appearance, or worsening, of clinical or radiographic manifestations consistent with an inflammatory process, (2) symptoms that cannot be explained by a newly acquired infection, and (3) negative culture results and/or a decrease in the fungal antigen level (BDG, galactomannan, histoplasma antigen, cryptococcal antigen . . .). Granuloma appear to be the histopathology hallmark, and hypercalcemia subsequent to endogenous production of 1,25 dihydroxyvitamin D by macrophages in granulomas should perhaps be sought more frequently [139]. Existing research bioassays need to be translated into clinical practice to support diagnosis. Until thorough diagnostic markers and a clear definition for fungal-associated IRIS are consensually acknowledged by the medical and scientific community, all studies included in this review ought to be considered with caution.

Regarding treatment, given the drawbacks of corticosteroid treatment, the benefit from such therapy might still be argued in cases where IRIS symptoms do not usually result in lethal complications. Furthermore, the optimal dose and length of treatment for a reasonable risk/benefit ratio need to be discussed. More studies, including randomized trials, are needed to evaluate the relevance of other anti-inflammatory drugs and to propose guidelines for the management of IRIS in fungal diseases.

The prevention of IRIS in HIV relies mostly on the timing of introduction of ART, which appears to be critical in IRIS involving the CNS compartment, such as cryptococcal IRIS resulting in significant morbidity and mortality. Other forms of IRIS, especially involving the lungs or skin, appear to be less life-threatening and have been set aside in research protocols. However, these forms of IRIS bear morbidity, a longer length of stay in the hospital, a high cost investigation, and unnecessary medications, thus requiring attention by the research community to improve the standard of care.

Our understanding of the pathogenesis of IRIS remains in its infancy. More data are available on TB-IRIS; however, additional research is needed to know if these results are applicable to fungi-associated IRIS. The heterogeneity of fungal infections and immunosuppression types contributes to the complexity of understanding IRIS occurring during fungal infections. Multifactorial approaches must be taken to understand its pathogenesis, including host genetics, fungal strain specificities, immunology, and histopathology, which could subsequently lead us to uncover new treatment options.

Author Contributions: O.L., R.G., and S.D. contributed to the conception and design of the review. O.L. and S.D. drafted the manuscript. All authors critically revised the entire manuscript, agreed to be fully accountable for ensuring the integrity and accuracy of the work, and read and approved the final submission.

Conflicts of Interest: The authors declare no conflict of interest.

References

1. Enoch, D.A.; Yang, H.; Aliyu, S.H.; Micallef, C. The changing epidemiology of invasive fungal infections. *Meth. Mol. Biol.* **2017**, *1508*, 17–65.

2. Singh, N.; Lortholary, O.; Alexander, B.D.; Gupta, K.L.; John, G.T.; Pursell, K.; Muñoz, P.; Klintmalm, G.B.; Stosor, V.; del Busto, R.; et al. An immune reconstitution syndrome-like illness associated with *Cryptococcus neoformans* infection in organ transplant recipients. *Clin. Infect. Dis.* **2005**, *40*, 1756–1761. [CrossRef] [PubMed]

3. Sun, H.Y.; Singh, N. Immune reconstitution inflammatory syndrome in non-HIV immunocompromised patients. *Curr. Opin. Infect. Dis.* **2009**, *22*, 394–402. [CrossRef]

4. Singh, N.; Perfect, J.R. Immune reconstitution syndrome and exacerbation of infections after pregnancy. *Clin. Infect. Dis.* **2007**, *45*, 1192–1199. [CrossRef] [PubMed]

5. French, M.A. Immune reconstitution inflammatory syndrome: A reappraisal. *Clin. Infect. Dis.* **2009**, *48*, 101–107. [CrossRef]

6. Haddow, L.J.; Colebunders, R.; Meintjes, G.; Lawn, S.D.; Elliott, J.H.; Manabe, Y.C.; Bohjanen, P.R.; Sungkanuparph, S.; Easterbrook, P.J.; French, M.A.; et al. Cryptococcal immune reconstitution inflammatory syndrome in HIV-1-infected individuals: Proposed clinical case definitions. *Lancet Infect. Dis.* **2010**, *10*, 791–802. [CrossRef]

7. Perfect, J.R.; Dismukes, W.E.; Dromer, F.; Goldman, D.L.; Graybill, J.R.; Hamill, R.J.; Harrison, T.S.; Larsen, R.A.; Lortholary, O.; Nguyen, M.H.; et al. Clinical practice guidelines for the management of cryptococcal disease: 2010 update by the Infectious Diseases Society of America. *Clin. Infect. Dis.* **2010**, *50*, 291–322. [CrossRef] [PubMed]

8. Beardsley, J.; Wolbers, M.; Kibengo, F.M.; Ggayi, A.-B.M.; Kamali, A.; Cuc, N.T.K.; Binh, T.Q.; Chau, N.V.V.; Farrar, J.; Merson, L.; et al. Adjunctive Dexamethasone in HIV-Associated Cryptococcal Meningitis. *N. Engl. J. Med.* **2016**, *374*, 542–554. [CrossRef] [PubMed]

9. Legris, T.; Massad, M.; Purgus, R.; Vacher-Coponat, H.; Ranque, S.; Girard, N.; Berland, Y.; Moal, V. Immune reconstitution inflammatory syndrome mimicking relapsing cryptococcal meningitis in a renal transplant recipient. *Transplant Infect. Dis.* **2010**, *13*, 303–308. [CrossRef] [PubMed]

10. Singh, N.; Sifri, C.D.; Silveira, F.P.; Miller, R.; Gregg, K.S.; Huprikar, S.; Lease, E.D.; Zimmer, A.; Dummer, J.S.; Spak, C.W.; et al. Cryptococcosis in patients with cirrhosis of the liver and posttransplant outcomes. *Transplantation* **2015**, *99*, 2132–2141. [CrossRef] [PubMed]

11. Sun, H.Y.; Alexander, B.D.; Huprikar, S.; Forrest, G.N.; Bruno, D.; Lyon, G.M.; Wray, D.; Johnson, L.B.; Sifri, C.D.; Razonable, R.R.; et al. Predictors of immune reconstitution syndrome in organ transplant recipients with cryptococcosis: Implications for the management of immunosuppression. *Clin. Infect. Dis.* **2014**, *60*, 36–44. [CrossRef] [PubMed]

12. Singh, N.; Lortholary, O.; Alexander, B.D.; Gupta, K.L.; John, G.T.; Pursell, K.; Muñoz, P.; Klintmalm, G.B.; Stosor, V.; Limaye, A.P.; et al. Allograft loss in renal transplant recipients with *Cryptococcus neoformans* associated immune reconstitution syndrome. *Transplantation* **2005**, *80*, 1131–1133. [CrossRef] [PubMed]

13. Lanternier, F.; Chandesris, M.O.; Poiree, S.; Bougnoux, M.E.; Mechai, F.; Mamzer-Bruneel, M.F.; Viard, J.P.; Galmiche-Rolland, L.; Lecuit, M.; Lortholary, O. Cellulitis revealing a Cryptococcosis-related immune reconstitution inflammatory syndrome in a renal allograft recipient. *Am. J. Transplant.* **2007**, *7*, 2826–2828. [CrossRef]

14. Rhein, J.; Bahr, N.C.; Morawski, B.M.; Schutz, C.; Zhang, Y.; Finkelman, M.; Meya, D.B.; Meintjes, G.; Boulware, D.R. Detection of high cerebrospinal fluid levels of (1→3)-β-d-Glucan in cryptococcal meningitis. *Open Forum Infect. Dis.* **2014**, *1*, 267. [CrossRef]

15. Rhein, J.; Bahr, N.C.; Hemmert, A.C.; Cloud, J.L.; Bellamkonda, S.; Oswald, C.; Lo, E.; Nabeta, H.; Kiggundu, R.; Akampurira, A.; et al. Diagnostic performance of a multiplex PCR assay for meningitis in an HIV-infected population in Uganda. *Diagn. Microbiol. Infect. Dis.* **2015**, *84*, 1–6. [CrossRef] [PubMed]

16. Ecevit, I.Z.; Clancy, C.J.; Schmalfuss, I.M.; Nguyen, M.H. The poor prognosis of central nervous system cryptococcosis among nonimmunosuppressed patients: A call for better disease recognition and evaluation of adjuncts to antifungal therapy. *Clin. Infect. Dis.* **2006**, *42*, 1443–1447. [CrossRef] [PubMed]

17. Somerville, L.K.; Henderson, A.P.; Chen, S.C.A.; Kok, J. Successful treatment of *Cryptococcus neoformans* immune reconstitution inflammatory syndrome in an immunocompetent host using thalidomide. *Med. Mycol. Case Rep.* **2015**, *7*, 12–14. [CrossRef]

18. Chen, S.C.A.; Meyer, W.; Sorrell, T.C. *Cryptococcus gattii* infections. *Clin. Microbiol. Rev.* **2014**, *27*, 980–1024. [CrossRef]

19. Rammaert, B.; Desjardins, A.; Lortholary, O. New insights into hepatosplenic candidosis, a manifestation of chronic disseminated candidosis. *Mycoses* **2012**, *55*, e74–e84. [CrossRef]

20. Anttila, V.J.; Lamminen, A.E.; Bondestam, S.; Korhola, O.; Färkkilä, M.; Sivonen, A.; Ruutu, T.; Ruutu, P. Magnetic resonance imaging is superior to computed tomography and ultrasonography in imaging infectious liver foci in acute leukaemia. *Eur. J. Haematol.* **1996**, *56*, 82–87. [CrossRef]

21. Hot, A.; Maunoury, C.; Poiree, S.; Lanternier, F.; Viard, J.P.; Loulergue, P.; Coignard, H.; Bougnoux, M.E.; Suarez, F.; Rubio, M.T.; et al. Diagnostic contribution of positron emission tomography with [18F]fluorodeoxyglucose for invasive fungal infections. *Clin. Microbiol. Infect.* **2011**, *17*, 409–417. [CrossRef] [PubMed]

22. Rammaert, B.; Candon, S.; Maunoury, C.; Bougnoux, M.-E.; Jouvion, G.; Braun, T.; Correas, J.-M.; Lortholary, O. Thalidomide for steroid-dependent chronic disseminated candidiasis after stem cell transplantation: A case report. *Transplant Infect. Dis.* **2016**, *19*, e12637. [CrossRef]

23. De Castro, N.; Mazoyer, E.; Porcher, R.; Raffoux, E.; Suarez, F.; Ribaud, P.; Lortholary, O.; Molina, J.M. Hepatosplenic candidiasis in the era of new antifungal drugs: A study in Paris 2000-2007. *Clin. Microbiol. Infect.* **2012**, *18*, E185–E187. [CrossRef]

24. Ellis, M.; Al-Ramadi, B.; Bernsen, R.; Kristensen, J.; Alizadeh, H.; Hedstrom, U. Prospective evaluation of mannan and anti-mannan antibodies for diagnosis of invasive *Candida* infections in patients with neutropenic fever. *J. Med. Microbiol.* **2009**, *58*, 606–615. [CrossRef] [PubMed]

25. Todeschini, G.; Murari, C.; Bonesi, R.; Pizzolo, G.; Verlato, G.; Tecchio, C.; Meneghini, V.; Franchini, M.; Giuffrida, C.; Perona, G.; Bellavite, P. Invasive aspergillosis in neutropenic patients: Rapid neutrophil recovery is a risk factor for severe pulmonary complications. *Eur. J. Clin. Investig.* **1999**, *29*, 453–457. [CrossRef]

26. Miceli, M.H.; Maertens, J.; Buvé, K.; Grazziutti, M.; Woods, G.; Rahman, M.; Barlogie, B.; Anaissie, E.J. Immune reconstitution inflammatory syndrome in cancer patients with pulmonary aspergillosis recovering from neutropenia: Proof of principle, description, and clinical and research implications. *Cancer* **2007**, *110*, 112–120. [CrossRef] [PubMed]

27. Singh, N.; Suarez, J.F.; Avery, R.; Lass-Flörl, C.; Geltner, C.; Pasqualotto, A.C.; Lyon, G.M.; Barron, M.; Husain, S.; Wagener, M.M.; Montoya, J.G. Immune reconstitution syndrome-like entity in lung transplant recipients with invasive aspergillosis. *Transplant Immunol.* **2013**, *29*, 109–113. [CrossRef] [PubMed]

28. Jung, J.; Hong, H.-L.; Lee, S.-O.; Choi, S.-H.; Kim, Y.S.; Woo, J.H.; Kim, S.-H. Immune reconstitution inflammatory syndrome in neutropenic patients with invasive pulmonary aspergillosis. *J. Infect.* **2015**, *70*, 1–9. [CrossRef] [PubMed]

29. Kiprono, S.K.; Masenga, J.E. Immune reconstitution inflammatory syndrome: Cutaneous and bone histoplasmosis mimicking leprosy after treatment. *J. Clin. Exp. Dermatol.* **2012**, *03*, 1–3. [CrossRef]

30. Kiggundu, R.; Nabeta, H.W.; Okia, R.; Rhein, J.; Lukande, R. Unmasking histoplasmosis immune reconstitution inflammatory syndrome in a patient recently started on antiretroviral therapy. *ACR* **2016**, *6*, 27–33. [CrossRef]

31. Passos, L.; Talhari, C.; Santos, M.; Ribeiro-Rodrigues, R.; Ferreira, L.C.D.L.; Talhari, S. Histoplasmosis-associated immune reconstitution inflammatory syndrome. *An Bras. Dermatol.* **2011**, *86*, S168–S172. [CrossRef]

32. Jazwinski, A.; Naggie, S.; Perfect, J. Immune reconstitution syndrome in a patient with disseminated histoplasmosis and steroid taper: Maintaining the perfect balance. *Mycoses* **2011**, *54*, 270–272. [CrossRef] [PubMed]

33. Vergidis, P.; Avery, R.K.; Wheat, L.J.; Dotson, J.L.; Assi, M.A.; Antoun, S.A.; Hamoud, K.A.; Burdette, S.D.; Freifeld, A.G.; McKinsey, D.S.; et al. Histoplasmosis complicating tumor necrosis factor–α blocker therapy: A retrospective analysis of 98 cases. *Clin. Infect. Dis.* **2015**, *61*, 409–417. [CrossRef]

34. Breton, G.; Adle-Biassette, H.; Therby, A.; Ramanoelina, J.; Choudat, L.; Bissuel, F.; Huerre, M.; Dromer, F.; Dupont, B.; Lortholary, O. Immune reconstitution inflammatory syndrome in HIV-infected patients with disseminated histoplasmosis. *AIDS* **2006**, *20*, 119–121. [CrossRef] [PubMed]

35. Edman, J.C.; Kovacs, J.A.; Masur, H.; Santi, D.V.; Elwood, H.J.; Sogin, M.L. Ribosomal RNA sequence shows *Pneumocystis carinii* to be a member of the fungi. *Nature* **1988**, *334*, 519–522. [CrossRef] [PubMed]

36. Thomas, C.F.; Limper, A.H. Current insights into the biology and pathogenesis of Pneumocystis pneumonia. *Nat. Rev. Micro.* **2007**, *5*, 298–308. [CrossRef]

37. Kim, H.-W.; Heo, J.Y.; Lee, Y.-M.; Kim, S.J.; Jeong, H.W. Unmasking granulomatous *Pneumocystis jirovecii* pneumonia with nodular opacity in an HIV-infected patient after initiation of antiretroviral therapy. *Yonsei Med. J.* **2016**, *57*, 1042–1045. [CrossRef]

38. Barry, S.M.; Lipman, M.C.I.; Deery, A.R.; Johnson, M.A.; Janossy, G. Immune reconstitution pneumonitis following *Pneumocystis carinii* pneumonia in HIV-infected subjects. *HIV Med.* **2002**, *3*, 207–211. [CrossRef]

39. Wu, A.K.; Cheng, V.C.; Tang, B.S.; Hung, I.F.; Lee, R.A.; Hui, D.S.; Yuen, K.Y. The unmasking of *Pneumocystis jiroveci* pneumonia during reversal of immunosuppression: Case reports and literature review. *BMC Infect. Dis.* **2004**, *4*, 543–548. [CrossRef]

40. Mok, H.P.; Hart, E.; Venkatesan, P. Early development of immune reconstitution inflammatory syndrome related to *Pneumocystis* pneumonia after antiretroviral therapy. *Int J STD AIDS* **2014**, *25*, 373–377. [CrossRef]

41. Godoy, M.C.B.; Silva, C.I.S.; Ellis, J.; Phillips, P.; Müller, N.L. Organizing pneumonia as a manifestation of *Pneumocystis jiroveci* immune reconstitution syndrome in HIV-positive patients: Report of 2 cases. *J. Thorac. Imaging* **2008**, *23*, 39–43. [CrossRef] [PubMed]

42. Wislez, M.; Bergot, E.; Antoine, M.; Parrot, A.; Carette, M.; Mayaud, C.; Cadranel, J. Acute Respiratory Failure Following HAART Introduction in Patients Treated for *Pneumocystis carinii* Pneumonia. *Am. J. Respir. Crit. Care Med.* **2001**, *164*, 847–851. [CrossRef] [PubMed]

43. Alanio, A.; Hauser, P.M.; Lagrou, K.; Melchers, W.J.G.; Helweg-Larsen, J.; Matos, O.; Cesaro, S.; Maschmeyer, G.; Einsele, H.; Donnelly, J.P.; et al. ECIL guidelines for the diagnosis of Pneumocystis jirovecii pneumonia in patients with haematological malignancies and stem cell transplant recipients. *J. Antimicrob. Chemother.* **2016**, *71*, 2386–2396. [CrossRef] [PubMed]

44. Mori, S.; Polatino, S.; Estrada-Y-Martin, R.M. *Pneumocystis*-associated organizing pneumonia as a manifestation of immune reconstitution inflammatory syndrome in an HIV-infected individual with a normal CD4+ T-cell count following antiretroviral therapy. *Int. J. STD AIDS* **2009**, *20*, 662–665. [CrossRef] [PubMed]

45. Kleindienst, R.; Fend, F.; Prior, C.; Margreiter, R.; Vogel, W. Bronchiolitis obliterans organizing pneumonia associated with Pneumocystis carinii infection in a liver transplant patient receiving tacrolimus. *Clin. Transplant.* **1999**, *13*, 65–67. [CrossRef] [PubMed]

46. Verma, N.; Soans, B. Cryptogenic organizing pneumonia associated with Pneumocystis carinii infection and sirolimus therapy in a renal transplant patient. *Australas. Radiol.* **2006**, *50*, 68–70. [CrossRef] [PubMed]

47. Hall, C.; Hajjawi, R.; Barlow, G.; Thaker, H.; Adams, K.; Moss, P. *Penicillium marneffei* presenting as an immune reconstitution inflammatory syndrome (IRIS) in a patient with advanced HIV. *BMJ Case Rep.* **2013**, *2013*, bcr2012007555. [CrossRef] [PubMed]

48. Sudjaritruk, T.; Sirisanthana, T.; Sirisanthana, V. Immune reconstitution inflammatory syndrome from *Penicillium marneffei* in an HIV-infected child: A case report and review of literature. *BMC Infect. Dis.* **2012**, *12*, 28–31. [CrossRef]

49. Saikia, L.; Nath, R.; Hazarika, D.; Mahanta, J. Atypical cutaneous lesions of *Penicillium marneffei* infection as a manifestation of the immune reconstitution inflammatory syndrome after highly active antiretroviral therapy. *Indian J. Dermatol. Venereol. Leprol.* **2010**, *76*, 45–48. [CrossRef]

50. Ho, A.; Shankland, G.S.; Seaton, R.A. *Penicillium marneffei* infection presenting as an immune reconstitution inflammatory syndrome in an HIV patient. *Int. J. STD AIDS* **2010**, *21*, 780–782. [CrossRef]

51. Thanh, N.T.; Vinh, L.D.; Liem, N.T.; Shikuma, C.; Day, J.N.; Thwaites, G.; Le, T. Clinical features of three patients with paradoxical immune reconstitution inflammatory syndrome associated with *Talaromyces marneffei* infection. *Med. Mycol. Case Rep.* **2018**, *19*, 33–37. [CrossRef] [PubMed]

52. Mortimer, R.B.; Libke, R.; Eghbalieh, B.; Bilello, J.F. Immune reconstitution inflammatory syndrome presenting as superior vena cava syndrome secondary to Coccidioides lymphadenopathy in an HIV-infected patient. *J. Int. Assoc. Physicians AIDS Care* **2008**, *7*, 283–285. [CrossRef] [PubMed]

53. Almeida, S.M.; Roza, T.H. HIV Immune Recovery Inflammatory Syndrome and Central Nervous System Paracoccidioidomycosis. *Mycopathologia* **2016**, *182*, 1–4. [CrossRef] [PubMed]

54. Galhardo, M.C.G.; Silva, M.T.T.; Lima, M.A.; Nunes, E.P.; Schettini, L.E.C.; de Freitas, R.F.; de Almeida Paes, R.; de Sousa Neves, E.; do Valle, A.C.F. *Sporothrix schenckii* meningitis in AIDS during immune reconstitution syndrome. *J. Neurol. Neurosurg. Psychiatry* **2010**, *81*, 696–699. [CrossRef]

55. Dony, A.; Perpoint, T.; Ducastelle, S.; Ferry, T. Disseminated fusariosis with immune reconstitution syndrome and cracking mycotic aortic aneurysm in a 55-year-old patient with acute myeloid leukaemia. *BMJ Case Rep.* **2013**. [CrossRef] [PubMed]

56. Crombie, K.; Spengane, Z.; Locketz, M.; Dlamini, S.; Lehloenya, R.; Wasserman, S.; Maphanga, T.G.; Govender, N.P.; Kenyon, C.; Schwartz, I.S. Paradoxical worsening of *Emergomyces africanus* infection in an HIV-infected male on itraconazole and antiretroviral therapy. *PLoS Negl. Trop. Dis.* **2018**, *12*, e0006173. [CrossRef] [PubMed]

57. Legrand, F.; Lecuit, M.; Dupont, B.; Bellaton, E.; Huerre, M.; Rohrlich, P.S.; Lortholary, O. Adjuvant corticosteroid therapy for chronic disseminated candidiasis. *Clin. Infect. Dis.* **2008**, *46*, 696–702. [CrossRef]

58. Lortholary, O.; Fontanet, A.; Mémain, N.; Martin, A.; Sitbon, K.; Dromer, F.; French Cryptococcosis Study Group. Incidence and risk factors of immune reconstitution inflammatory syndrome complicating HIV-associated cryptococcosis in France. *AIDS* **2005**, *19*, 1043–1049. [CrossRef]

59. Bettelli, E.; Korn, T.; Oukka, M.; Kuchroo, V.K. Induction and effector functions of TH17 cells. *Nature* **2008**, *453*, 1051–1057. [CrossRef]

60. Weaver, C.T.; Hatton, R.D. Interplay between the TH17 and TReg cell lineages: A (co-)evolutionary perspective. *Nat. Rev. Immunol.* **2009**, *9*, 883–889. [CrossRef]

61. Weaver, C.T.; Hatton, R.D.; Mangan, P.R.; Harrington, L.E. IL-17 Family Cytokines and the Expanding Diversity of Effector T Cell Lineages. *Ann. Rev. Immunol.* **2007**, *25*, 821–852. [CrossRef] [PubMed]

62. Martinez, F.O.; Gordon, S. The M1 and M2 paradigm of macrophage activation: Time for reassessment. *F1000Prime Rep* **2014**, *6*, 13–19. [CrossRef] [PubMed]

63. Shearer, G.M. HIV-Induced immunopathogenesis. *Immunity* **1998**, *9*, 587–593. [CrossRef]

64. Becker, Y. The changes in the T helper 1 (Th1) and T helper 2 (Th2) cytokine balance during HIV-1 infection are indicative of an allergic response to viral proteins that may be reversed by Th2 cytokine inhibitors and immune response modifiers-a review and hypothesis. *Virus Genes* **2004**, *28*, 5–18. [CrossRef]

65. Zhang, Y.-H.; He, M.; Wang, Y.; Liao, A.-H. Modulators of the Balance between M1 and M2 Macrophages during Pregnancy. *Front. Immunol.* **2017**, *8*, 120. [CrossRef] [PubMed]

66. Tsiodras, S.; Samonis, G.; Boumpas, D.T.; Kontoyiannis, D.P. Fungal Infections Complicating Tumor Necrosis Factor α Blockade Therapy. *Mayo Clin. Proc.* **2008**, *83*, 181–194. [CrossRef]

67. Yasunami, Y.; Kojo, S.; Kitamura, H.; Toyofuku, A.; Satoh, M.; Nakano, M.; Nabeyama, K.; Nakamura, Y.; Matsuoka, N.; Ikeda, S.; et al. Vα14 NK T cell–triggered IFN-γ production by Gr-1 +CD11b +cells mediates early graft loss of syngeneic transplanted islets. *J. Exp. Med.* **2005**, *202*, 913–918. [CrossRef] [PubMed]

68. Flynn, J.L.; Chan, J.; Lin, P.L. Macrophages and control of granulomatous inflammation in tuberculosis. *Mucosal Immunol.* **2011**, *4*, 271–278. [CrossRef] [PubMed]

69. Das, P.; Rampal, R.; Udinia, S.; Kumar, T.; Pilli, S.; Wari, N.; Ahmed, I.K.; Kedia, S.; Gupta, S.D.; Kumar, D.; et al. Selective M1 macrophage polarization in granuloma-positive and granuloma-negative Crohn's disease, in comparison to intestinal tuberculosis. *Intest. Res.* **2018**, *16*, 426–435. [CrossRef] [PubMed]

70. Gordon, S. Alternative activation of macrophages. *Nat. Rev. Immunol.* **2003**, *3*, 23–35. [CrossRef] [PubMed]

71. Martin-Blondel, G.; Mars, L.T.; Liblau, R.S. Pathogenesis of the immune reconstitution inflammatory syndrome in HIV-infected patients. *Curr. Opin. Infect. Dis.* **2012**, *25*, 312–320. [CrossRef]

72. Nilsson, J.; Boasso, A.; Velilla, P.A.; Zhang, R.; Vaccari, M.; Franchini, G.; Shearer, G.M.; Andersson, J.; Chougnet, C. HIV-1-driven regulatory T-cell accumulation in lymphoid tissues is associated with disease progression in HIV/AIDS. *Blood* **2006**, *108*, 3808–3817. [CrossRef] [PubMed]

73. Montes, M.; Sanchez, C.; Lewis, D.E.; Graviss, E.A.; Seas, C.; Gotuzzo, E.; White, A.C. Jr. Normalization of FoxP3+ Regulatory T Cells in Response to Effective Antiretroviral Therapy. *J. Infect. Dis.* **2010**, *203*, 496–499. [CrossRef] [PubMed]

74. Seddiki, N.; Sasson, S.C.; Santner-Nanan, B.; Munier, M.; van Bockel, D.; Ip, S.; Marriott, D.; Pett, S.; Nanan, R.; Cooper, D.A.; et al. Proliferation of weakly suppressive regulatory CD4+ T cells is associated with over-active CD4+ T-cell responses in HIV-positive patients with mycobacterial immune restoration disease. *Eur. J. Immunol.* **2009**, *39*, 391–403. [CrossRef] [PubMed]

75. Ingram, P.R.; Howman, R.; Leahy, M.F.; Dyer, J.R. Cryptococcal Immune Reconstitution Inflammatory Syndrome following Alemtuzumab Therapy. *Clin. Infect. Dis.* **2007**, *44*, e115–e117. [CrossRef] [PubMed]

76. Otahbachi, M.; Nugent, K. Granulomatous *Pneumocystis jiroveci* pneumonia in a patient with chronic lymphocytic leukemia: A literature review and hypothesis on pathogenesis. *Phytochemistry.* **2007**, *333*, 131–135. [CrossRef]

77. Li, L.; Huang, L.; Sung, S.-S.J.; Lobo, P.I.; Brown, M.G.; Gregg, R.K.; Engelhard, V.H.; Okusa, M.D. NKT cell activation mediates neutrophil IFN-gamma production and renal ischemia-reperfusion injury. *J. Immunol.* **2007**, *178*, 5899–5911. [CrossRef]

78. Mantovani, A.; Cassatella, M.A.; Costantini, C.; Jaillon, S. Neutrophils in the activation and regulation of innate and adaptive immunity. *Nat. Rev. Immunol.* **2011**, *11*, 519–531. [CrossRef]

79. Pelletier, M.; Maggi, L.; Micheletti, A.; Lazzeri, E.; Tamassia, N.; Costantini, C.; Cosmi, L.; Lunardi, C.; Annunziato, F.; Romagnani, S.; et al. Evidence for a cross-talk between human neutrophils and Th17 cells. *Blood* **2010**, *115*, 335–343. [CrossRef]

80. Elenkov, I.J.; Wilder, R.L.; Bakalov, V.K.; Link, A.A.; Dimitrov, M.A.; Fisher, S.; Crane, M.; Kanik, K.S.; Chrousos, G.P. IL-12, TNF-alpha, and hormonal changes during late pregnancy and early postpartum: Implications for autoimmune disease activity during these times. *J. Clin. Endocrinol. Metab.* **2001**, *86*, 4933–4938.

81. Bellocchio, S.; Gaziano, R.; Bozza, S.; Rossi, G.; Montagnoli, C.; Perruccio, K.; Calvitti, M.; Pitzurra, L.; Romani, L. Liposomal amphotericin B activates antifungal resistance with reduced toxicity by diverting Toll-like receptor signalling from TLR-2 to TLR-4. *J. Antimicrob. Chemother.* **2005**, *55*, 214–222. [CrossRef] [PubMed]

82. Ben-Ami, R.; Lewis, R.E.; Kontoyiannis, D.P. Immunocompromised Hosts: Immunopharmacology of Modern Antifungals. *Clin. Infect. Dis.* **2008**, *47*, 226–235. [CrossRef]

83. Gil-Lamaignere, C.; Roilides, E.; Maloukou, A.; Georgopoulou, I.; Petrikkos, G.; Walsh, T.J. Amphotericin B lipid complex exerts additive antifungal activity in combination with polymorphonuclear leucocytes against *Scedosporium prolificans* and *Scedosporium apiospermum*. *J. Antimicrob. Chemother.* **2002**, *50*, 1027–1030. [CrossRef]

84. Roilides, E.; Lyman, C.A.; Filioti, J.; Akpogheneta, O.; Sein, T.; Lamaignere, C.G.; Petraitiene, R.; Walsh, T.J. Amphotericin B Formulations Exert Additive Antifungal Activity in Combination with Pulmonary Alveolar Macrophages and Polymorphonuclear Leukocytes against Aspergillus fumigatus. *Antimicrob. Agents Chemother.* **2002**, *46*, 1974–1976. [CrossRef]

85. Simitsopoulou, M.; Roilides, E.; Maloukou, A.; Gil-Lamaignere, C.; Walsh, T.J. Interaction of amphotericin B lipid formulations and triazoles with human polymorphonuclear leucocytes for antifungal activity against Zygomycetes. *Mycoses* **2008**, *51*, 147–154. [CrossRef] [PubMed]

86. Voelz, K.; May, R.C. Cryptococcal Interactions with the Host Immune System. *Eukaryotic Cell* **2010**, *9*, 835–846. [CrossRef] [PubMed]

87. Wiesner, D.L.; Moskalenko, O.; Corcoran, J.M.; McDonald, T.; Rolfes, M.A.; Meya, D.B.; Kajumbula, H.; Kambugu, A.; Bohjanen, P.R.; Knight, J.F.; et al. Cryptococcal genotype influences immunologic response and human clinical outcome after meningitis. *mBio* **2012**, *3*, e00196-12. [CrossRef] [PubMed]

88. Desnos-Ollivier, M.; Patel, S.; Spaulding, A.R.; Charlier, C.; Garcia-Hermoso, D.; Nielsen, K.; Dromer, F. Mixed infections and *in vivo* evolution in the human fungal pathogen *Cryptococcus neoformans*. *mBio* **2010**, *1*, 1–9. [CrossRef] [PubMed]

89. Alanio, A.; Gits-Muselli, M.; Mercier-Delarue, S.; Dromer, F.; Bretagne, S. Diversity of *Pneumocystis jirovecii* during infection revealed by ultra-deep pyrosequencing. *Front. Microbiol.* **2016**, *7*, 733. [CrossRef] [PubMed]

90. Price, P.; Morahan, G.; Huang, D.; Stone, E.; Cheong, K.Y.M.; Castley, A.; Rodgers, M.; McIntyre, M.Q.; Abraham, L.J.; French, M.A. Polymorphisms in cytokine genes define subpopulations of HIV-1 patients who experienced immune restoration diseases. *AIDS* **2002**, *16*, 2043–2047. [CrossRef]

91. Ogola, G.O.; Ouma, C.; Jura, W.G.; Muok, E.O.; Colebunders, R.; Mwinzi, P.N. A non-synonymous polymorphism in IL-23R Gene (rs1884444) is associated with reduced risk to schistosomiasis-associated Immune Reconstitution Inflammatory Syndrome in a Kenyan population. *BMC Infect. Dis.* **2014**, *14*, 1–7. [CrossRef] [PubMed]

92. Choi, E.H.; Foster, C.B.; Taylor, J.G.; Erichsen, H.C.; Chen, R.A.; Walsh, T.J.; Anttila, V.-J.; Ruutu, T.; Palotie, A.; Chanock, S.J. Association between chronic disseminated candidiasis in adult acute leukemia and common IL4 promoter haplotypes. *J. Infect. Dis.* **2003**, *187*, 1153–1156. [CrossRef] [PubMed]

93. Narendran, G.; Kavitha, D.; Karunaianantham, R.; Gil-Santana, L.; Almeida-Junior, J.L.; Reddy, S.D.; Kumar, M.M.; Hemalatha, H.; Jayanthi, N.N.; Ravichandran, N.; et al. Role of LTA4H Polymorphism in Tuberculosis-Associated Immune Reconstitution Inflammatory Syndrome Occurrence and Clinical Severity in Patients Infected with HIV. *PLoS ONE* **2016**, *11*, e0163298. [CrossRef] [PubMed]

94. Meintjes, G.; Scriven, J.; Marais, S. Management of the Immune Reconstitution Inflammatory Syndrome. *Curr HIV/AIDS Rep* **2012**, *9*, 238–250. [CrossRef] [PubMed]

95. Rhen, T.; Cidlowski, J.A. Anti-inflammatory action of glucocorticoids—New mechanisms for old drugs. *N. Engl. J. Med.* **2005**, *353*, 1711–1723. [CrossRef] [PubMed]

96. Nahid, P.; Dorman, S.E.; Alipanah, N.; Barry, P.M.; Brozek, J.L.; Cattamanchi, A.; Chaisson, L.H.; Chaisson, R.E.; Daley, C.L.; Grzemska, M.; et al. Official American Thoracic Society/Centers for Disease Control and Prevention/Infectious Diseases Society of America Clinical Practice Guidelines: Treatment of Drug-Susceptible Tuberculosis. *Clin. Infect. Dis.* **2016**, *63*, e147–e195. [CrossRef] [PubMed]

97. Meintjes, G.; Wilkinson, R.J.; Morroni, C.; Pepper, D.J.; Rebe, K.; Rangaka, M.X.; Oni, T.; Maartens, G. Randomized placebo-controlled trial of prednisone for paradoxical tuberculosis-associated immune reconstitution inflammatory syndrome. *AIDS* **2010**, *24*, 1–19. [CrossRef] [PubMed]

98. Caillot, D.; Couaillier, J.F.; Bernard, A.; Casasnovas, O.; Denning, D.W.; Mannone, L.; Lopez, J.; Couillault, G.; Piard, F.; Vagner, O.; et al. Increasing volume and changing characteristics of invasive pulmonary aspergillosis on sequential thoracic computed tomography scans in patients with neutropenia. *J. Clin. Oncol.* **2001**, *19*, 253–259. [CrossRef] [PubMed]

99. Crespo, G.; Cervera, C.; Michelena, J.; Marco, F.; Moreno, A.; Navasa, M. Immune reconstitution syndrome after voriconazole treatment for cryptococcal meningitis in a liver transplant recipient. *Liver Transplant.* **2008**, *14*, 1671–1674. [CrossRef] [PubMed]

100. Montaner, J.S.G. Corticosteroids Prevent Early Deterioration in Patients with Moderately Severe *Pneumocystis carinii* Pneumonia and the Acquired Immunodeficiency Syndrome (AIDS). *Ann. Intern. Med.* **1990**, *113*, 14–20. [CrossRef] [PubMed]
101. Chabria, S.; Barakat, L.; Ogbuagu, O. Steroid-exacerbated HIV-associated cutaneous Kaposi's sarcoma immune reconstitution inflammatory syndrome: "Where a good intention turns bad." *Int. J. STD AIDS* **2016**, *27*, 1026–1029. [CrossRef] [PubMed]
102. Fernández-Sánchez, M.; Iglesias, M.C.; Ablanedo-Terrazas, Y.; Ormsby, C.E.; Alvarado-De La Barrera, C.; Reyes-Terán, G. Steroids are a risk factor for Kaposi's sarcoma-immune reconstitution inflammatory syndrome and mortality in HIV infection. *AIDS* **2016**, *30*, 909–914. [CrossRef]
103. Blackmore, T.K.; Manning, L.; Taylor, W.J.; Wallis, R.S. Therapeutic Use of Infliximab in Tuberculosis to Control Severe Paradoxical Reaction of the Brain and Lymph Nodes. *Clin. Infect. Dis.* **2008**, *47*, e83–e85. [CrossRef] [PubMed]
104. Hsu, D.C.; Faldetta, K.F.; Pei, L.; Sheikh, V.; Utay, N.S.; Roby, G.; Rupert, A.; Fauci, A.S.; Sereti, I. A paradoxical treatment for a paradoxical condition: Infliximab use in three cases of mycobacterial IRIS. *Clin. Infect. Dis.* **2015**, *62*, 258–261. [CrossRef] [PubMed]
105. Lwin, N.; Boyle, M.; Davis, J.S. Adalimumab for Corticosteroid and Infliximab-Resistant Immune Reconstitution Inflammatory Syndrome in the Setting of TB/HIV Coinfection. *Open Forum Infect. Dis.* **2018**, *5*, ofy027. [CrossRef] [PubMed]
106. Gaube, G.; De Castro, N.; Gueguen, A.; Lascoux, C.; Zagdanski, A.M.; Alanio, A.; Molina, J.M. Treatment with adalimumab for severe immune reconstitution inflammatory syndrome in an HIV-infected patient presenting with cryptococcal meningitis. *Medecine et Maladies Infectieuses* **2016**, *46*, 154–156. [CrossRef] [PubMed]
107. Sitapati, A.M.; Kao, C.L.; Cachay, E.R.; Masoumi, H.; Wallis, R.S.; Mathews, W.C. Treatment of HIV-Related Inflammatory Cerebral Cryptococcoma with Adalimumab. *Clin. Infect. Dis.* **2010**, *50*, e7–e10. [CrossRef]
108. Brunel, A.-S.; Reynes, J.; Tuaillon, E.; Rubbo, P.-A.; Lortholary, O.; Montes, B.; Le Moing, V.; Makinson, A. Thalidomide for steroid-dependent immune reconstitution inflammatory syndromes during AIDS. *AIDS* **2012**, *26*, 2110–2112. [CrossRef]
109. Kaplan, J.E.; Benson, C.; Holmes, K.K.; Brooks, J.T.; Pau, A.; Masur, H. Guidelines for prevention and treatment of opportunistic infections in HIV-infected adults and adolescents: Recommendations from CDC, the National Institutes of Health, and the HIV Medicine Association of the Infectious Diseases Society of America. *MMR Recomm. Rep.* **2009**, *58*, 1–207.
110. Bergman, P.W.; Björkhem-Bergman, L. Is there a role for statins in fungal infections? *Expert Rev. Anti-Infect. Ther.* **2014**, *11*, 1391–1400. [CrossRef]
111. Sun, H.Y.; Singh, N. Potential role of statins for the management of immune reconstitution syndrome. *Med. Hypotheses* **2011**, *76*, 307–310. [CrossRef] [PubMed]
112. Tleyjeh, I.M.; Kashour, T.; Hakim, F.A.; Zimmerman, V.A.; Erwin, P.J.; Sutton, A.J.; Ibrahim, T. Statins for the prevention and treatment of infections: A systematic review and meta-analysis. *Arch. Intern. Med.* **2009**, *169*, 1658–1667. [CrossRef] [PubMed]
113. Zorzou, M.P.; Chini, M.; Lioni, A.; Tsekes, G.; Nitsotolis, T.; Tierris, I.; Panagiotou, N.; Rontogianni, D.; Harhalakis, N.; Lazanas, M. Successful treatment of immune reconstitution inflammatory syndrome-related hemophagocytic syndrome in an HIV patient with primary effusion lymphoma. *Hematol. Rep.* **2016**, *8*, 1–4. [CrossRef] [PubMed]
114. Calic, Z.; Cappelen-Smith, C.; Hodgkinson, S.J.; McDougall, A.; Cuganesan, R.; Brew, B.J. Treatment of progressive multifocal leukoencephalopathy immune reconstitution inflammatory syndrome with intravenous immunoglobulin in a patient with multiple sclerosis treated with fingolimod after discontinuation of natalizumab. *J. Clin. Neurosci.* **2015**, *22*, 598–600. [CrossRef] [PubMed]
115. Müller, M.; Wandel, S.; Colebunders, R.; Attia, S.; Furrer, H.; Egger, M. Immune reconstitution inflammatory syndrome in patients starting antiretroviral therapy for HIV infection: A systematic review and meta-analysis. *Lancet Infect. Dis.* **2010**, *10*, 251–261. [CrossRef]
116. Manabe, Y.C.; Campbell, J.D.; Sydnor, E.; Moore, R.D. Immune reconstitution inflammatory syndrome: Risk factors and treatment implications. *J. Acquir. Immune Defic. Syndr.* **2007**, *46*, 456–462. [CrossRef] [PubMed]

117. Chang, C.C.; Lim, A.; Omarjee, S.; Levitz, S.M.; Gosnell, B.I.; Spelman, T.; Elliott, J.H.; Carr, W.H.; Moosa, M.-Y.S.; Ndung'u, T.; et al. Cryptococcosis-IRIS is associated with lower *Cryptococcus*-specific IFN-γ responses before antiretroviral therapy but not higher T-cell responses during therapy. *J. Infect. Dis.* **2013**, *208*, 898–906. [CrossRef]

118. Chang, C.C.; Dorasamy, A.A.; Gosnell, B.I.; Elliott, J.H.; Spelman, T.; Omarjee, S.; Naranbhai, V.; Coovadia, Y.; Ndung'u, T.; Moosa, M.-Y.S.; et al. Clinical and mycological predictors of cryptococcosis-associated immune reconstitution inflammatory syndrome. *AIDS* **2013**, *27*, 2089–2099. [CrossRef]

119. Yoon, H.A.; Nakouzi, A.; Chang, C.C.; Kuniholm, M.H.; Carreño, L.J.; Wang, T.; Ndung'u, T.; Lewin, S.R.; French, M.A.; Pirofski, L.-A. Association between plasma antibody responses and risk for Cryptococcus-associated immune reconstitution inflammatory syndrome. *J. Infect. Dis.* **2018**, *17*, 873–880. [CrossRef]

120. Boulware, D.R.; Meya, D.B.; Bergemann, T.L.; Wiesner, D.L.; Rhein, J.; Musubire, A.; Lee, S.J.; Kambugu, A.; Janoff, E.N.; Bohjanen, P.R. Clinical features and serum biomarkers in HIV immune reconstitution inflammatory syndrome after cryptococcal meningitis: A prospective cohort study. *PLoS Med.* **2010**, *7*, e1000384. [CrossRef]

121. Boulware, D.R.; Bonham, S.C.; Meya, D.B.; Wiesner, D.L.; Park, G.S.; Kambugu, A.; Janoff, E.N.; Bohjanen, P.R. Paucity of initial cerebrospinal fluid inflammation in cryptococcal meningitis is associated with subsequent immune reconstitution inflammatory syndrome. *J. Infect. Dis.* **2010**, *202*, 962–970. [CrossRef] [PubMed]

122. Chang, C.C.; Omarjee, S.; Lim, A.; Spelman, T.; Gosnell, B.I.; Carr, W.H.; Elliott, J.H.; Moosa, M.-Y.S.; Ndung'u, T.; French, M.A.; Lewin, S.R. Chemokine levels and chemokine receptor expression in the blood and the cerebrospinal fluid of HIV-infected patients with cryptococcal meningitis and cryptococcosis-associated immune reconstitution inflammatory syndrome. *J. Infect. Dis.* **2013**, *208*, 1604–1612. [CrossRef] [PubMed]

123. Wilkinson, R.J.; Walker, N.F.; Scriven, J.; Meintjes, G. Immune reconstitution inflammatory syndrome in HIV-infected patients. *HIV* **2015**, *7*, 49–64. [CrossRef] [PubMed]

124. Bicanic, T.; Meintjes, G.; Rebe, K.; Williams, A.; Loyse, A.; Wood, R.; Hayes, M.; Jaffar, S.; Harrison, T. Immune reconstitution inflammatory syndrome in HIV-associated cryptococcal meningitis: A prospective study. *J. Acquir. Immune Defic. Syndr.* **2009**, *51*, 130–134. [CrossRef] [PubMed]

125. Shelburne, S.A.; Darcourt, J.; White, A.C.; Greenberg, S.B.; Hamill, R.J.; Atmar, R.L.; Visnegarwala, F. The role of immune reconstitution inflammatory syndrome in AIDS-related *Cryptococcus neoformans* disease in the era of highly active antiretroviral therapy. *Clin. Infect. Dis.* **2005**, *40*, 1049–1052. [CrossRef] [PubMed]

126. Vitoria, M.; Ford, N.; Clayden, P.; Pozniak, A.L.; Hill, A.M. When could new antiretrovirals be recommended for national treatment programmes in low-income and middle-income countries. *Curr. Opin. HIV AIDS* **2017**, *12*, 414–422. [CrossRef] [PubMed]

127. Psichogiou, M.; Basoulis, D.; Tsikala-Vafea, M.; Vlachos, S.; Kapelios, C.J.; Daikos, G.L. Integrase Strand Transfer inhibitors and the emergence of Immune Reconstitution Inflammatory Syndrome (IRIS). *CHR* **2018**, *15*, 1–6. [CrossRef]

128. Dutertre, M.; Cuzin, L.; Demonchy, E.; Puglièse, P.; Joly, V.; Valantin, M.-A.; Cotte, L.; Huleux, T.; Delobel, P.; Martin-Blondel, G. Initiation of antiretroviral therapy containing integrase inhibitors increases the risk of IRIS requiring hospitalization. *J. Acquir. Immune Defic. Syndr.* **2017**, *76*, e23–e26. [CrossRef]

129. Chandesris, M.-O.; Kelaidi, C.; Méchaï, F.; Bougnoux, M.-E.; Brousse, N.; Viard, J.-P.; Poirée, S.; Lecuit, M.; Hermine, O.; Lortholary, O. Granulocyte Colony Stimulating Factor-induced exacerbation of fungus-related Immune Restoration Inflammatory Syndrome: A case of chronic disseminated candidiasis exacerbation. *J. Microbiol. Immunol. Infect.* **2010**, *43*, 339–343. [CrossRef]

130. Blanc, F.-X.; Sok, T.; Laureillard, D.; Borand, L.; Rekacewicz, C.; Nerrienet, E.; Madec, Y.; Marcy, O.; Chan, S.; Prak, N.; et al. Earlier versus later start of antiretroviral therapy in HIV-infected adults with tuberculosis. *N. Engl. J. Med.* **2011**, *365*, 1471–1481. [CrossRef]

131. Torok, M.E.; Yen, N.T.B.; Chau, T.T.H.; Mai, N.T.H.; Phu, N.H.; Mai, P.P.; Dung, N.T.; Chau, N.V.V.; Bang, N.D.; Tien, N.A.; et al. Timing of Initiation of antiretroviral therapy in human immunodeficiency virus (HIV)-associated tuberculous meningitis. *Clin. Infect. Dis.* **2011**, *52*, 1374–1383. [CrossRef] [PubMed]

132. Zolopa, A.R.; Andersen, J.; Komarow, L.; Sanne, I.; Sanchez, A.; Hogg, E.; Suckow, C.; Powderly, W. Early antiretroviral therapy reduces AIDS progression/death in individuals with acute opportunistic infections: A multicenter randomized strategy trial. *PLoS ONE* **2009**, *4*, e5575. [CrossRef] [PubMed]

133. Makadzange, A.T.; Ndhlovu, C.E.; Takarinda, K.; Reid, M.; Kurangwa, M.; Gona, P.; Hakim, J.G. Early versus delayed initiation of antiretroviral therapy for concurrent HIV infection and cryptococcal meningitis in sub-saharan Africa. *Clin. Infect. Dis.* **2010**, *50*, 1532–1538. [CrossRef] [PubMed]

134. Bisson, G.P.; Molefi, M.; Bellamy, S.; Thakur, R.; Steenhoff, A.; Tamuhla, N.; Rantleru, T.; Tsimako, I.; Gluckman, S.; Ravimohan, S.; et al. Early versus delayed antiretroviral therapy and cerebrospinal fluid fungal clearance in adults with HIV and cryptococcal meningitis. *Clin. Infect. Dis.* **2013**, *56*, 1165–1173. [CrossRef] [PubMed]

135. Boulware, D.R.; Meya, D.B.; Muzoora, C.; Rolfes, M.A.; Huppler Hullsiek, K.; Musubire, A.; Taseera, K.; Nabeta, H.W.; Schutz, C.; Williams, D.A.; et al. Timing of antiretroviral therapy after diagnosis of cryptococcal meningitis. *N. Engl. J. Med.* **2014**, *370*, 2487–2498. [CrossRef] [PubMed]

136. Scriven, J.E.; Graham, L.M.; Schutz, C.; Scriba, T.J.; Wilkinson, K.A.; Wilkinson, R.J.; Boulware, D.R.; Urban, B.C.; Meintjes, G.; Lalloo, D.G. The CSF Immune Response in HIV-1-Associated Cryptococcal Meningitis: Macrophage Activation, Correlates of Disease Severity, and Effect of Antiretroviral Therapy. *J. Acquir. Imm. Defic. Syndr.* **2017**, *75*, 299–307. [CrossRef]

137. Southern African HIV Clinicians Society, T. Guideline for the prevention, diagnosis and management of cryptococcal meningitis among HIV-infected persons: 2013 update. *S. Afr. J. HIV Med.* **2013**, *14*, 76–86. [CrossRef]

138. Abassi, M.; Boulware, D.R.; Rhein, J. Cryptococcal Meningitis: Diagnosis and Management Update. *Curr. Trop. Med. Rep.* **2015**, *2*, 90–99. [CrossRef]

139. Singh, N. Hypercalcemia Related to Immune Reconstitution in Organ Transplant Recipients with Granulomatous Opportunistic Infections. *Transplantation* **2006**, *82*, 986. [CrossRef]

Journal of
Fungi

MDPI

Review

Pre-Existing Liver Disease and Toxicity of Antifungals

Nikolaos Spernovasilis [1] and Diamantis P. Kofteridis [1,2,]*

[1] Department of Internal Medicine and Infectious Diseases, University Hospital of Heraklion, P.O. Box 1352, 71110 Heraklion, Crete, Greece; nikspe@hotmail.com
[2] School of Medicine, University of Crete, 71110 Heraklion, Crete, Greece
* Correspondence: kofterid@med.uoc.gr; Tel.: +30-2810-392688; Fax: +30-2810-392359

Received: 30 October 2018; Accepted: 7 December 2018; Published: 10 December 2018

Abstract: Pre-existing liver disease in patients with invasive fungal infections further complicates their management. Altered pharmacokinetics and tolerance issues of antifungal drugs are important concerns. Adjustment of the dosage of antifungal agents in these cases can be challenging given that current evidence to guide decision-making is limited. This comprehensive review aims to evaluate the existing evidence related to antifungal treatment in individuals with liver dysfunction. This article also provides suggestions for dosage adjustment of antifungal drugs in patients with varying degrees of hepatic impairment, after accounting for established or emerging pharmacokinetic–pharmacodynamic relationships with regard to antifungal drug efficacy in vivo.

Keywords: liver disease; hepatic impairment; invasive fungal infection; antifungal agent; antifungal drug; toxicity

1. Introduction

Invasive fungal infection (IFI) is a leading cause of morbidity and mortality among immunocompromised and critically ill patients [1,2]. Although antifungal drug options have increased in recent years, effective management of IFI depends mainly on early and appropriate individualized treatment that optimizes efficacy and safety based on local epidemiology, drug spectrum of activity, pharmacokinetic (PK) and pharmacodynamic (PD) properties of the antifungal agent, and patient related factors [3].

Pre-existing liver disease in patients with IFIs raises significant concern about the safety of antifungal agent administration. The liver is the primary site of drug metabolism, and hepatic disease can significantly alter the PKs of antifungal drugs, mainly through impaired clearance [4]. Moreover, other variables that affect PKs such as liver blood flow, biliary excretion and plasma protein binding may be altered in patients with pre-existing hepatic dysfunction [4]. These patients may also tolerate drug-induced liver injury (DILI) more poorly than healthy individuals [5]. Furthermore, in the cirrhotic patients, drug-related extrahepatic effects, such as renal failure, gastrointestinal bleeding and hepatic encephalopathy, are more likely to occur [6]. Hepatic functional status is also an important determinant of the drug–drug interaction (DDI) magnitude due to enzyme inhibition or induction in the liver [7].

It is important to distinguish isolated biochemical injury from hepatic dysfunction [8]. In general, DILI is characterized by elevations in hepatic enzymes, resulting from the effect of an active drug or its metabolites to the liver [9]. This biochemical abnormality is not necessarily accompanied by clinically significant liver dysfunction, since liver has a notable healing capacity [8]. However, DILI can be the cause of hepatic dysfunction, manifested by hyperbilirubinemia and coagulopathy [10], or even acute liver failure, presented with jaundice and hepatic encephalopathy [11].

Liver injury induced by a drug is generally classified as either intrinsic, which is predictable, dose-dependent and reproducible in preclinical models, or idiosyncratic, which is unpredictable and dose-independent [12–14]. An international expert group of clinicians and scientists comprehensibly proposed the clinical chemistry criteria for the diagnosis of DILI, taking also into account the possibility of pre-existing liver enzymes abnormalities (Table 1) [15]. Furthermore, the ratio of serum alanine aminotransferase (ALT) to alkaline phosphatase (ALP), expressed as multiples of upper limit of normal (ULN), is called R ratio or value, and is used to classify DILI in individuals with previous normal liver tests into three categories: hepatocellular (*R* > 5), cholestatic (*R* < 2) and mixed (*R* of 2–5) [16]. Bilirubin, although not incorporated into the R ratio, remains an essential marker in calculating the Model for End-Stage Liver Disease (MELD) score and the Child–Pugh score [17,18]. Both these prognostic models are also used to assess hepatic function, with the Child–Pugh score being the most commonly used method in cirrhotic patients among studies submitted to the US Food and Drug Administration (FDA) although it is not associated directly with PK changes [19] and does not represent a reliable estimator of liver function [20].

Table 1. Clinical chemistry criteria for DILI.

Anyone of the Following *:
ALT elevation $\geq$ 5 $\times$ ULN [¶]
ALP elevation $\geq$ 2 $\times$ ULN [¶], especially with accompanying elevations in concentrations of 5'-NT or GGT
ALT elevation $\geq$ 3 $\times$ ULN [¶] and simultaneous TB elevation $\geq$ 2 $\times$ ULN [¶]

DILI: drug-induced liver injury; ALT: alanine transaminase; ULN: upper limit of normal; AST: aspartate transaminase; ALP: alkaline phosphatase; 5'-NT: 5'-nucleotidase; GGT: γ-glutamyl transpeptidase; TB: total bilirubin. * After other causes have been ruled-out [15]. [¶] In cases of pre-existing abnormal biochemistry before the administration of the implicated drug, ULN is replaced by the mean baseline values obtained prior to drug exposure [15].

The risk of developing liver injury and possible hepatic dysfunction by an antifungal agent depends on several factors. The chemical properties of the agent, demographics, genetic predisposition, comorbidities including underlying hepatic disease, concomitant hepatotoxic drugs and DDIs, severity of the illness, and liver involvement by the fungal infection, all affect the possibility for hepatotoxicity [21]. Under these circumstances, it can be difficult to attribute DILI due to antifungals to only one factor.

In general, published literature regarding the use of antifungal agents in patients with pre-existing liver disease is somewhat inconclusive. A clear understanding of antifungal-caused liver injury in patients with underlying hepatic impairment is lacking, and recommendations for dosage adjustments in these cases are not straightforward [3,22]. Most of the information about antifungal dosing regimens is derived from clinical trials and PK studies, in which only few patients with a varying level of liver impairment were included [20]. For some antifungals, a dose reduction is recommended in the manufacturers' product characteristics in cases of pre-existing hepatic dysfunction, while for other antifungal agents no dosage adjustment is required or recommended [22].

The aim of the present review is to provide an overview of the safety profile of the various antifungal agents in patients with underlying liver disease. The intention is to summarize current data on the PKs of antifungals in these individuals and to increase clinical awareness of how various antifungal compounds should be used under these circumstances.

2. Antifungal Agents

The current antifungal armory for IFIs includes polyenes (amphotericin B-based preparations), flucytosine, triazoles (fluconazole, itraconazole, voriconazole, posaconazole, and isavuconazole), and echinocandins (caspofungin, micafungin, and anidulafungin) [23]. These compounds differ from each other in their spectrum of activity, pharmacokinetics/pharmacodynamics (PK/PD) properties, indications, dosing, safety profile, cost, and ease of use [3,24,25].

2.1. Polyenes

Amphotericin was introduced in therapy in 1958 as amphotericin B deoxycholate (AmBD), but its clinical usefulness is limited because of nephrotoxicity and infusion-related reactions [24,26]. Three lipid formulations of amphotericin B (AmB), liposomal amphotericin B (LAmB), amphotericin B lipid complex (ABLC), and amphotericin colloidal dispersion (ABCD; discontinued in most countries) were developed in the 1990s to reduce the toxicity observed with AmBD [24]. AmB interacts with ergosterol in the fungal membranes leading to the formation of membrane-spanning pores, ion leakage, and ultimately fungal cell death [27]. Additional cytotoxic mechanisms of AmB are inhibition of the fungal proton-ATPase and lipid peroxidation [28]. It is eliminated unchanged mainly via urine and feces [29]. Because of its broad antimycotic spectrum, AmB is a cornerstone in the treatment of serious and life-threatening fungal infections. The daily dose for AmBD ranges from 0.3 to 1.5 mg/kg, while the recommended standard doses for the lipid formulations of AmB are much higher [29,30]. Specifically, for LAmB the usual daily dose ranges from 3 to 5 mg/kg, but doses up to 10 mg/kg/d can be administered in cases of rhino-orbital-cerebral mucormycosis [29]. For ABLC the usual dose is 5mg/kg/d, while for ABCD the daily dose ranges from 3 to 4 mg/kg [30].

Generally, lipid-based formulations of AmB present at least the same efficacy as AmBD and are even superior in the treatment of certain fungal infections, such as mild to moderate disseminated histoplasmosis in patients with acquired immunodeficiency syndrome (AIDS), while they are associated with a safer profile [30–32]. Notably, in some studies, the administration of LAmB was associated with lower toxicity rates, namely infusional and kidney toxicity, compared to other lipid formulations [33–35]. However, differences in drug-induced nephrotoxicity between lipid-based formulations of AmB continue to be a subject of debate [36,37]. Other commonly encountered adverse effects of AmB preparations, apart from nephrotoxicity and infusion reactions, include hypokalemia, hypomagnesemia, and anemia [27,38]. Liver injury due to AmB therapy is relatively subtle and reversible, with its incidence reaching 32% for LAmB and 41% for ABLC in some clinical studies [21,39,40]. Interestingly, lipid formulations of AmB, mainly LAmB, seem to have a stronger association with DILI than AmBD, probably due to the carriers of these formulations [24,33,40,41]. In any case, clinically evident liver injury and treatment discontinuation due to AmB preparations are rare [21,27].

No specific recommendations are available for AmB preparations in the case of pre-existing hepatic impairment, but considering their limited hepatic metabolism, dosage adjustment is unlikely to be necessary [22]. Data on the PKs of AmB in pre-existing liver disease are sparse and clinical studies are lacking so far. In a retrospective single-center non-randomized autopsy-controlled study, Chamilos et al. compared hepatic enzymes elevations and histopathological findings in the livers of 64 patients with hematologic malignancies who had received LAmB or ABLC for at least 7 days, as a treatment for IFIs [42]. Among these patients, there were 22 patients with elevated liver enzymes at baseline, more than five times the ULN. None of the patients with acute liver injury, including those with abnormal baseline hepatic biochemical parameters, showed the histopathological changes induced by liposomal formulations of AmB that have been reported in animal studies [42]. Another study assessed the PK properties of ABCD in 11 patients with cholestatic liver disease compared to 9 subjects with normal liver enzymes [43]. Pre-existing cholestatic liver disease had no significant influence on steady state PKs of liberated AmB, and the authors concluded that the standard dosage of ABCD is probably appropriate for these patients [43].

2.2. Flucytosine

Flucytosine became available in 1968 [44]. It is taken up by fungal cells by cytosine permease and converted intracellularly into fluorouracil, which is further metabolized into 5-fluorouridine triphosphate and 5-fluorodeoxyuridine monophosphate, resulting in inhibition of fungal protein and DNA synthesis [45]. It is mainly eliminated by the kidneys, while it is minimally metabolized in the liver [46]. The high occurrence of resistance precludes its use as a single agent. Nowadays, flucytosine is used in combination therapy with AmB as first-line therapy in cryptococcal meningoencephalitis [47]. Furthermore, it may be added to other regimens for the treatment of severe pulmonary cryptococcosis, central nervous system candidiasis, *Candida* endocarditis, and *Candida* urinary tract infections [47–49]. Flucytosine's recommended dosage in individuals with normal renal function ranges from 50 to 150 mg/kg/d divided in four doses for both oral and intravenous formulation, while dosages up to 200 mg/kg/d can be administered [29,50].

Flucytosine's most significant adverse effects is myelotoxicity, mainly neutropenia and thrombocytopenia, and hepatotoxicity, and both are thought to be due to the effects of fluorouracil [46,51]. Because human intestinal flora is capable of converting flucytosine into fluorouracil in vitro, oral administration of the drug might be associated with more side effects than intravenous administration [51]. Liver injury is frequently encountered during treatment with flucytosine and the incidence varies from 0% to 41%, probably due to the different definition of liver injury in different studies [24]. The elevation in liver enzymes is usually mild to moderate and reversible on discontinuation, while two cases of severe liver necrosis have been reported in patients who received flucytosine for candidal endocarditis [46,52]. Both myelotoxicity and liver toxicity have been associated with high flucytosine concentrations in the blood. Therapeutic drug monitoring (TDM) is advisable 3–5 days after initiating therapy and after any changes in the glomerular filtration rate (GFR) to keep the 2 h flucytosine post-dose levels between 30 to 80 mg/L [53]. DDIs involving the cytochrome P450 (CYP450) pose a minor concern for flucytosine administration [29].

For patients with pre-existing hepatic impairment, limited data are available regarding the PK properties and the safety of flucytosine. In 1973, Block studied for the first time the effect of hepatic insufficiency on flucytosine concentrations in the serum of rabbits with chemically induced acute hepatitis [54]. No influence of the hepatic function on serum concentration of the drug was observed. In the same paper, a single patient with biopsy-proven cirrhosis was described as treated with flucytosine for cryptococcal meningitis. Drug concentrations in serum were measured at 1, 2, and 6 h after a dose and did not differ from concentrations determined simultaneously in 10 patients with cryptococcal infection and normal liver function being treated with the same dose of flucytosine [54]. However, given the fact that liver injury due to flucytosine treatment is a common adverse effect in many studies, this antifungal agent should be used with extreme caution or even be avoided in this patient population, although there are no dosage adjustments provided in the manufacturer's labeling [29,50]. Combined treatment with AmB may lead to the accumulation of flucytosine because of AmB-induced nephrotoxicity, further complicating the matter [55]. In addition, a recent study examining the hepatotoxicity induced by combined therapy of flucytosine and AmB in animal models showed a synergistic inflammatory activation in a dose-dependent manner, through the NF-κB pathway, which promoted an inflammatory cascade in the liver. The authors suggested that the combination of flucytosine and AmB for the treatment of IFIs in patients with hepatic dysfunction requires careful clinical, biochemical, and drug monitoring [56].

2.3. Azoles

The azole antifungals are synthetic compounds that can be divided into two subclasses, the imidazoles and the triazoles, according to the number of nitrogen atoms in the five-membered azole ring [29]. The imidazoles include ketoconazole, miconazole, and clotrimazole [21]. Miconazole was at one time administered intravenously for the treatment of certain IFIs, but soon this formulation was withdrawn due to toxicity associated with drug solvent [57]. Ketoconazole was frequently applied for systemic mycoses in the past, but it is now avoided due to its liver and hormonal toxicity [23]. The triazoles consist of fluconazole, itraconazole, voriconazole, posaconazole, and isavuconazole [29]. Azole antifungals inhibit the synthesis of ergosterol in the fungal cell membrane [29]. Despite this mechanism of action, azoles are generally fungistatic against yeasts, while the newer members of this subclass possess fungicidal activity against certain molds [23,48]. At present, these agents are considered the backbone of IFI therapy [23,49,58].

The most common adverse events (AEs) with all the triazoles, and especially with oral itraconazole, are nausea, vomiting, diarrhea, and abdominal pain [23,59]. Liver injury has been described also with all triazoles, ranging from mild elevations in transaminases to fatal hepatic failure [60–62]. Generally, in most cases of hepatic injury due to triazoles, normalization of the liver enzymes and resolution of the clinical symptoms occurred gradually after the discontinuation of the drug [21,63]. Additionally, triazoles are involved in numerous DDIs because they are substrates and inhibitors of CYP450 isoenzymes [63,64].

2.3.1. Fluconazole

Fluconazole, unlike the other triazoles, is characterized by high water solubility and approximately 60–80% of the drug is eliminated by the kidneys, while hepatic metabolism does not play an important role in the elimination of the drug [29]. The fluconazole dosage regimen for IFIs is guided by the indication, and the daily dose recommended by the manufacturer is up to 400 mg, but in clinical practice it usually ranges from 400 to 800 mg [49,65]. It is well tolerated, even in cases requiring long-term administration of the drug [21]. Nevertheless, up to 10% of patients treated with fluconazole developed asymptomatic liver injury, with those with AIDS or bone marrow transplantation being at greater risk [40,66–69]. Hepatic injury was typically transient and usually resolved despite drug continuation [21]. Cholestatic and mixed patterns of hepatic injury have been reported, and reinstitution of fluconazole resulted in recurrences in many cases [67,70–72]. Furthermore, there are some limited data to suggest that liver injury is dose-related [67,73]. In a large meta-analysis of antifungals tolerability and hepatotoxicity, the risk of liver injury with standard dose of fluconazole not requiring treatment discontinuation was 9.3%, while the risk of drug discontinuation due to elevated liver enzymes was 0.7% [74]. Despite the fact that the risk of acute liver failure due to fluconazole treatment is minimal [74,75], there are some case reports describing deaths attributable to liver dysfunction [66,76–78].

Few reports exist regarding the use of fluconazole in patients with pre-existing liver disease. Ruhnke et al. evaluated the PKs of a single 100 mg dose of fluconazole in 9 patients with cirrhosis, classified as group B or group C according to Child-Pugh score, compared with 10 healthy subjects [79]. They found that in cirrhotic patients the terminal elimination constant for fluconazole was lower, and that the total plasma clearance was reduced and the mean residence time increased. The authors assumed that this may be due to kidney dysfunction not reflected in creatinine clearance or the DDIs between fluconazole and diuretics that cirrhotic individuals were receiving. Nevertheless, the authors argued that dosage adjustment of fluconazole in patients with liver impairment is unnecessary, because of the wide range of values they found and the known low toxicity of fluconazole [79]. At the clinical level, Gearhart first described a 50-year old woman with hepatitis who received fluconazole for *Candida* infection and experienced worsening of liver function, which returned to baseline after discontinuation of the drug [80].

A population-based study by Lo Re et al. assessed the risk of acute liver injury with oral azole antifungals in the outpatient setting [81]. Liver aminotransferase levels and development of hepatic dysfunction were examined in 195,334 new initiators of these drugs, for a period of 182 days after the last day's supply. Fluconazole initiators were 178,879 and, among them, 7073 individuals had pre-existing liver disease. The authors found that the risk of transaminitis (liver aminotransferases > 200 U/L) and severe liver injury [international normalized ratio (INR) $\geq$ 1.5 and total bilirubin (TB) > 2× ULN] in patients without history of chronic liver disease was lower among users of fluconazole, compared to other azoles. Nevertheless, it should be taken into account that, with the exception of itraconazole, patients administered other azoles were probably of worse health status compared to those administered fluconazole. More interestingly, compared to patients without chronic liver disease who received fluconazole, patients with pre-existing liver disease who were treated with the same drug had higher absolute risk and incidence rate of transaminitis (p value interaction < 0.001) and of severe liver injury (p value interaction < 0.001) [81]. Whether this observation was due to fluconazole, the natural history of the disease, or both, is unclear [81]. However, no dosage adjustment is provided by the manufacturer for patients with liver impairment, although prescribing information includes a warning that fluconazole should be administered with caution to patients with hepatic dysfunction [65].

2.3.2. Itraconazole

Itraconazole is highly lipophilic, undergoes extensive hepatic metabolism, and is eliminated mostly via feces and urine [29]. It is available as capsule, oral solution, and intravenous formulation [82]. The oral solution has higher bioavailability than capsule formulation, and thus they should not be used interchangeably [83]. The adults recommended by the manufacturer dosage depends on the drug formulation and the indication, usually ranging from 200 mg to 400 mg per day, and doses above 200 mg should be divided [82,83]. However, for the treatment of certain fungal infections, such as blastomycosis and histoplasmosis, doses of 200 mg t.i.d. for 3 days and then 200 mg q.d. or b.i.d. as long-term therapy are recommended, while for coccidioidal meningitis doses up to 800 mg per day can be administrated [84–86]. Itraconazole-induced liver injury is not uncommon, and the pattern is typically cholestatic, although hepatocellular injury has been described in cases of acute liver failure [21]. In a large meta-analysis, 31.5% of patients treated with itraconazole developed hepatotoxicity, but a great variability of hepatotoxicity definition was noted in the included studies and many patients may have developed liver injury owing to the underlying IFI itself, limiting the validity of these results [87]. Treatment discontinuation due to itraconazole-induced liver injury was observed in 1.6% of patients [87]. In a more recent meta-analysis, Wang et al. estimated the risk of elevation of liver enzymes not requiring discontinuation of therapy at 17.4% among itraconazole recipients, while the respective risk of treatment discontinuation due to liver injury was 1.5% [74].

The use of itraconazole in patients with liver disease is not well studied. In a PK study, a single 100 mg dose of itraconazole was administered in 12 cirrhotic and 6 healthy individuals [88]. Compared with healthy volunteers, a statistically significant reduction in C_{max} and an increase in the elimination half-time of the drug were observed in patients with cirrhosis. Nevertheless, based on the area under the curve (AUC), cirrhotic and healthy individuals had comparable overall exposure to the drug [88]. In the already mentioned observational study of Lo Re et al., 55 patients with chronic liver disease received itraconazole, and onychomycosis was the most common indication for treatment initiation [81]. Interestingly, none of them developed transaminitis or severe acute liver injury [81]. The fact that, in this study, itraconazole was prescribed mainly for a less severe condition such as onychomycosis and probably in lower doses than those recommended for severe IFIs treatment, may be the reasons for its decreased hepatotoxic potential, compared with what has been observed in other studies which included patients with severe fungal infections and multiple comorbidities. No dose adjustment is available for patients with hepatic impairment, but it is recommended that these patients should be carefully monitored when treated with itraconazole [83]. Apart from the periodic assessment of a patient's liver enzymes levels while on itraconazole, TDM is generally recommended, in order to assure adequate exposure and to minimize potential toxicities [55,58,82,89,90].

2.3.3. Voriconazole

Voriconazole's chemical structure is similar to fluconazole, but its spectrum of activity is much broader [48]. It is metabolized by CYP450, mainly CYP2C19, which exhibit significant genetic polymorphism, and it is involved in many DDIs. In addition, recent data suggest that voriconazole metabolism can be inhibited in cases of severe inflammation [91]. It is available as tablet, oral suspension, and intravenous solution [92]. The manufacturer's recommended dose of intravenous formulation for most IFIs is 6 mg/kg b.i.d. on day 1 as a loading dose, followed by 4 mg/kg b.i.d. as a maintenance dose [92,93]. The oral dose for adult patients is 400 mg b.i.d. on the first day followed by 200 mg b.i.d., while if patient response is inadequate, the maintenance dose may be increased from 200 mg b.i.d. to 300 mg b.i.d. [92,93]. A 50% reduction of both loading and maintenance oral doses is recommended for adult patients with a body weight less than 40 kg [92,93]. The incidence of liver injury in patients treated with voriconazole varies significantly among studies, depending mostly on the characteristics of the study population, while the pattern of liver enzyme abnormality is not uniform [94–97]. Wang et al. found in their meta-analysis that 19.7% of 881 patients who received voriconazole developed elevation of liver enzymes without the need for treatment discontinuation [74]. A more recent meta-analysis of the utility of voriconazole's TDM included 11 studies and reported a pooled incidence rate of liver injury among voriconazole recipients at 5.7% [98].

Compared with other triazoles, more data exist regarding the use of voriconazole in patients with underlying hepatic impairment. After a single oral dose of 200 mg of voriconazole in 12 patients with mild to moderate hepatic impairment (Child–Pugh Classes A and B), AUC was 3.2-fold higher than in age and weight matched controls with normal liver function [92]. In an oral multiple-dose PK study, AUC at steady state (AUCτ) was similar in individuals with Child–Pugh Class B cirrhosis given a maintenance dose of 100 mg twice daily and individuals with normal liver function given 200 mg twice daily [99]. Based on the aforementioned data, the medication label of voriconazole recommends that individuals with mild to moderate cirrhosis (Child–Pugh Class A and B) receive the same loading dose as individuals with hepatic function, but half the maintenance dose, while no recommendation is given for individuals with Child–Pugh Class C cirrhosis [92].

In a cohort study of 29 patients with severe liver dysfunction, defined as MELD score > 9, who received at least four doses of voriconazole, a deterioration of hepatic biochemistry was observed in 69% of them [100]. The pattern of the liver injury was mixed; hepatocellular and cholestatic in 45%, 35% and 15% of patients, respectively. None of them developed clinical or laboratory signs of worsening hepatic function. The biochemical parameters returned to baseline levels in all patients after the cessation of voriconazole treatment [100]. Lo Re et al included in their study 97 patients with pre-existing liver disease who received oral voriconazole. Among them, 4 developed ALT or AST > 200 U/L and 2 developed severe liver injury (INR > 1.5 and TB > 2 × ULN), but none of them experienced acute liver failure. Individuals with pre-existing liver disease treated with voriconazole had higher rates of severe liver injury than recipients of voriconazole without underlying hepatic disease [81]. A recent single-center retrospective study compared 6 patients with severe liver cirrhosis (Child–Pugh Class C) who were treated with oral voriconazole based on TDM, with 56 individuals without severe liver cirrhosis who received voriconazole in the recommended dosage for IFIs, also under TDM [101]. The daily maintenance doses of voriconazole of the severe cirrhotic patients were in the range of 50 to 200 mg, with a median daily dose at one-third of the median daily dose of the individuals without severe cirrhosis. The median trough serum concentration of the drug was within recommended levels in both groups of patients. Thus, the authors argued that a dose reduction to about one-third that of the standard maintenance dose is required in patients with Child–Pugh Class C cirrhosis [101].

A multicenter retrospective study aimed to investigate the voriconazole trough concentrations and safety in cirrhotic patients receiving the drug [102]. Seventy-eight patients with Child–Pugh Class B or C cirrhosis who had been treated with voriconazole under TDM were allocated to two groups, according to the dosage regimen they had received. Patients in the first group had received the recommended dosage by the manufacturer or a fixed dose of 200 mg twice daily. Patients in the second group had received a loading dose of 200 mg twice daily on day 1, followed by 100 mg twice daily, or a fixed dose of 100 mg twice daily. The steady-state trough concentration of voriconazole was measured in all patients and its relationship with AEs was analyzed. Voriconazole C_{min} values were significantly different between the two groups, and the proportion of C_{min} higher than the super-therapeutic concentration (defined as 5 mg/L) was 63% in the first group and 28% in the second group of patients. While no statistically significant differences were observed in the incidence of AEs between the two groups, these incidences were considered excessively high (26.5% of patients in the first group and 15.9% of patients in the second group). Interestingly, voriconazole C_{min} between patients with an AE and those without AEs in both groups was similar. However, based on the high C_{min} and incidence of AEs in these patients, both the recommended maintenance dose and halved maintenance dose were considered as inappropriately high [102].

The same authors conducted another study including solely patients with Child–Pugh Class C cirrhosis [103]. Patients were allocated to two groups, according to the dosage schedule of voriconazole's maintenance dose. The first group included those who received 100 mg of voriconazole twice daily, while the second group included those who received 200 mg of voriconazole once daily. There was no significant difference in voriconazole C_{min} between the two groups. However, the proportion of voriconazole C_{min} higher than the upper limit of therapeutic level (defined again as 5 mg/L) in the first and second groups was 34% and 48%, respectively. The incidence of AEs was 21% in the first group and 27% in the second group, with no statistically significant difference. Further analysis revealed that the increasing C_{min} of voriconazole was associated with increasing incidence of AEs, although no statistical significance was found. It was suggested that in patients with Child–Pugh Class C cirrhosis the halved maintenance dose is probably inappropriate, and that lower dosage should be considered in conjunction with early TDM [103].

Voriconazole TDM is generally recommended because of its highly variable PKs, in order to enhance efficacy, to evaluate therapeutic failure due to possible suboptimal drug exposure, and to avoid associated toxicity due to increased serum drug levels [55,58,104]. It is well established in the literature that an elevated drug's level in the serum is correlated with increased risk of toxicity [104–106]. Thus, voriconazole TDM is of paramount importance in patients with pre-existing liver disease, since the drug is extensively metabolized by the liver and this population is more difficult to tolerate a deterioration of hepatic function due to voriconazole-induced liver injury [101–103,107]. Various target trough concentrations associated with efficacy and safety have been reported, and most experts aim for voriconazole trough serum concentration of more than 1–1.5 µg/mL for efficacy but less than 5–6 µg/mL for avoiding toxicity [58,89,98,104].

2.3.4. Posaconazole

Posaconazole's chemical structure resembles that of itraconazole, but it has a wider antimycotic spectrum [29]. Initially, posaconazole was available only as an oral suspension which displays poor and highly variable absorption [108]. Recently, tablet and intravenous formulations with improved bioavailability were approved [109–111]. Posaconazole is metabolized in the liver by UDP-glucuronic-transferase, usually without previous oxidation by CYP450, and is eliminated mainly in the feces and, secondarily, in the urine [112]. Noticeably, posaconazole is a potent inhibitor of CYP3A4, thus clinically relevant DDIs may occur [29]. Regarding IFIs, the adult recommended therapeutic dose for oral suspension is 200 mg q.i.d., while the prophylactic dose is 200 mg t.i.d. [113,114]. In addition, for both tablet and intravenous formulation a loading dose of 300 mg b.i.d. on day 1, followed by a maintenance dose of 300 mg once daily, is recommended as prophylactic as well as therapeutic dosage regimen for several IFIs [113,114]. Liver injury occurs in up to 25% of patients receiving posaconazole regardless of the formulation, but this may be multifactorial and not only attributable to the drug [81,115–118]. The dominant pattern of hepatic injury varies among studies, partly depending on the studied population [21,115–117]. In addition, hepatic failure due to posaconazole treatment is generally uncommon [81,110,111,115–117].

Regarding the use of posaconazole in individuals with pre-existing hepatic impairment, Moton et al. conducted a PK study to evaluate the need of posaconazole dose adjustment in this population [119]. In their single-center study, the researchers aimed to compare the PKs of a single dose 400 mg of posaconazole oral suspension in 19 patients with varying degrees of hepatic dysfunction with 18 matched healthy individuals who received the same regimen. No clear trend was observed of an increase or decrease in posaconazole exposure linked with increasing degrees of hepatic dysfunction. The detected differences of PKs between healthy individuals and those with hepatic dysfunction were not clinically significant, and the authors suggested that posaconazole dosage adjustment may not be required in individuals with hepatic impairment [119]. A case-report also described a patient with Child–Pugh Class B cirrhosis suffering from maxillary mucormycosis who, after surgical debridement and initial treatment with AmB followed by itraconazole, was successfully treated with oral posaconazole suspension 400 mg twice daily for nine months without hepatic decompensation [120]. In addition, Lo Re et al included in their observational study 9 patients with chronic liver disease who received posaconazole, and only one of them developed severe acute liver injury (INR > 1.5 and TB > 2× ULN) [81].

In a recent single-center retrospective cohort study, Tverdek et al. assessed the real-life safety and effectiveness of primary antifungal prophylaxis with new tablet and intravenous posaconazole formulations in high-risk patients with leukemia and/or hematopoietic stem cell transplantation (HSCT) [116]. A total of 343 patients were included, 62% of whom received 300 mg of posaconazole twice daily on day 1, while 99% received the maintenance dose of 300 mg per day. Among them, 316 patients had baseline liver assessment, including 144 patients with baseline elevations of ALT, ALP, or/and TB, of which 23 had grade 3 or 4 liver injury [121]. Concerning the 121 patients with baseline liver injury but no grade 3 or 4 abnormalities, 34 (28%) of them developed grade 3 or 4 liver injury. Liver abnormalities were developed in nearly 20% of all patients, primarily manifested as hyperbilirubinemia. These abnormalities were more frequent in individuals with pre-existing liver injury, but this may not be solely due to DILI, as the underlying disease and concomitant drugs may also have contributed [116].

Noticeably, in patients with new-onset hepatotoxicity due to voriconazole administration for IFIs, sequential use of posaconazole seems to be safe and effective, with favorable outcomes and improvement of liver biochemistry in most of the cases [122–124]. Independently of the acute or chronic nature of pre-existing liver injury, no dosage adjustments are recommended for individuals with hepatic impairment treated with posaconazole [113]. In addition, while many guidelines recommend TDM in patients receiving posaconazole oral suspension for IFI prophylaxis or treatment to confirm adequate absorption and ensure efficacy [58,89], PK/PD analyses conducted with oral posaconazole suspension do not support a relationship between plasma concentrations and toxicity [125,126]. On the contrary, Tverdek et al identified a potential association between elevated serum posaconazole levels and hepatotoxicity in patients treated with the new tablet and intravenous formulations of the drug, but further evaluation is needed [116].

2.3.5. Isavuconazole

Isavuconazole is the newest member of triazoles antifungals. In both oral and intravenous formulations, it is administered as a water-soluble prodrug, isavuconazonium sulfate [127]. After intravenous administration, the prodrug is rapidly hydrolyzed to isavuconazole by plasma esterases, while oral formulation of isavuconazonium sulfate sustains chemical hydrolysis in the gastrointestinal lumen [112]. Metabolism of isavuconazole takes place in the liver by CYP450 isoenzymes, with subsequent glucuronidation by uridine diphosphate-glucuronosyl transferase (UGT) [127]. Isavuconazole is generally well tolerated and safe, and has fewer DDIs compared with voriconazole and posaconazole, but clinical experience is still limited [60,61]. It is approved by the FDA and the European Medicine Agency (EMA) for the treatment of adult patients with invasive aspergillosis or invasive mucormycosis, with a loading dose of 200 mg t.i.d. for the first two days, followed by a maintenance dose of 200 mg q.d., via oral or intravenous administration [127,128]. Elevations in liver enzymes have been reported in clinical trials but they are generally reversible and rarely only require treatment discontinuation [129–131]. However, cases of severe liver injury have occurred during treatment with this antifungal agent [127,129]. In a phase 3 comparative study evaluating isavuconazole versus voriconazole for the treatment of invasive aspergillosis, there were significantly higher liver disorders in the voriconazole arm (p value = 0.016), but the protocol of the study did not allow TDM [131]. Since voriconazole displays highly variable non-linear pharmacokinetics in adults and, thus, TDM is recommended, these results should be interpreted with caution, and further research is needed.

An initial single-dose PK study aimed to assess the effect of mild to moderate hepatic impairment due to alcoholic cirrhosis on the disposition of isavuconazole [132]. Clearance values of isavuconazole were significantly decreased and half-life values were significantly increased in cirrhotic patients compared with healthy individuals, leading the authors to recommend a 50% decrease in the maintenance dose of the drug for patients with mild or moderate liver disease [132]. However, a subsequent population PK analysis used data from the aforementioned study and from another study and reported different results [133]. The PK and safety results showed that dose adjustment appears to be unnecessary for patients with Child–Pugh Class A or Class B cirrhosis treated with isavuconazole, since there was a less than twofold increase in trough concentrations for those compared with healthy subjects, while the AEs profile was similar between cirrhotic and healthy individuals [133].

Notwithstanding, both these PK studies did not take PD into consideration, which may affect the dose of isavuconazole against different fungi in this population of patients. In a recently published PK/PD study, Zheng et al. examined the efficacy of various isavuconazole dosing regimens for healthy individuals and patients with renal and hepatic impairment, namely Child-Pugh Class A or B cirrhosis, against *Aspergillus* spp. and other fungi [134]. The Monte Carlo simulation was used in each scenario to calculate target attainment and cumulative fractions of response probabilities. The clinically recommended dose of 200 mg isavuconazole per day was effective for all individuals against *A. fumigatus*, *A. flavus*, *A. nidulans*, *A. terreus*, and *A. versicolor*. [134].

In the manufacturer's labeling, the standard dose of isavuconazole is recommended for patients with mild or moderate liver dysfunction, while the drug has not been studied in patients with Child–Pugh Class C hepatic impairment, and should be used in these individuals only when the benefits outweigh the risks [127]. Although TDM of isavuconazole may be considered in selected patients, such as those with severe hepatic impairment, routine TDM for isavuconazole is not recommended [135].

2.4. Echinocandins

Echinocandins inhibit the synthesis of 1,3-β-D-glucan, a fungal cell wall component, resulting in instability of the cell wall, cell lysis, and death [136]. The fact that this class of antifungals agents targets the fungal cell wall and not the cell membrane explains the absence of cross-reactivity with mammalian cells and the excellent tolerability of this class of compounds in humans [48]. They are fungicidal to *Candida*, including several non-albicans strains, and fungistatic to *Aspergilli*, thus they are considered the first-line treatment for *Candida* spp. infections [29,49]. At present, the available agents of this class include caspofungin, micafungin, and anidulafungin [23]. Common AEs related with echinocandins treatment include phlebitis, nausea, diarrhea, headache and pruritus, but also other drug reactions such as leukopenia, anemia, hypokalemia, and liver injury have been reported [29]. Noticeably, the echinocandins have less than half the likelihood of discontinuation of therapy due to AEs, compared with triazoles [137].

2.4.1. Caspofungin

Caspofungin bounds to plasma proteins at 95%; it is transformed in the liver but only minimally undergoes degradation by CYP450 isoenzymes, and the metabolites are eliminated via urine [138,139]. The recommended dosage for adults is 70 mg as a single loading dose on day 1, followed by a maintenance dose of 50 mg once daily [140,141]. The EMA recommends an increase of maintenance dose to 70 mg daily when patient's body weight exceeds 80 kg [140]. Generally, hepatic abnormalities related to caspofungin treatment are uncommon and severe hepatic AEs are rare [21]. In most studies, elevated hepatic enzymes were observed in up to 9% of patients, and they were often clinically irrelevant [24].

Regarding patients with pre-existing liver disease treated with caspofungin, Mistry et al. conducted single- and multiple-dose open-label studies to assess dosage and safety of caspofungin in hepatic impairment [142]. Patients with Child–Pugh score 5–6 or 7–9 hepatic impairment were matched with healthy individuals. Patients with Child–Pugh score 5–6 hepatic impairment had a mild elevation in caspofungin serum concentration, which was considered as clinically irrelevant. Patients with Child–Pugh score 7–9 hepatic impairment needed a reduced maintenance dose of caspofungin in order to achieve drug concentrations similar with the healthy individuals in the control group [142]. Based mainly on these data, a reduction of caspofungin maintenance dose from 50 mg to 35 mg per day is recommended for patients with Child–Pugh Class 7–9 hepatic impairment, while no recommendation is given for patients with Child–Pugh score 10–15 hepatic impairment [141].

However, Spriet et al. initially described a patient with Child-Pugh Score 9 cirrhosis diagnosed with acute myeloid leukemia, who was treated for a severe IFI with a full dose of caspofungin 70 mg per day, since his body weight was over 80 kg [143]. The PK data of this case-report indicated that if the reduced dose of caspofungin had been used, it would probably have resulted in a low caspofungin systemic exposure and a possible therapeutic failure [143]. A subsequent population PK analysis concluded that a reduction of caspofungin maintenance dose in non-cirrhotic intensive-care unit (ICU) patients, who are misclassified due to hypoalbuminemia as with Child–Pugh Class B hepatic impairment, is not recommended, because it may result in significantly lower drug exposure and possible therapeutic failure [144]. On the contrary, authors suggested that, depending on pathogens MIC, a caspofungin maintenance dose of 70–100 mg daily may be reasonable in many cases [144].

Furthermore, data from the aforementioned population PK analysis in non-cirrhotic ICU patients were used in another PK study of a single-dose of 70 mg of caspofungin in patients with decompensated Child–Pugh Class B or C cirrhosis to evaluate the impact of cirrhosis and hepatic impairment severity on the PK of the drug [145]. Remarkably, their data showed that cirrhosis had a limited impact on clearance of caspofungin. Also, it was the first study providing PK data of caspofungin for patients with Child–Pugh Class C cirrhosis and compared with patients with Child–Pugh Class B cirrhosis, no further decrease of caspofungin clearance was observed in the former group of individuals. Thus, the researchers concluded that reducing the dose of caspofungin in patients with Child–Pugh Class B or C cirrhosis leads to a decrease in exposure and this may result in a suboptimal clinical outcome [145]. In another recent PK study for general patients, ICU patients, and patients with hepatic impairment receiving caspofungin, a whole-body physiology-based PK model was developed and was combined with Monte Carlo stimulation to optimize dosage regimen of the drug in patients with different characteristics [146]. The results of this study indicated that the caspofungin maintenance dose should not be reduced to 35 mg per day for ICU patients classified as Child–Pugh Class B when this classification is driven by hypoalbuminemia, as lower drug exposure occurs. On the contrary, authors argued that, in any other case, a reduction of caspofungin maintenance dose to 35 mg per day for patients with moderate hepatic impairment classified as Child–Pugh Class B, may be reasonable [146].

2.4.2. Micafungin

Micafungin is highly bound to proteins, it is metabolized in the liver by enzymes unrelated to CYP450, and the metabolites are excreted primarily via feces [147]. The recommended dosage for patients weighing greater than 40 kg is 100 once daily for the treatment of invasive candidiasis, and 150 mg once daily for the treatment of *Candida* esophagitis [148,149]. It is a well-tolerated antifungal agent with few AEs requiring cessation of the drug [21]. Mild elevations of hepatic enzymes may occur, but clinically overt liver toxicity is rare [23,150]. Nevertheless, rat models demonstrated an association between micafungin and foci of altered hepatocytes and hepatocellular tumors when this was given for more than 3 months, but this finding has not been replicated in humans [23,29].

Micafungin has a low hepatic extraction ratio with high protein binding in plasma, and while its total plasma concentration may decrease in some clinical cases, the unbound fraction of the drug is likely to remain stable [151,152]. A phase I parallel group open-label PK study of a single-dose of micafungin included 8 patients with Child–Pugh Score 7–9 hepatic dysfunction and did not find significant difference in unbound plasma concentration of the drug compared with healthy controls, while a lower AUC was found in the patients with hepatic impairment [153]. The latter was attributed to the differences in body weight among patients, and no dose adjustment was recommended [153]. In an another open-label single-dose PK study, 8 patients with Child–Pugh score 10-12 hepatic impairment and 8 healthy individuals received 100 mg of micafungin [154]. Compared with healthy subjects, patients with hepatic dysfunction had lower C_{max} and AUC values, but the magnitude of differences was considered as clinically meaningless and no dose reduction was recommended in patients with severe hepatic impairment [154]. In addition, Luque et al. conducted a prospective observational study to assess the possibility of DILI due to micafungin use in daily practice including 12 patients, 8 of whom had elevated liver enzymes at the beginning of the treatment [155]. The daily dose of micafungin was 100 mg for 10 patients and 150 mg for the remaining two. There was no correlation between the degree of the pre-existing liver injury and micafungin levels. In steady state, C_{max} and C_{min} were similar in subjects with and without initial liver abnormalities. Hepatic enzymes levels remained stable or even improved in all but one patient. These results further support the safety of micafungin in patients with pre-existing liver injury and IFIs [155]. Based on most of the aforementioned studies, the summary of manufacturers' product characteristics approved by the FDA recommends that no dosage adjustment is required in patients with hepatic impairment [149]. Contrarily, EMA recommends avoidance of micafungin use in patients with severe hepatic impairment, while it has issued a black-box warning for hepatotoxicity and potential for liver tumors [148].

2.4.3. Anidulafungin

Anidulafungin has a very high protein binding of 99%; it is degraded non-hepatically in the blood, and the metabolites are eliminated via feces [156]. The recommended adult dosage for invasive candidiasis is a single loading dose of 200 mg on day 1, followed by a maintenance dose of 100 mg once daily [157,158]. Anidulafungin AEs, including DILI, are generally infrequent [159,160]. With regard to patients with pre-existing hepatic disease treated with this antifungal agent, Dowel et al. conducted a phase I, open-label, single-dose study including 20 patients with varying degrees of hepatic impairment and 7 healthy controls [161]. No statistically significant differences in PK parameters were observed between healthy controls and patients with mild or moderate hepatic impairment. However, compared with healthy controls, subjects with severe hepatic impairment (Child–Pugh Class C) showed statistically significant decreases in C_{max} and AUC values, most likely secondary to ascites and edema, but anidulafungin exposure remained significantly above MIC_{90} of many common fungal pathogens. Additionally, the values of all PK parameters still remained within the range that had been previously reported in healthy subjects. No evidence of dose-depended toxicity or serious AEs was observed. Thus, the authors suggested that anidulafungin can be safely used in patients with hepatic dysfunction without dosage adjustment [161].

In a retrospective cohort study, Verma et al. assessed the safety and efficacy of anidulafungin in the treatment of IFIs in patients with hepatic impairment or multiorgan failure [162]. Fifty patients were included, among them 30 with a calculated baseline MELD score, of whom 13 had a score $\geq$ 30. A dose of 200 mg was given to all patients on day 1, followed by 100 mg per day onwards. Before initiation of treatment with anidulafungin, at least one abnormal liver function test (LFT) was observed in 49 of 50 patients (98%). During treatment, LFTs worsened in many patients, but fewer patients had elevated LFTs at the completion of treatment than at the beginning. A favorable outcome was seen in more than 75% of patients. The latter further supports indications that anidulafungin is efficacious and safe in patients with decompensated hepatic disease and, in agreement with package insert recommendations, no dosage reduction is needed in patients with any degree of hepatic impairment [157,162].

3. Clinical Implications and Future Directions

Patients treated with antifungal agents for IFIs may have underlying hepatic impairment of varying degrees and origin. Clinicians should be aware of that, since it further complicates management with regard to efficacy and safety of the antifungal therapy. Firstly, metabolism and elimination of many antifungals are significantly altered by hepatic dysfunction, while DDIs are somewhat unpredictable compared to individuals with intact liver function. Moreover, it may be difficult to attribute further deterioration of liver biochemistry or function only to antifungals in patients with severe comorbidities and concomitant administration of other hepatotoxic drugs. In addition, precise estimates of hepatic function are currently unavailable. The Child–Pugh system, on which most dosage modifications in hepatic impairment are based, was initially developed to assess the prognosis of chronic liver disease and not the degree of hepatic dysfunction [20]. For all the above reasons, the optimal use of antifungals in patients with pre-existing liver disease with IFIs is still unfolding. Data discussed in the present review give rise to useful clinical suggestions for the optimization of treatment. Table 2 summarizes the dosage adjustments of antifungal agents that are approved and recommended by FDA and/or EMA for patients with hepatic impairment treated for IFIs, and also presents the recommendations included in many guidelines regarding TDM of certain antifungal drugs for optimizing efficacy and safety.

Table 2. Antifungal agent dosage adjustment for patients with hepatic impairment.

Antifungal Agent	Severity of Hepatic Impairment by Child–Pugh Score		
	Score 5–6 (Class A)	Score 7–9 (Class B)	Score 10–15 (Class C)
AmB preparations	No recommendations available		
Flucytosine	No recommendations available, use with caution, TDM recommended Authors' comment: extra caution when combined with AmB preparations		
Fluconazole	No recommendations available, use with caution		
Itraconazole	No recommendations available, strongly discouraged unless benefit exceeds risk, use with caution and under close monitoring, TDM is recommended		
Voriconazole	50% reduction of maintenance dosage and TDM are recommended		No recommendations available, use only if benefit outweighs risk, close monitoring and TDM are recommended Authors' comment: reduction of maintenance dosage to about one-third may be considered
Posaconazole	No dosage adjustment is recommended, TDM when oral suspension is used Authors' comment: TDM may also be considered when tablet or intravenous drug formulation is used		
Isavuconazole	No dosage adjustment is recommended		No recommendations available, use only if benefit outweighs risk
Caspofungin	No dosage adjustment is recommended	Reduced maintenance dose from 50 mg to 35 mg daily Authors' comment: in critically ill patients, reduced dosage may lead to decreased drug exposure	No recommendations available
Micafungin	No dosage adjustment is recommended		US FDA recommends no dosage adjustment, EMA recommends avoidance of its use
Anidulafungin	No dosage adjustment is recommended		

AmB: amphotericin B; TDM: therapeutic drug monitoring; US FDA: United States Food and Drug Administration; EMA: European Medicines Agency.

With regard to AmB, to date few data exist on the necessity for dosage adjustment of any AmB formulations in patients with hepatic impairment. However, the lipid formulations of the drug seem to have a higher potential for hepatotoxicity compared to AmBD. In addition, AmB formulations combined with flucytosine for the treatment of certain fungal infections may lead to increased flucytosine serum levels due to kidney injury and accumulation of the renally eliminated drug. Flucytosine TDM is of clinical importance generally, in order to assure efficacy and to prevent AEs, including hepatotoxicity.

Fluconazole dosage modification for hepatic impairment per se is not required. Nevertheless, it should be used cautiously in this subset of patients due to the increased risk of further deterioration of hepatic enzymes levels and/or hepatic function compared to subjects with normal liver function. For itraconazole there are no dosage adjustment recommendations available for patients with hepatic dysfunction, however its use is discouraged in this subset of patients unless benefit exceeds risk. In the latter case, close monitoring, including TDM, is recommended, but further work is necessary for establishing clear drug target levels.

Use of voriconazole has also an increased risk for severe live injury in patients with chronic liver disease. While reduction of voriconazole's maintenance dose by 50% is recommended in patients with Child–Pugh Class A or B cirrhosis, data for patients with more severe hepatic impairment were lacking until recently. New evidence suggests that dose should be lowered more than 50% in patients with Child–Pugh Class C hepatic dysfunction, and always under TDM for safety and efficacy enhancement [101–103,163]. However, optimal dosage in this setting has not formally been defined and this is a noteworthy area of active research. Likewise, posaconazole and isavuconazole have not been studied sufficiently in patients with severe hepatic impairment and more research on that topic is of paramount importance. Furthermore, only recently a possible relationship between increased posaconazole serum levels and liver toxicity was identified in patients receiving the new intravenous and tablet drug formulations, thus more PK studies are needed, especially in patients with underlying liver disease [116]. Regarding isavuconazole, generally it demonstrates a favorable safety profile in relation to DDIs and hepatotoxicity. Nevertheless, compared with other triazoles, published clinical experience and post-marketing data, including its use in special patient populations, are still limited.

Compared with triazoles, echinocandin use in patients with underlying hepatic impairment is considered relatively safe. A reduction to caspofungin maintenance dose is recommended for patients classified with Child–Pugh Score 7–9 hepatic dysfunction, yet this has been challenged recently and clinicians should be aware of that, since it may result in suboptimal exposure in critically ill patients [144–146]. With regard to micafungin, no dosage modification is recommended in mild and moderate hepatic insufficiency, but additional research seems necessary for patients with severe hepatic impairment. Among this class of antifungal agents, anidulafungin may have an advantage for use in cirrhotic patients due to its non-hepatic metabolism, more predictable PK, and favorable tolerability. However, this remains to be further evaluated with future comparative studies in this subset of patients.

4. Conclusions

Treatment of IFIs in patients with pre-existing liver disease poses a significant challenge for clinicians. These patients are often more vulnerable to the hepatotoxic potential of many antifungal agents, while possible alterations of the PKs of these drugs may trigger adverse effects not localized only to the liver. Current evidence from PK studies and safety data from the existing clinical trials and post-marketing studies can help physicians optimize IFIs treatment in this special group of patients. However, most of the existing evidence is limited to subjects with mild to moderate hepatic disease, and clear recommendations for dosage adjustments in cases of severe hepatic impairment are not yet available for the majority of antifungal agents. This raises the need for more PK and clinical studies in this subset of patients. Furthermore, additional attention should be paid to future pharmacovigilance monitoring of antifungal agent use in patients with liver disease of any degree. In any case, close clinical and laboratory monitoring, including TDM for specific antifungal drugs, is essential in the majority of these patients in order to prevent or promptly recognize further deterioration of the hepatic function, thus avoiding unfavorable outcomes.

Funding: This research received no external funding.

Conflicts of Interest: The authors declare no conflict of interest.

References

1. Limper, A.H.; Adenis, A.; Le, T.; Harrison, T.S. Fungal infections in HIV/AIDS. *Lancet Infect. Dis.* **2017**, *17*, e334–e343. [CrossRef]
2. Colombo, A.L.; de Almeida Júnior, J.N.; Slavin, M.A.; Chen, S.C.A.; Sorrell, T.C. Candida and invasive mould diseases in non-neutropenic critically ill patients and patients with haematological cancer. *Lancet Infect. Dis.* **2017**, *17*, e344–e356. [CrossRef]

3.   Kontoyiannis, D.P. Invasive mycoses: Strategies for effective management. *Am. J. Med.* **2012**, *125*, S25–S38. [CrossRef] [PubMed]

4.   Rodighiero, V. Effects of liver disease on pharmacokinetics. An update. *Clin. Pharmacokinet.* **1999**, *37*, 399–431. [CrossRef] [PubMed]

5.   Gupta, N.K.; Lewis, J.H. Review article: The use of potentially hepatotoxic drugs in patients with liver disease. *Aliment. Pharmacol. Ther.* **2008**, *28*, 1021–1041. [CrossRef] [PubMed]

6.   Lewis, J.H.; Stine, J.G. Review article: Prescribing medications in patients with cirrhosis—A practical guide. *Aliment. Pharmacol. Ther.* **2013**, *37*, 1132–1156. [CrossRef] [PubMed]

7.   Palatini, P.; De Martin, S. Pharmacokinetic drug interactions in liver disease: An update. *World J. Gastroenterol.* **2016**, *22*, 1260–1278. [CrossRef]

8.   Navarro, V.J.; Senior, J.R. Drug-related hepatotoxicity. *New Eng. J. Med.* **2006**, *354*, 731–739. [CrossRef]

9.   Lee, W.M. Drug-induced hepatotoxicity. *New Eng. J. Med.* **2003**, *349*, 474–485. [CrossRef]

10.   Lo Re, V.; Haynes, K.; Goldberg, D.; Forde, K.A.; Carbonari, D.M.; Leidl, K.B.F.; Hennessy, S.; Reddy, K.R.; Pawloski, P.A.; Daniel, G.W.; et al. Validity of diagnostic codes to identify cases of severe acute liver injury in the U.S. Food and Drug Administration's Mini-Sentinel Distributed Database. *Pharmacoepidemiol. Drug Saf.* **2013**, *22*, 861–872. [CrossRef]

11.   Bernal, W.; Wendon, J. Acute Liver Failure. *New Eng. J. Med.* **2013**, *369*, 2525–2534. [CrossRef] [PubMed]

12.   Ortega-Alonso, A.; Stephens, C.; Lucena, M.I.; Andrade, R.J. Case Characterization, Clinical Features and Risk Factors in Drug-Induced Liver Injury. *Int. J. Mol. Sci.* **2016**, *17*, 714. [CrossRef] [PubMed]

13.   Kullak-Ublick, G.A.; Andrade, R.J.; Merz, M.; End, P.; Benesic, A.; Gerbes, A.L.; Aithal, G.P. Drug-induced liver injury: Recent advances in diagnosis and risk assessment. *Gut* **2017**, *66*, 1154–1164. [CrossRef] [PubMed]

14.   Alempijevic, T.; Zec, S.; Milosavljevic, T. Drug-induced liver injury: Do we know everything? *World J. Hepatol.* **2017**, *9*, 491–502. [CrossRef] [PubMed]

15.   Aithal, G.P.; Watkins, P.B.; Andrade, R.J.; Larrey, D.; Molokhia, M.; Takikawa, H.; Hunt, C.M.; Wilke, R.A.; Avigan, M.; Kaplowitz, N.; et al. Case definition and phenotype standardization in drug-induced liver injury. *Clin. Pharmacol. Ther.* **2011**, *89*, 806–815. [CrossRef]

16.   Ahmad, J.; Odin, J.A. Epidemiology and Genetic Risk Factors of Drug Hepatotoxicity. *Clin. Liver Dis.* **2017**, *21*, 55–72. [CrossRef]

17.   Temple, R. Hy's law: Predicting serious hepatotoxicity. *Pharmacoepidemiol. Drug Saf.* **2006**, *15*, 241–243. [CrossRef]

18.   Lewis, J.H. The Art and Science of Diagnosing and Managing Drug-induced Liver Injury in 2015 and Beyond. *Clin. Gastroenterol. Hepatol.* **2015**, *13*, 2173–2189. [CrossRef]

19.   Pena, M.A.; Horga, J.F.; Zapater, P. Variations of pharmacokinetics of drugs in patients with cirrhosis. *Expert Rev. Clin. Pharmacol.* **2016**, *9*, 441–458. [CrossRef]

20.   Cota, J.M.; Burgess, D.S. Antifungal Dose Adjustment in Renal and Hepatic Dysfunction: Pharmacokinetic and Pharmacodynamic Considerations. *Curr. Fungal Infect. Rep.* **2010**, *4*, 120–128. [CrossRef]

21.   Tverdek, F.P.; Kofteridis, D.; Kontoyiannis, D.P. Antifungal agents and liver toxicity: A complex interaction. *Expert Rev. Anti Infect. Ther.* **2016**, *14*, 765–776. [CrossRef] [PubMed]

22.   Pea, F.; Lewis, R.E. Overview of antifungal dosing in invasive candidiasis. *J. Antimicrob. Chemother.* **2018**, *73*, i33–i43. [PubMed]

23.   Mourad, A.; Perfect, J.R. Tolerability profile of the current antifungal armoury. *J. Antimicrob. Chemother.* **2018**, *73*, i26–i32. [CrossRef] [PubMed]

24.   Kyriakidis, I.; Tragiannidis, A.; Munchen, S.; Groll, A.H. Clinical hepatotoxicity associated with antifungal agents. *Expert Opin. Drug Saf.* **2017**, *16*, 149–165. [CrossRef]

25.   Bader, J.C.; Bhavnani, S.M.; Andes, D.R.; Ambrose, P.G. We can do better: A fresh look at echinocandin dosing. *J. Antimicrob. Chemother.* **2018**, *73*, i44–i50. [CrossRef] [PubMed]

26.   Utz, J.P.; Treger, A.; Mc, C.N.; Emmons, C.W. Amphotericin B: Intravenous use in 21 patients with systemic fungal diseases. *Antibiot. Annu.* **1958**, *6*, 628–634.

27.   Loo, A.S.; Muhsin, S.A.; Walsh, T.J. Toxicokinetic and mechanistic basis for the safety and tolerability of liposomal amphotericin B. *Expert Opin. Drug Saf.* **2013**, *12*, 881–895. [CrossRef]

28.   Brajtburg, J.; Bolard, J. Carrier effects on biological activity of amphotericin B. *Clin. Microbiol. Rev.* **1996**, *9*, 512–531. [CrossRef]

29. Bellmann, R.; Smuszkiewicz, P. Pharmacokinetics of antifungal drugs: Practical implications for optimized treatment of patients. *Infection* **2017**, *45*, 737–779. [CrossRef]

30. Steimbach, L.M.; Tonin, F.S.; Virtuoso, S.; Borba, H.H.; Sanches, A.C.; Wiens, A.; Fernandez-Llimos, F.; Pontarolo, R. Efficacy and safety of amphotericin B lipid-based formulations—A systematic review and meta-analysis. *Mycoses* **2017**, *60*, 146–154. [CrossRef]

31. Hamill, R.J. Amphotericin B formulations: A comparative review of efficacy and toxicity. *Drugs* **2013**, *73*, 919–934. [CrossRef] [PubMed]

32. Johnson, P.C.; Wheat, L.J.; Cloud, G.A.; Goldman, M.; Lancaster, D.; Bamberger, D.M.; Powderly, W.G.; Hafner, R.; Kauffman, C.A.; Dismukes, W.E. Safety and efficacy of liposomal amphotericin B compared with conventional amphotericin B for induction therapy of histoplasmosis in patients with AIDS. *Ann. Intern. Med.* **2002**, *137*, 105–109. [CrossRef] [PubMed]

33. Fleming, R.V.; Kantarjian, H.M.; Husni, R.; Rolston, K.; Lim, J.; Raad, I.; Pierce, S.; Cortes, J.; Estey, E. Comparison of amphotericin B lipid complex (ABLC) vs. ambisome in the treatment of suspected or documented fungal infections in patients with leukemia. *Leuk. Lymphoma* **2001**, *40*, 511–520. [CrossRef] [PubMed]

34. Wade, R.L.; Chaudhari, P.; Natoli, J.L.; Taylor, R.J.; Nathanson, B.H.; Horn, D.L. Nephrotoxicity and other adverse events among inpatients receiving liposomal amphotericin B or amphotericin B lipid complex. *Diagn. Microbiol. Infect. Dis.* **2013**, *76*, 361–367. [CrossRef] [PubMed]

35. Wingard, J.R.; White, M.H.; Anaissie, E.; Raffalli, J.; Goodman, J.; Arrieta, A. A randomized, double-blind comparative trial evaluating the safety of liposomal amphotericin B versus amphotericin B lipid complex in the empirical treatment of febrile neutropenia. L Amph/ABLC Collaborative Study Group. *Clin. Infect. Dis.* **2000**, *31*, 1155–1163. [CrossRef]

36. Safdar, A.; Ma, J.; Saliba, F.; Dupont, B.; Wingard, J.R.; Hachem, R.Y.; Mattiuzzi, G.N.; Chandrasekar, P.H.; Kontoyiannis, D.P.; Rolston, K.V.; et al. Drug-induced nephrotoxicity caused by amphotericin B lipid complex and liposomal amphotericin B: A review and meta-analysis. *Medicine* **2010**, *89*, 236–244. [CrossRef]

37. Stone, N.R.; Bicanic, T.; Salim, R.; Hope, W. Liposomal Amphotericin B (AmBisome((R))): A Review of the Pharmacokinetics, Pharmacodynamics, Clinical Experience and Future Directions. *Drugs* **2016**, *76*, 485–500. [CrossRef]

38. Shigemi, A.; Matsumoto, K.; Ikawa, K.; Yaji, K.; Shimodozono, Y.; Morikawa, N.; Takeda, Y.; Yamada, K. Safety analysis of liposomal amphotericin B in adult patients: Anaemia, thrombocytopenia, nephrotoxicity, hepatotoxicity and hypokalaemia. *Int. J. Antimicrob. Agents* **2011**, *38*, 417–420. [CrossRef]

39. Inselmann, G.; Inselmann, U.; Heidemann, H.T. Amphotericin B and liver function. *Eur. J. Int. Med.* **2002**, *13*, 288–292. [CrossRef]

40. Fischer, M.A.; Winkelmayer, W.C.; Rubin, R.H.; Avorn, J. The hepatotoxicity of antifungal medications in bone marrow transplant recipients. *Clin. Infect. Dis.* **2005**, *41*, 301–307. [CrossRef]

41. Patel, G.P.; Crank, C.W.; Leikin, J.B. An evaluation of hepatotoxicity and nephrotoxicity of liposomal amphotericin B (L-AMB). *J. Med. Toxicol.* **2011**, *7*, 12–15. [CrossRef] [PubMed]

42. Chamilos, G.; Luna, M.; Lewis, R.E.; Chemaly, R.; Raad, I.I.; Kontoyiannis, D.P. Effects of liposomal amphotericin B versus an amphotericin B lipid complex on liver histopathology in patients with hematologic malignancies and invasive fungal infections: A retrospective, nonrandomized autopsy study. *Clin. Ther.* **2007**, *29*, 1980–1986. [CrossRef] [PubMed]

43. Weiler, S.; Überlacher, E.; Schöfmann, J.; Stienecke, E.; Dunzendorfer, S.; Joannidis, M.; Bellmann, R. Pharmacokinetics of Amphotericin B Colloidal Dispersion in Critically Ill Patients with Cholestatic Liver Disease. *Antimicrob. Agents Chemother.* **2012**, *56*, 5414–5418. [CrossRef]

44. Tassel, D.; Madoff, M.A. Treatment of Candida sepsis and Cryptococcus meningitis with 5-fluorocytosine. A new antifungal agent. *JAMA* **1968**, *206*, 830–832. [CrossRef]

45. Waldorf, A.R.; Polak, A. Mechanisms of action of 5-fluorocytosine. *Antimicrob. Agents Chemother.* **1983**, *23*, 79–85. [CrossRef] [PubMed]

46. Vermes, A.; Guchelaar, H.J.; Dankert, J. Flucytosine: A review of its pharmacology, clinical indications, pharmacokinetics, toxicity and drug interactions. *J. Antimicrob. Chemother.* **2000**, *46*, 171–179. [CrossRef]

47. Maziarz, E.K.; Perfect, J.R. Cryptococcosis. *Infect. Dis. Clin. N. Am.* **2016**, *30*, 179–206. [CrossRef] [PubMed]

48. Ashley, E.S.D.; Lewis, R.; Lewis, J.S.; Martin, C.; Andes, D. Pharmacology of Systemic Antifungal Agents. *Clin. Infect. Dis.* **2006**, *43*, S28–S39. [CrossRef]

49. Pappas, P.G.; Kauffman, C.A.; Andes, D.R.; Clancy, C.J.; Marr, K.A.; Ostrosky-Zeichner, L.; Reboli, A.C.; Schuster, M.G.; Vazquez, J.A.; Walsh, T.J.; et al. Clinical Practice Guideline for the Management of Candidiasis: 2016 Update by the Infectious Diseases Society of America. *Clin. Infect. Dis.* **2016**, *62*, e1–e50. [CrossRef] [PubMed]

50. *Ancobon*; Valeant Pharmaceuticals: Bridgewater, NJ, USA, 2017.

51. Brouwer, A.E.; van Kan, H.J.; Johnson, E.; Rajanuwong, A.; Teparrukkul, P.; Wuthiekanun, V.; Chierakul, W.; Day, N.; Harrison, T.S. Oral versus intravenous flucytosine in patients with human immunodeficiency virus-associated cryptococcal meningitis. *Antimicrob. Agents Chemother.* **2007**, *51*, 1038–1042. [CrossRef]

52. Record, C.O.; Skinner, J.M.; Sleight, P.; Speller, D.C. Candida endocarditis treated with 5-fluorocytosine. *Br. Med. J.* **1971**, *1*, 262–264. [CrossRef] [PubMed]

53. Pasqualotto, A.C.; Howard, S.J.; Moore, C.B.; Denning, D.W. Flucytosine therapeutic monitoring: 15 years experience from the UK. *J. Antimicrob. Chemother.* **2007**, *59*, 791–793. [CrossRef] [PubMed]

54. Block, E.R. Effect of hepatic insufficiency on 5-fluorocytosine concentrations in serum. *Antimicrob. Agents Chemother.* **1973**, *3*, 141–142. [CrossRef] [PubMed]

55. Ashbee, H.R.; Barnes, R.A.; Johnson, E.M.; Richardson, M.D.; Gorton, R.; Hope, W.W. Therapeutic drug monitoring (TDM) of antifungal agents: Guidelines from the British Society for Medical Mycology. *J. Antimicrob. Chemother.* **2014**, *69*, 1162–1176. [CrossRef] [PubMed]

56. Folk, A.; Cotoraci, C.; Balta, C.; Suciu, M.; Herman, H.; Boldura, O.M.; Dinescu, S.; Paiusan, L.; Ardelean, A.; Hermenean, A. Evaluation of Hepatotoxicity with Treatment Doses of Flucytosine and Amphotericin B for Invasive Fungal Infections. *BioMed Res. Int.* **2016**, *2016*, 9. [CrossRef] [PubMed]

57. Fothergill, A.W. Miconazole: A historical perspective. *Expert Rev. Anti Infect. Ther* **2006**, *4*, 171–175. [CrossRef] [PubMed]

58. Patterson, T.F.; Thompson, G.R., III; Denning, D.W.; Fishman, J.A.; Hadley, S.; Herbrecht, R.; Kontoyiannis, D.P.; Marr, K.A.; Morrison, V.A.; Nguyen, M.H.; et al. Practice Guidelines for the Diagnosis and Management of Aspergillosis: 2016 Update by the Infectious Diseases Society of America. *Clin. Infect. Dis.* **2016**, *63*, e1–e60. [CrossRef]

59. Tucker, R.M.; Haq, Y.; Denning, D.W.; Stevens, D.A. Adverse events associated with itraconazole in 189 patients on chronic therapy. *J. Antimicrob. Chemother.* **1990**, *26*, 561–566. [CrossRef]

60. Natesan, S.K.; Chandrasekar, P.H. Isavuconazole for the treatment of invasive aspergillosis and mucormycosis: Current evidence, safety, efficacy, and clinical recommendations. *Infect. Drug Resist.* **2016**, *9*, 291–300. [CrossRef]

61. Wilson, D.T.; Dimondi, V.P.; Johnson, S.W.; Jones, T.M.; Drew, R.H. Role of isavuconazole in the treatment of invasive fungal infections. *Ther. Clin. Risk Manag.* **2016**, *12*, 1197–1206. [CrossRef]

62. Raschi, E.; Poluzzi, E.; Koci, A.; Caraceni, P.; Ponti, F.D. Assessing liver injury associated with antimycotics: Concise literature review and clues from data mining of the FAERS database. *World J. Hepatol.* **2014**, *6*, 601–612. [CrossRef] [PubMed]

63. Song, J.C.; Deresinski, S. Hepatotoxicity of antifungal agents. *Curr. Opin. Investig. Drugs* **2005**, *6*, 170–177. [PubMed]

64. Bruggemann, R.J.; Alffenaar, J.W.; Blijlevens, N.M.; Billaud, E.M.; Kosterink, J.G.; Verweij, P.E.; Burger, D.M. Clinical relevance of the pharmacokinetic interactions of azole antifungal drugs with other coadministered agents. *Clin. Infect. Dis.* **2009**, *48*, 1441–1458. [CrossRef] [PubMed]

65. *Diflucan*; Pfizer Inc.: New York, NY, USA, 2018.

66. Muñoz, P.P.; Moreno, S.S.; Berenguer, J.J.; de Quirós, J.; Bouza, E.E. Fluconazole-related hepatotoxicity in patients with acquired immunodeficiency syndrome. *Arch. Intern. Med.* **1991**, *151*, 1020–1021. [CrossRef] [PubMed]

67. Wells, C.; Lever, A.M. Dose-dependent fluconazole hepatotoxicity proven on biopsy and rechallenge. *J. Infect.* **1992**, *24*, 111–112. [CrossRef]

68. Hay, R.J. Risk/benefit ratio of modern antifungal therapy: Focus on hepatic reactions. *J. Am. Acad. Dermatol.* **1993**, *29*, S50–S54. [CrossRef]

69. Como, J.A.; Dismukes, W.E. Oral azole drugs as systemic antifungal therapy. *N. Engl. J. Med.* **1994**, *330*, 263–272. [PubMed]

70. Franklin, I.M.; Elias, E.; Hirsch, C. Fluconazole-induced jaundice. *Lancet* **1990**, *336*, 565. [CrossRef]

71. Trujillo, M.A.; Galgiani, J.N.; Sampliner, R.E. Evaluation of hepatic injury arising during fluconazole therapy. *Arch. Intern. Med.* **1994**, *154*, 102–104. [CrossRef]
72. Guillaume, M.P.; De Prez, C.; Cogan, E. Subacute mitochondrial liver disease in a patient with AIDS: Possible relationship to prolonged fluconazole administration. *Am. J. Gastroenterol.* **1996**, *91*, 165–168.
73. Anaissie, E.J.; Kontoyiannis, D.P.; Huls, C.; Vartivarian, S.E.; Karl, C.; Prince, R.A.; Bosso, J.; Bodey, G.P. Safety, plasma concentrations, and efficacy of high-dose fluconazole in invasive mold infections. *J. Infect. Dis.* **1995**, *172*, 599–602. [CrossRef] [PubMed]
74. Wang, J.L.; Chang, C.H.; Young-Xu, Y.; Chan, K.A. Systematic review and meta-analysis of the tolerability and hepatotoxicity of antifungals in empirical and definitive therapy for invasive fungal infection. *Antimicrob. Agents Chemother.* **2010**, *54*, 2409–2419. [CrossRef] [PubMed]
75. Garcia Rodriguez, L.A.; Duque, A.; Castellsague, J.; Perez-Gutthann, S.; Stricker, B.H. A cohort study on the risk of acute liver injury among users of ketoconazole and other antifungal drugs. *Br. J. Clin. Pharmacol.* **1999**, *48*, 847–852. [CrossRef] [PubMed]
76. Jacobson, M.A.; Hanks, D.K.; Ferrell, L.D. Fatal acute hepatic necrosis due to fluconazole. *Am. J. Med.* **1994**, *96*, 188–190. [CrossRef]
77. Chmel, H. Fatal acute hepatic necrosis due to fluconazole. *Am. J. Med.* **1995**, *99*, 224–225. [CrossRef]
78. Bronstein, J.A.; Gros, P.; Hernandez, E.; Larroque, P.; Molinie, C. Fatal acute hepatic necrosis due to dose-dependent fluconazole hepatotoxicity. *Clin. Infect. Dis.* **1997**, *25*, 1266–1267. [CrossRef]
79. Ruhnke, M.; Yeates, R.A.; Pfaff, G.; Sarnow, E.; Hartmann, A.; Trautmann, M. Single-dose pharmacokinetics of fluconazole in patients with liver cirrhosis. *J. Antimicrob. Chemother.* **1995**, *35*, 641–647. [CrossRef]
80. Gearhart, M.O. Worsening of Liver Function with Fluconazole and Review of Azole Antifungal Hepatotoxicity. *Ann. Pharmacother.* **1994**, *28*, 1177–1181. [CrossRef]
81. Lo Re, V., 3rd; Carbonari, D.M.; Lewis, J.D.; Forde, K.A.; Goldberg, D.S.; Reddy, K.R.; Haynes, K.; Roy, J.A.; Sha, D.; Marks, A.R.; et al. Oral Azole Antifungal Medications and Risk of Acute Liver Injury, Overall and by Chronic Liver Disease Status. *Am. J. Med.* **2016**, *129*, 283–291. [CrossRef]
82. Lestner, J.; Hope, W.W. Itraconazole: An update on pharmacology and clinical use for treatment of invasive and allergic fungal infections. *Expert Opin. Drug Metab. Toxicol.* **2013**, *9*, 911–926. [CrossRef]
83. *Sporanox*; Janssen Pharmaceuticals: Titusville, FL, USA, 2017.
84. Chapman, S.W.; Dismukes, W.E.; Proia, L.A.; Bradsher, R.W.; Pappas, P.G.; Threlkeld, M.G.; Kauffman, C.A. Clinical Practice Guidelines for the Management of Blastomycosis: 2008 Update by the Infectious Diseases Society of America. *Clin. Infect. Dis.* **2008**, *46*, 1801–1812. [CrossRef] [PubMed]
85. Galgiani, J.N.; Ampel, N.M.; Blair, J.E.; Catanzaro, A.; Geertsma, F.; Hoover, S.E.; Johnson, R.H.; Kusne, S.; Lisse, J.; MacDonald, J.D.; et al. 2016 Infectious Diseases Society of America (IDSA) Clinical Practice Guideline for the Treatment of Coccidioidomycosis. *Clin. Infect. Dis.* **2016**, *63*, e112–e146. [CrossRef] [PubMed]
86. Wheat, L.J.; Freifeld, A.G.; Kleiman, M.B.; Baddley, J.W.; McKinsey, D.S.; Loyd, J.E.; Kauffman, C.A. Clinical practice guidelines for the management of patients with histoplasmosis: 2007 Update by the Infectious Diseases Society of America. *Clin. Infect. Dis.* **2007**, *45*, 807–825. [CrossRef] [PubMed]
87. Girois, S.B.; Chapuis, F.; Decullier, E.; Revol, B.G. Adverse effects of antifungal therapies in invasive fungal infections: Review and meta-analysis. *Eur. J. Clin. Microbiol. Infect. Dis.* **2006**, *25*, 138–149. [CrossRef]
88. Levron, J.C.; Chwetzoff, E.; Perrichon, P.; Autic, A.; Berthelot, P.; Boboc, D. *Pharmacokinetics of Itraconazole in Cirrhotic Patients*; Clinical Research Report R 51211; Laboratoires Janssen: Val-de-Reuil, France, 1987.
89. Ullmann, A.J.; Aguado, J.M.; Arikan-Akdagli, S.; Denning, D.W.; Groll, A.H.; Lagrou, K.; Lass-Flörl, C.; Lewis, R.E.; Munoz, P.; Verweij, P.E.; et al. Diagnosis and management of Aspergillus diseases: Executive summary of the 2017 ESCMID-ECMM-ERS guideline. *Clin. Microbiol. Infect.* **2018**, *24*, e1–e38. [CrossRef]
90. Lestner, J.M.; Roberts, S.A.; Moore, C.B.; Howard, S.J.; Denning, D.W.; Hope, W.W. Toxicodynamics of itraconazole: Implications for therapeutic drug monitoring. *Clin. Infect. Dis.* **2009**, *49*, 928–930. [CrossRef]
91. Veringa, A.; Ter Avest, M.; Span, L.F.; van den Heuvel, E.R.; Touw, D.J.; Zijlstra, J.G.; Kosterink, J.G.; van der Werf, T.S.; Alffenaar, J.C. Voriconazole metabolism is influenced by severe inflammation: A prospective study. *J. Antimicrob. Chemother.* **2017**, *72*, 261–267. [CrossRef]
92. *Vfend*; Pfizer Inc.: New York, NY, USA, 2018.

93. European Medicines Agency. Summary of Product Characteristics: Vfend. Available online: https://www.ema.europa.eu/documents/product-information/vfend-epar-product-information_en.pdf (accessed on 22 November 2018).

94. Denning, D.W.; Ribaud, P.; Milpied, N.; Caillot, D.; Herbrecht, R.; Thiel, E.; Haas, A.; Ruhnke, M.; Lode, H. Efficacy and safety of voriconazole in the treatment of acute invasive aspergillosis. *Clin. Infect. Dis.* **2002**, *34*, 563–571. [CrossRef]

95. Zonios, D.; Yamazaki, H.; Murayama, N.; Natarajan, V.; Palmore, T.; Childs, R.; Skinner, J.; Bennett, J.E. Voriconazole metabolism, toxicity, and the effect of cytochrome P450 2C19 genotype. *J. Infect. Dis.* **2014**, *209*, 1941–1948. [CrossRef]

96. Amigues, I.; Cohen, N.; Chung, D.; Seo, S.; Plescia, C.; Jakubowski, A.; Barker, J.; Papanicolaou, G.A. Hepatic Safety of Voriconazole after Allogeneic Hematopoietic Stem Cell Transplantation. *Biol. Blood Marrow Transplant.* **2010**, *16*, 46–52. [CrossRef]

97. Saito, T.; Fujiuchi, S.; Tao, Y.; Sasaki, Y.; Ogawa, K.; Suzuki, K.; Tada, A.; Kuba, M.; Kato, T.; Kawabata, M.; et al. Efficacy and safety of voriconazole in the treatment of chronic pulmonary aspergillosis: Experience in Japan. *Infection* **2012**, *40*, 661–667. [CrossRef] [PubMed]

98. Luong, M.L.; Al-Dabbagh, M.; Groll, A.H.; Racil, Z.; Nannya, Y.; Mitsani, D.; Husain, S. Utility of voriconazole therapeutic drug monitoring: A meta-analysis. *J. Antimicrob. Chemother.* **2016**, *71*, 1786–1799. [CrossRef] [PubMed]

99. Tan, K.K.C.; Wood, N.; Weil, A. Multiple-dose pharmacokinetics of voriconazole in chronic hepatic impairment. In Proceedings of the 41st Interscience Conference on Antimicrobial Agents and Chemotherapy, Chicago, IL, USA, 16–19 December 2001.

100. Solis-Munoz, P.; Lopez, J.C.; Bernal, W.; Willars, C.; Verma, A.; Heneghan, M.A.; Wendon, J.; Auzinger, G. Voriconazole hepatotoxicity in severe liver dysfunction. *J. Infect.* **2013**, *66*, 80–86. [CrossRef] [PubMed]

101. Yamada, T.; Imai, S.; Koshizuka, Y.; Tazawa, Y.; Kagami, K.; Tomiyama, N.; Sugawara, R.; Yamagami, A.; Shimamura, T.; Iseki, K. Necessity for a Significant Maintenance Dosage Reduction of Voriconazole in Patients with Severe Liver Cirrhosis (Child-Pugh Class C). *Biol. Pharm. Bull.* **2018**, *41*, 1112–1118. [CrossRef] [PubMed]

102. Wang, T.; Yan, M.; Tang, D.; Xue, L.; Zhang, T.; Dong, Y.; Zhu, L.; Wang, X.; Dong, Y. Therapeutic drug monitoring and safety of voriconazole therapy in patients with Child-Pugh class B and C cirrhosis: A multicenter study. *Int. J. Infect. Dis.* **2018**, *72*, 49–54. [CrossRef]

103. Wang, T.; Yan, M.; Tang, D.; Xue, L.; Zhang, T.; Dong, Y.; Zhu, L.; Wang, X.; Dong, Y. A retrospective, multicenter study of voriconazole trough concentrations and safety in patients with Child-Pugh class C cirrhosis. *J. Clin. Pharm. Ther.* **2018**. [CrossRef]

104. Hashemizadeh, Z.; Badiee, P.; Malekhoseini, S.A.; Raeisi Shahraki, H.; Geramizadeh, B.; Montaseri, H. Observational Study of Associations between Voriconazole Therapeutic Drug Monitoring, Toxicity, and Outcome in Liver Transplant Patients. *Antimicrob. Agents Chemother.* **2017**, *61*. [CrossRef] [PubMed]

105. Pascual, A.; Calandra, T.; Bolay, S.; Buclin, T.; Bille, J.; Marchetti, O. Voriconazole therapeutic drug monitoring in patients with invasive mycoses improves efficacy and safety outcomes. *Clin. Infect. Dis.* **2008**, *46*, 201–211. [CrossRef] [PubMed]

106. Pasqualotto, A.C.; Xavier, M.O.; Andreolla, H.F.; Linden, R. Voriconazole therapeutic drug monitoring: Focus on safety. *Expert Opin. Drug Saf.* **2010**, *9*, 125–137. [CrossRef]

107. Liu, X.; Su, H.; Tong, J.; Chen, J.; Yang, H.; Xiao, L.; Hu, J.; Zhang, L. Significance of monitoring plasma concentration of voriconazole in a patient with liver failure: A case report. *Medicine* **2017**, *96*, e8039. [CrossRef]

108. Courtney, R.; Pai, S.; Laughlin, M.; Lim, J.; Batra, V. Pharmacokinetics, safety, and tolerability of oral posaconazole administered in single and multiple doses in healthy adults. *Antimicrob. Agents Chemother.* **2003**, *47*, 2788–2795. [CrossRef] [PubMed]

109. Sime, F.B.; Stuart, J.; Butler, J.; Starr, T.; Wallis, S.C.; Pandey, S.; Lipman, J.; Roberts, J.A. Pharmacokinetics of Intravenous Posaconazole in Critically Ill Patients. *Antimicrob. Agents Chemother.* **2018**, *62*. [CrossRef] [PubMed]

110. Strommen, A.; Hurst, A.L.; Curtis, D.; Abzug, M.J. Use of Intravenous Posaconazole in Hematopoietic Stem Cell Transplant Patients. *J. Pediatr. Hematol. Oncol.* **2018**, *40*, e203–e206. [CrossRef]

111. Wiederhold, N.P. Pharmacokinetics and safety of posaconazole delayed-release tablets for invasive fungal infections. *Clin. Pharmacol.* **2016**, *8*, 1–8. [CrossRef]

112. Jovic, Z.; Jankovic, S.M.; Ruzic Zecevic, D.; Milovanovic, D.; Stefanovic, S.; Folic, M.; Milovanovic, J.; Kostic, M. Clinical Pharmacokinetics of Second-Generation Triazoles for the Treatment of Invasive Aspergillosis and Candidiasis. *Eur. J. Drug Metab. Pharmacokinet.* **2018**. [CrossRef]

113. *Noxafil*; Merk & Co., Inc.: Whitehouse Station, NJ, USA, 2017.

114. European Medicines Agency. Summary of Product Characteristics: Noxafil. Available online: https://www.ema.europa.eu/documents/product-information/noxafil-epar-product-information_en.pdf (accessed on 22 November 2018).

115. Cornely, O.A.; Duarte, R.F.; Haider, S.; Chandrasekar, P.; Helfgott, D.; Jimenez, J.L.; Candoni, A.; Raad, I.; Laverdiere, M.; Langston, A.; et al. Phase 3 pharmacokinetics and safety study of a posaconazole tablet formulation in patients at risk for invasive fungal disease. *J. Antimicrob. Chemother.* **2016**, *71*, 718–726. [CrossRef]

116. Tverdek, F.P.; Heo, S.T.; Aitken, S.L.; Granwehr, B.; Kontoyiannis, D.P. Real-Life Assessment of the Safety and Effectiveness of the New Tablet and Intravenous Formulations of Posaconazole in the Prophylaxis of Invasive Fungal Infections via Analysis of 343 Courses. *Antimicrob. Agents Chemother.* **2017**, *61*. [CrossRef] [PubMed]

117. Boglione-Kerrien, C.; Picard, S.; Tron, C.; Nimubona, S.; Gangneux, J.P.; Lalanne, S.; Lemaitre, F.; Bellissant, E.; Verdier, M.C.; Petitcollin, A. Safety study and therapeutic drug monitoring of the oral tablet formulation of posaconazole in patients with haematological malignancies. *J. Cancer Res. Clin. Oncol.* **2018**, *144*, 127–134. [CrossRef] [PubMed]

118. Zhang, S.; He, Y.; Jiang, E.; Wei, J.; Yang, D.; Zhang, R.; Zhai, W.; Zhang, G.; Wang, Z.; Zhang, L.; et al. Efficacy and safety of posaconazole in hematopoietic stem cell transplantation patients with invasive fungal disease. *Future Microbiol.* **2017**, *12*, 1371–1379. [CrossRef] [PubMed]

119. Moton, A.; Krishna, G.; Ma, L.; O'Mara, E.; Prasad, P.; McLeod, J.; Preston, R.A. Pharmacokinetics of a single dose of the antifungal posaconazole as oral suspension in subjects with hepatic impairment. *Curr. Med. Res. Opin.* **2010**, *26*, 1–7. [CrossRef]

120. Lin, S.Y.; Lu, P.L.; Tsai, K.B.; Lin, C.Y.; Lin, W.R.; Chen, T.C.; Chang, Y.T.; Huang, C.H.; Chen, C.Y.; Lai, C.C.; et al. A mucormycosis case in a cirrhotic patient successfully treated with posaconazole and review of published literature. *Mycopathologia* **2012**, *174*, 499–504. [CrossRef] [PubMed]

121. National Cancer Institute. *Common Terminology Criteria for Adverse Events*; Version 4.0; NIH, U.S. Department of Health and Human Services: Washington, DC, USA, 2009.

122. Heinz, W.J.; Egerer, G.; Lellek, H.; Boehme, A.; Greiner, J. Posaconazole after previous antifungal therapy with voriconazole for therapy of invasive aspergillus disease, a retrospective analysis. *Mycoses* **2013**, *56*, 304–310. [CrossRef]

123. Foo, H.; Gottlieb, T. Lack of Cross-Hepatotoxicity between Voriconazole and Posaconazole. *Clin. Infect. Dis.* **2007**, *45*, 803–805. [CrossRef] [PubMed]

124. Martinez-Casanova, J.; Carballo, N.; Luque, S.; Sorli, L.; Grau, S. Posaconazole achieves prompt recovery of voriconazole-induced liver injury in a case of invasive aspergillosis. *Infect. Drug Resist.* **2018**, *11*, 317–321. [CrossRef] [PubMed]

125. Jang, S.H.; Colangelo, P.M.; Gobburu, J.V. Exposure-response of posaconazole used for prophylaxis against invasive fungal infections: Evaluating the need to adjust doses based on drug concentrations in plasma. *Clin. Pharmacol. Ther.* **2010**, *88*, 115–119. [CrossRef] [PubMed]

126. Catanzaro, A.; Cloud, G.A.; Stevens, D.A.; Levine, B.E.; Williams, P.L.; Johnson, R.H.; Rendon, A.; Mirels, L.F.; Lutz, J.E.; Holloway, M.; et al. Safety, tolerance, and efficacy of posaconazole therapy in patients with nonmeningeal disseminated or chronic pulmonary coccidioidomycosis. *Clin. Infect. Dis.* **2007**, *45*, 562–568. [CrossRef]

127. *Cresemba*; Astellas Pharma Inc.: Northbrook, IL, USA, 2018.

128. European Medicines Agency. Summary of Product Characteristics: Cresemba. Available online: https://www.ema.europa.eu/documents/product-information/cresemba-epar-product-information_en.pdf (accessed on 22 November 2018).

129. Marty, F.M.; Ostrosky-Zeichner, L.; Cornely, O.A.; Mullane, K.M.; Perfect, J.R.; Thompson, G.R., 3rd; Alangaden, G.J.; Brown, J.M.; Fredricks, D.N.; Heinz, W.J.; et al. Isavuconazole treatment for mucormycosis: A single-arm open-label trial and case-control analysis. *Lancet Infect. Dis.* **2016**, *16*, 828–837. [CrossRef]

130. Jenks, J.D.; Salzer, H.J.; Prattes, J.; Krause, R.; Buchheidt, D.; Hoenigl, M. Spotlight on isavuconazole in the treatment of invasive aspergillosis and mucormycosis: Design, development, and place in therapy. *Drug Des. Devel. Ther.* **2018**, *12*, 1033–1044. [CrossRef]

131. Maertens, J.A.; Raad, I.I.; Marr, K.A.; Patterson, T.F.; Kontoyiannis, D.P.; Cornely, O.A.; Bow, E.J.; Rahav, G.; Neofytos, D.; Aoun, M.; et al. Isavuconazole versus voriconazole for primary treatment of invasive mould disease caused by Aspergillus and other filamentous fungi (SECURE): A phase 3, randomised-controlled, non-inferiority trial. *Lancet* **2016**, *387*, 760–769. [CrossRef]

132. Schmitt-Hoffmann, A.; Roos, B.; Spickermann, J.; Heep, M.; Peterfai, E.; Edwards, D.J.; Stoeckel, K. Effect of mild and moderate liver disease on the pharmacokinetics of isavuconazole after intravenous and oral administration of a single dose of the prodrug BAL8557. *Antimicrob. Agents Chemother.* **2009**, *53*, 4885–4890. [CrossRef]

133. Desai, A.; Schmitt-Hoffmann, A.H.; Mujais, S.; Townsend, R. Population Pharmacokinetics of Isavuconazole in Subjects with Mild or Moderate Hepatic Impairment. *Antimicrob. Agents Chemother.* **2016**, *60*, 3025–3031. [CrossRef] [PubMed]

134. Zheng, X.; Xu, G.; Zhu, L.; Fang, L.; Zhang, Y.; Ding, H.; Tong, Y.; Sun, J.; Huang, P. Pharmacokinetic/Pharmacodynamic Analysis of Isavuconazole against *Aspergillus* spp. and *Candida* spp. in Healthy Subjects and Patients With Hepatic or Renal Impairment by Monte Carlo Simulation. *J. Clin. Pharmacol.* **2018**, *58*, 1266–1273. [CrossRef] [PubMed]

135. Desai, A.V.; Kovanda, L.L.; Hope, W.W.; Andes, D.; Mouton, J.W.; Kowalski, D.L.; Townsend, R.W.; Mujais, S.; Bonate, P.L. Exposure-Response Relationships for Isavuconazole in Patients with Invasive Aspergillosis and Other Filamentous Fungi. *Antimicrob. Agents Chemother.* **2017**, *61*. [CrossRef] [PubMed]

136. Wiederhold, N.P.; Lewis, R.E. The echinocandin antifungals: An overview of the pharmacology, spectrum and clinical efficacy. *Expert Opin. Investig. Drugs* **2003**, *12*, 1313–1333. [CrossRef] [PubMed]

137. Wang, J.F.; Xue, Y.; Zhu, X.B.; Fan, H. Efficacy and safety of echinocandins versus triazoles for the prophylaxis and treatment of fungal infections: A meta-analysis of RCTs. *Eur. J. Clin. Microbiol. Infect. Dis.* **2015**, *34*, 651–659. [CrossRef]

138. Dekkers, B.G.J.; Veringa, A.; Marriott, D.J.E.; Boonstra, J.M.; van der Elst, K.C.M.; Doukas, F.F.; McLachlan, A.J.; Alffenaar, J.C. Invasive Candidiasis in the Elderly: Considerations for Drug Therapy. *Drugs Aging* **2018**, *35*, 781–789. [CrossRef]

139. Balani, S.K.; Xu, X.; Arison, B.H.; Silva, M.V.; Gries, A.; DeLuna, F.A.; Cui, D.; Kari, P.H.; Ly, T.; Hop, C.E.; et al. Metabolites of caspofungin acetate, a potent antifungal agent, in human plasma and urine. *Drug Metab. Dispos.* **2000**, *28*, 1274–1278.

140. European Medicines Agency. Summary of Product Characteristics: Cancidas. Available online: https://www.ema.europa.eu/documents/product-information/cancidas-epar-product-information_en.pdf (accessed on 22 November 2018).

141. *Cancidas*; Merk & Co., Inc.: Whitehouse Station, NJ, USA, 2018.

142. Mistry, G.C.; Migoya, E.; Deutsch, P.J.; Winchell, G.; Hesney, M.; Li, S.; Bi, S.; Dilzer, S.; Lasseter, K.C.; Stone, J.A. Single- and multiple-dose administration of caspofungin in patients with hepatic insufficiency: Implications for safety and dosing recommendations. *J. Clin. Pharmacol.* **2007**, *47*, 951–961. [CrossRef]

143. Spriet, I.; Meersseman, W.; Annaert, P.; de Hoon, J.; Willems, L. Pharmacokinetics of caspofungin in a critically ill patient with liver cirrhosis. *Eur. J. Clin. Pharmacol.* **2011**, *67*, 753–755. [CrossRef]

144. Martial, L.C.; Bruggemann, R.J.; Schouten, J.A.; van Leeuwen, H.J.; van Zanten, A.R.; de Lange, D.W.; Muilwijk, E.W.; Verweij, P.E.; Burger, D.M.; Aarnoutse, R.E.; et al. Dose Reduction of Caspofungin in Intensive Care Unit Patients with Child Pugh B Will Result in Suboptimal Exposure. *Clin. Pharmacokinet.* **2016**, *55*, 723–733. [CrossRef]

145. Gustot, T.; Ter Heine, R.; Brauns, E.; Cotton, F.; Jacobs, F.; Bruggemann, R.J. Caspofungin dosage adjustments are not required for patients with Child-Pugh B or C cirrhosis. *J. Antimicrob. Chemother.* **2018**, *73*, 2493–2496. [CrossRef]

146. Yang, Q.T.; Zhai, Y.J.; Chen, L.; Zhang, T.; Yan, Y.; Meng, T.; Liu, L.C.; Chen, L.M.; Wang, X.; Dong, Y.L. Whole-body physiology-based pharmacokinetics of caspofungin for general patients, intensive care unit patients and hepatic insufficiency patients. *Acta Pharmacol. Sin.* **2018**, *39*, 1533–1543. [CrossRef] [PubMed]

147. Kofla, G.; Ruhnke, M. Pharmacology and metabolism of anidulafungin, caspofungin and micafungin in the treatment of invasive candidosis: Review of the literature. *Eur. J. Med. Res.* **2011**, *16*, 159–166. [CrossRef] [PubMed]

148. European Medicines Agency. Summary of Product Characteristics: Mycamine. Available online: https://www.ema.europa.eu/documents/product-information/mycamine-epar-product-information_en.pdf (accessed on 15 October 2018).

149. *Mycamine*; Astellas Pharma Inc.: Northbrook, IL, USA, 2018.

150. Lee, C.H.; Lin, J.C.; Ho, C.L.; Sun, M.; Yen, W.T.; Lin, C. Efficacy and safety of micafungin versus extensive azoles in the prevention and treatment of invasive fungal infections for neutropenia patients with hematological malignancies: A meta-analysis of randomized controlled trials. *PLoS ONE* **2017**, *12*, e0180050. [CrossRef] [PubMed]

151. Yeoh, S.F.; Lee, T.J.; Chew, K.L.; Lin, S.; Yeo, D.; Setia, S. Echinocandins for management of invasive candidiasis in patients with liver disease and liver transplantation. *Infect. Drug Resist.* **2018**, *11*, 805–819. [CrossRef] [PubMed]

152. Wasmann, R.E.; Muilwijk, E.W.; Burger, D.M.; Verweij, P.E.; Knibbe, C.A.; Bruggemann, R.J. Clinical Pharmacokinetics and Pharmacodynamics of Micafungin. *Clin. Pharmacokinet.* **2018**, *57*, 267–286. [CrossRef] [PubMed]

153. Hebert, M.F.; Smith, H.E.; Marbury, T.C.; Swan, S.K.; Smith, W.B.; Townsend, R.W.; Buell, D.; Keirns, J.; Bekersky, I. Pharmacokinetics of micafungin in healthy volunteers, volunteers with moderate liver disease, and volunteers with renal dysfunction. *J. Clin. Pharmacol.* **2005**, *45*, 1145–1152. [CrossRef]

154. Undre, N.; Pretorius, B.; Stevenson, P. Pharmacokinetics of micafungin in subjects with severe hepatic dysfunction. *Eur. J. Drug Metab. Pharmacokinet.* **2015**, *40*, 285–293. [CrossRef]

155. Luque, S.; Campillo, N.; Alvarez-Lerma, F.; Ferrandez, O.; Horcajada, J.P.; Grau, S. Pharmacokinetics of micafungin in patients with pre-existing liver dysfunction: A safe option for treating invasive fungal infections. *Enferm. Infecc. Microbiol. Clin.* **2016**, *34*, 652–654. [CrossRef]

156. Damle, B.D.; Dowell, J.A.; Walsky, R.L.; Weber, G.L.; Stogniew, M.; Inskeep, P.B. In vitro and in vivo studies to characterize the clearance mechanism and potential cytochrome P450 interactions of anidulafungin. *Antimicrob. Agents Chemother.* **2009**, *53*, 1149–1156. [CrossRef]

157. *Eraxis*; Pfizer Inc.: New York, NY, USA, 2018.

158. European Medicines Agency. Summary of Product Characteristics: Ecalta. Available online: https://www.ema.europa.eu/documents/product-information/ecalta-epar-product-information_en.pdf (accessed on 22 November 2018).

159. Reboli, A.C.; Rotstein, C.; Pappas, P.G.; Chapman, S.W.; Kett, D.H.; Kumar, D.; Betts, R.; Wible, M.; Goldstein, B.P.; Schranz, J.; et al. Anidulafungin versus fluconazole for invasive candidiasis. *N. Eng. J. Med.* **2007**, *356*, 2472–2482. [CrossRef] [PubMed]

160. Aguado, J.M.; Varo, E.; Usetti, P.; Pozo, J.C.; Moreno, A.; Catalan, M.; Len, O.; Blanes, M.; Sole, A.; Munoz, P.; et al. Safety of anidulafungin in solid organ transplant recipients. *Liver Transplant.* **2012**, *18*, 680–685. [CrossRef]

161. Dowell, J.A.; Stogniew, M.; Krause, D.; Damle, B. Anidulafungin does not require dosage adjustment in subjects with varying degrees of hepatic or renal impairment. *J. Clin. Pharmacol.* **2007**, *47*, 461–470. [CrossRef] [PubMed]

162. Verma, A.; Auzinger, G.; Kantecki, M.; Campling, J.; Spurden, D.; Percival, F.; Heaton, N. Safety and Efficacy of Anidulafungin for Fungal Infection in Patients With Liver Dysfunction or Multiorgan Failure. *Open Forum Infect. Dis.* **2017**, *4*. [CrossRef]

163. Weiler, S.; Zoller, H.; Graziadei, I.; Vogel, W.; Bellmann-Weiler, R.; Joannidis, M.; Bellmann, R. Altered Pharmacokinetics of Voriconazole in a Patient with Liver Cirrhosis. *Antimicrob. Agents Chemother.* **2007**, *51*, 3459–3460. [CrossRef] [PubMed]

MDPI

St. Alban-Anlage 66

4052 Basel

Switzerland

Tel. +41 61 683 77 34

Fax +41 61 302 89 18

www.mdpi.com

Journal of Fungi Editorial Office

E-mail: jof@mdpi.com

www.mdpi.com/journal/jof

www.ingramcontent.com/pod-product-compliance
Lightning Source LLC
Chambersburg PA
CBHW051847210326
41597CB00033B/5805